Studies in Systems, Decision and Control

Volume 422

Series Editor

Janusz Kacprzyk, Systems Research Institute, Polish Academy of Sciences, Warsaw, Poland

The series "Studies in Systems, Decision and Control" (SSDC) covers both new developments and advances, as well as the state of the art, in the various areas of broadly perceived systems, decision making and control–quickly, up to date and with a high quality. The intent is to cover the theory, applications, and perspectives on the state of the art and future developments relevant to systems, decision making, control, complex processes and related areas, as embedded in the fields of engineering, computer science, physics, economics, social and life sciences, as well as the paradigms and methodologies behind them. The series contains monographs, textbooks, lecture notes and edited volumes in systems, decision making and control spanning the areas of Cyber-Physical Systems, Autonomous Systems, Sensor Networks, Control Systems, Energy Systems, Automotive Systems, Biological Systems, Vehicular Networking and Connected Vehicles, Aerospace Systems, Automation, Manufacturing, Smart Grids, Nonlinear Systems, Power Systems, Robotics, Social Systems, Economic Systems and other. Of particular value to both the contributors and the readership are the short publication timeframe and the world-wide distribution and exposure which enable both a wide and rapid dissemination of research output.

Indexed by SCOPUS, DBLP, WTI Frankfurt eG, zbMATH, SCImago.

All books published in the series are submitted for consideration in Web of Science.

More information about this series at https://link.springer.com/bookseries/13304

Roshni Maiti · Kaushik Das Sharma ·
Gautam Sarkar

Hybrid L_1 Adaptive Control

Applications of Fuzzy Modeling, Stochastic Optimization and Metaheuristics

Springer

Roshni Maiti
Department of Applied Physics
University of Calcutta
Kolkata, West Bengal, India

Kaushik Das Sharma
Department of Applied Physics
University of Calcutta
Kolkata, West Bengal, India

Gautam Sarkar
Department of Applied Physics
University of Calcutta
Kolkata, West Bengal, India

ISSN 2198-4182 ISSN 2198-4190 (electronic)
Studies in Systems, Decision and Control
ISBN 978-3-030-97101-4 ISBN 978-3-030-97102-1 (eBook)
https://doi.org/10.1007/978-3-030-97102-1

This Springer imprint is published by the registered company Springer Nature Switzerland AG
The registered company address is: Gewerbestrasse 11, 6330 Cham, Switzerland

Dedicated to my Parents Smt. Kajal Maiti and Mr. Asit Kumar Maiti. Also dedicated to the beloved students.

—Roshni Maiti

Preface

Conventional control methodologies, viz., Proportional Integral Derivative (PID) controller, Linear Quadratic Regulator (LQR), etc., can control Linear Time Invariant (LTI) systems. Though, their performances degrade when nonlinearities, time varying uncertainties, time varying disturbances, delays, etc., present into the system. Adaptive controllers arise to tackle time varying uncertainties and disturbances. The most extensively used adaptive control scheme, viz., Model Reference Adaptive Controller (MRAC) utilizes low adaptation gain to maintain robustness. Hence, MRAC possess sluggish transient performance. With the aim to overcome such problem, L_1 adaptive controller was introduced to provide fast transient performance with high robustness. However, the performance of basic L_1 adaptive controller degrades when nonlinearities, delays, etc., are present in the system. On the other hand, nonlinearities can be properly modelled through universal approximator fuzzy logic. Moreover, the parameters of the controllers can be chosen by employing different stochastic optimization and metaheuristics techniques to assure optimal performance. A number of controller designing schemes can be hybridized to control practical systems consist of nonlinearities, time varying uncertainties, disturbances, cross-couplings, unmodelled dynamics, delays, etc., simultaneously.

This book deals with the designing of hybrid control strategies to control practical systems containing time varying uncertainties, disturbances, nonlinearities, unknown parameters, unmodelled dynamics, delays, etc., concurrently. In this book, the advantages of different controllers are brought together to produce superior control performance for the practical systems. Being aware of the advantages of adaptive controller to tackle unknown constant, time varying uncertainties and time varying disturbances, a newly invented adaptive controller, namely L_1 adaptive controller is hybridized with other strategies. The parameters of L_1 adaptive controller should be chosen sensibly to maintain proper balance between good transient performance and high robustness.

In this book, to facilitate optimal parameter setting of the basic L_1 adaptive controller, stochastic optimization and metaheuristics techniques are hybridized with it. A variant of Harmony Search (HS) algorithm, viz., Local Best Harmony Search (lbestHS) algorithm, which is a metaheuristic technique, is employed to tune the parameter values of L_1 adaptive controller. This method is termed as lbest HS based

L_1 (lbest HS-L_1) adaptive controller. At first, the parameter values of L_1 adaptive controller are tuned utilizing lbest HS algorithm within the ranges obtained from mathematical calculation of L_1 norm condition. Then, the unknown constant, time varying uncertainties and time varying disturbances are adapted concurrently following adaptation laws to obtain fine-tuned values. The stability of the metaheuristic technique along with the controller is guaranteed analytically with the help of spectral radius convergence. This method exhibits satisfactory exploration and exploitation capabilities. Its results are compared with stochastic optimization, viz., PSO-based L_1 adaptive controller.

Again, this book throws light on tackling nonlinearities along with uncertainties and disturbances by hybridized fuzzy logic with L_1 adaptive controller. In this case, the nonlinear system is designed by combining finite number of linear fuzzy systems. Fuzzy logic-based L_1 adaptive controller is implemented for each linearized zone with the same premise part utilized to design the system. The conventional state feedback controller of the basic L_1 adaptive controller is substituted with fuzzy Parallel Distributed Compensation (PDC) controller which is a nonlinear state feedback controller. The adaptive fuzzy logic systems are designed from the fuzzy Lyapunov function which is the combination of zone-wise Lyapunov functions. The overall stability of the nonlinear system with this controller is guaranteed with the help of fuzzy Lyapunov function to retain the zonal behaviour of the system. The fuzzy PDC-L_1 adaptive controller is efficient to tackle nonlinearities and at the same time cancels unknown constant, time varying uncertainties and disturbances adequately.

The performances of these two controllers are compared with different control methodologies to validate their effectiveness. At first, two classical controllers, viz., Proportional Integral Derivative (PID) controller, Linear Quadratic Gaussian (LQG) are examined. Then, conventional adaptive controller, viz., Model Reference Adaptive Controller (MRAC) and basic L_1 adaptive controller are tested. After those, a stochastic optimization-based L_1 adaptive controller, viz., Particle Swarm Optimization (PSO) based L_1 (PSO-L_1) adaptive, metaheuristic HS-based L_1 (HS-L_1) adaptive and metaheuristiclbest HS-based L_1 (lbest HS-L_1) adaptive controller are employed. Finally, fuzzy PDC-L_1 adaptive controller is evaluated. In simulation environment, two Single-Input Single-Output (SISO) systems, viz., Duffing's oscillatory system and nonlinear spring mass damper system are examined. Then, one Single-Input Multi-Output (SIMO) system, viz., 4th order inverted pendulum with cart system and one Multi-Input Multi-Output (MIMO) system, viz., Twin Rotor MIMO System (TRMS) are investigated. In experimental case studies, speed control of an electrical actuator, angular position control of a Two Link Robot Manipulator (TLRM) and temperature control of a delay dependent air heater system are performed. The results show that, the disturbance rejection phenomenon of basic L_1 adaptive controller is better than the classical controllers and Model Reference Adaptive Controller (MRAC). Though, the tracking performance of basic L_1 adaptive controller is not satisfactory. In case of the lbest HS-L_1 adaptive controller, transient as well as tracking performances improve due to the optimal parameter setting of L_1 adaptive controller. The fuzzy PDC-L_1 adaptive controller provides better steady state tracking performance by tackling nonlinearities through fuzzy logic-based PDC controller as well as fast transient performance by properly eradicating uncertainties and disturbances by means of fuzzy logic-based L_1 adaptive controller.

Therefore, the salient features of the methods presented in this book can be summarized as follows.

I Designing of local best harmony search-based L_1 (lbest HS-L_1) adaptive controller.

(a) A newly developed adaptive controller, viz., L_1 adaptive controller is hybridized with a metaheuristic local neighbourhood variant of HS algorithm, viz., local best harmony search (lbest HS) algorithm to obtain optimal balance between fast transient performance and high robustness by eliminating uncertainties and disturbances.
(b) Satisfactory exploration phenomenon of the lbest HS-L_1 adaptive controller is proved analytically by means of increasing population variance with iterations.
(c) Stability of the lbest HS-L_1 adaptive controller is guaranteed through mathematical analysis of exploitation phenomenon in terms of spectral radius convergence of iterative matrix.
(d) Satisfactory transient and steady state performance of this method are guaranteed.

II Designing of fuzzy parallel distributed compensation type L_1 (fuzzy PDC-L_1) adaptive controller.

(a) Fuzzy PDC strategy is augmented with L_1 adaptive controller to tackle nonlinearities, unmodelled dynamics, delays, as well as uncertainties and disturbances, present in the system.
(b) Fuzzy adaptive rules are formulated to design different components of fuzzy PDC-L_1 adaptive controller, viz., predictor, L_1 adaptation laws, L_1 control law, PDC control law and low-pass filter.
(c) The overall stability of the fuzzy PDC-L_1 adaptive controller is assured with the help of fuzzy Lyapunov function.
(d) The transient state and steady state stable performance of the fuzzy PDC-L_1 adaptive controller are guaranteed analytically.

III Evaluation of the control strategies on four simulation case studies and three experimental case studies.

(a) Different types of controllers, viz., PID, LQG, MRAC, basic L_1 adaptive, PSO-L_1 adaptive, HS-L_1 adaptive are compared with lbest HS-L_1 adaptive and fuzzy PDC-L_1 adaptive controller in simulation case studies. In simulation, at first, two SISO systems, viz., chaotic Duffing's oscillatory system and nonlinear spring mass damper system are examined. Then, a SIMO, non-minimum phase, unstable 4th order inverted pendulum with cart system is investigated. After that, a nonlinear, cross-coupled twin rotor MIMO system is considered.

(b) The control strategies are employed efficiently on three experimental case studies, viz., speed control of electrical actuator, angular position control of two link robot manipulator and temperature control of delay dependent air heater system. The performances of the lbest HS-L_1 adaptive and fuzzy PDC-L_1 adaptive control methodologies are compared with PID, LQG, MRAC, basic L_1 adaptive, PSO-L_1 adaptive and HS-L_1 adaptive controller.

(c) The controllers are at first tuned for systems with no disturbance or small disturbance and then subjected to the systems with large disturbance without further tuning to examine the robustness of the controllers. The results demonstrate that, the lbest HS-L_1 adaptive controller provides better tracking and disturbance rejection phenomenon than the classical controllers, MRAC, basic L_1 adaptive controller as well as other optimization- based L_1 adaptive controllers. The fuzzy PDC-L_1 adaptive controller provides fast transient performance, better steady state tracking performance and high robustness than all other control strategies tested.

This book is composed of total eight number of chapters. This book is organized as follows.

Part-I: Prologue

Prologue contains introduction of the book in Chap. 1. In this chapter, the journey towards modern control theory is portrayed. The state of the art of the modern control theories are elaborated next with three sub-sections describing stochastic optimization and metaheuristics techniques; fuzzy logic systems; and L_1 adaptive controller. Then, the literatures of hybrid L_1 adaptive controller are articulated. The literatures of stability analysis of the controllers, optimization techniques are provided in the next section. The motivations of this book from the drawbacks of existing literatures are discussed next. Then, the proposals of the book and main contributions are discussed. At last, the structure of the book is provided followed by the summary of the chapter.

Part-II: Preliminaries

In preliminary part, Chap. 2 explains the motivations of designing basic L_1 adaptive controller. The architecture and formulation of basic L_1 adaptive controller are elaborated with its stability analysis. The controller is implemented and then the designed controller is subjected to unknown disturbances. Four simulation case studies are performed employing basic L_1 adaptive controller to show its effectiveness in elimination of unknown time varying uncertainties and time varying disturbances, compared to the classical controllers and MRAC.

Part-III: Design of Hybrid L_1 Adaptive Controller

This part comprises of two chapters containing the theoretical formulations of the designed methodologies, viz., stochastic optimization and metaheuristics-based L_1 adaptive controller and fuzzy PDC-L_1 adaptive controller. The effectiveness of these methodologies are demonstrated through simulation case studies in these chapters.

Stochastic optimization and metaheuristics-based design of L_1 adaptive controller is provided in Chap. 3. In this chapter, the lbest HS algorithm is utilized to tune the parameters of L_1 adaptive controller. The values of unknown constant, time varying uncertainties and time varying disturbances are first obtained from the lbest HS optimization technique. Then, their values are concurrently adapted following L_1 adaptation laws. The superior exploration capability of the lbest HS-L_1 adaptive controller is demonstrated analytically. The stability of the lbest HS-L_1 adaptive controller is also guaranteed in terms of exploitation through spectral radius convergence. This method is employed on four simulation case studies and compared with basic L_1 adaptive controller to show its fruitfulness.

Chapter 4 describes the designing of fuzzy PDC-L_1 adaptive controller. L_1 adaptive controller can handle system with uncertainties and disturbances very well. On the other hand, fuzzy PDC logic can handle nonlinear system well enough. Augmenting the advantages of L_1 adaptive controller to handle uncertainties, disturbances and fuzzy PDC logic to tackle nonlinearities present in the system, fuzzy PDC-L_1 adaptive controller is designed. The nonlinear system is considered as the fuzzy blending of finite number of linear fuzzy systems. For each linearized zone, L_1 adaptive controller and fuzzy PDC controller are developed with the same premise part, used to design the system. Utilization of same premise part keeps the zonal behaviour of the system in controller design which provides better controlling for nonlinear system. The stability of this method is guaranteed by means of fuzzy Lyapunov function. This method is employed on four simulation case studies and compared with basic L_1 adaptive controller and lbest HS-L_1 adaptive controller to demonstrate its superiority.

Part-IV: Applications

This part contains the validation of the lbest HS-L_1 adaptive and fuzzy PDC-L_1 adaptive control methods on three experimental case studies, viz., speed control of electrical actuator, angular position control of two link robot manipulator and temperature control of delay dependent air heater system.

In Chap. 5, the lbest HS-L_1 adaptive and fuzzy PDC-L_1 adaptive control methods are employed to control the speed of an electrical actuator, viz., DC motor experimental setup. The controllers are designed as a way that the DC motor can track a variable step trajectory and then the tuned controllers are subjected to the 100% load disturbance without further tuning. The experimental results show that, the lbest HS-L_1 adaptive controller performs better than the basic L_1 adaptive controller. The fuzzy PDC-L_1 adaptive controller performs better than both basic L_1 adaptive controller and lbest HS-L_1 adaptive controller.

Angular position control of a Two Link Robot Manipulator (TLRM) experimental setup is examined in Chap. 6. This laboratory setup of TLRM is the emulated version of large scale industrial TLRM system. The system modelling and controller designing for this laboratory scale setup is performed as a way that it can be replicated in case of large scale TLRM system. The parameter values of industrial large scale TLRM may be unknown or unmeasurable which leads to unknown dynamics of it. Therefore, the nonlinear TLRM system is modelled utilizing fuzzy PDC logic. Then, different controllers are developed for this TLRM system. Experimental results show that, the lbest HS-L_1 adaptive controller performs better than the basic L_1 adaptive controller and the fuzzy PDC-L_1 adaptive controller outperforms the basic L_1 adaptive controller as well as lbest HS-L_1 adaptive controller.

Chapter 7 portrays the control of an air heater system which contains time varying uncertainties, time varying disturbances, nonlinearities and large delay. lbest HS-L_1 adaptive and fuzzy PDC-L_1 adaptive control methodologies show better performances than other control methodologies tested in this book. In this chapter, the obtained results show that, these two designed controllers are very much efficient in tackling system with nonlinearities, delays, time varying uncertainties, time varying disturbances simultaneously.

Part-IV: Epilogue

In this part, concluding remarks, significant features of the book are summarized with the future research directions.

Chapter 8 draws the conclusion with brief description of obtained results. The effectiveness of the lbest HS-L_1 adaptive and fuzzy PDC-L_1 adaptive control methods and key features of the book are summarized. The future prospective of the current works are also discussed in this chapter.

Appendices and index terms are provided thereafter.

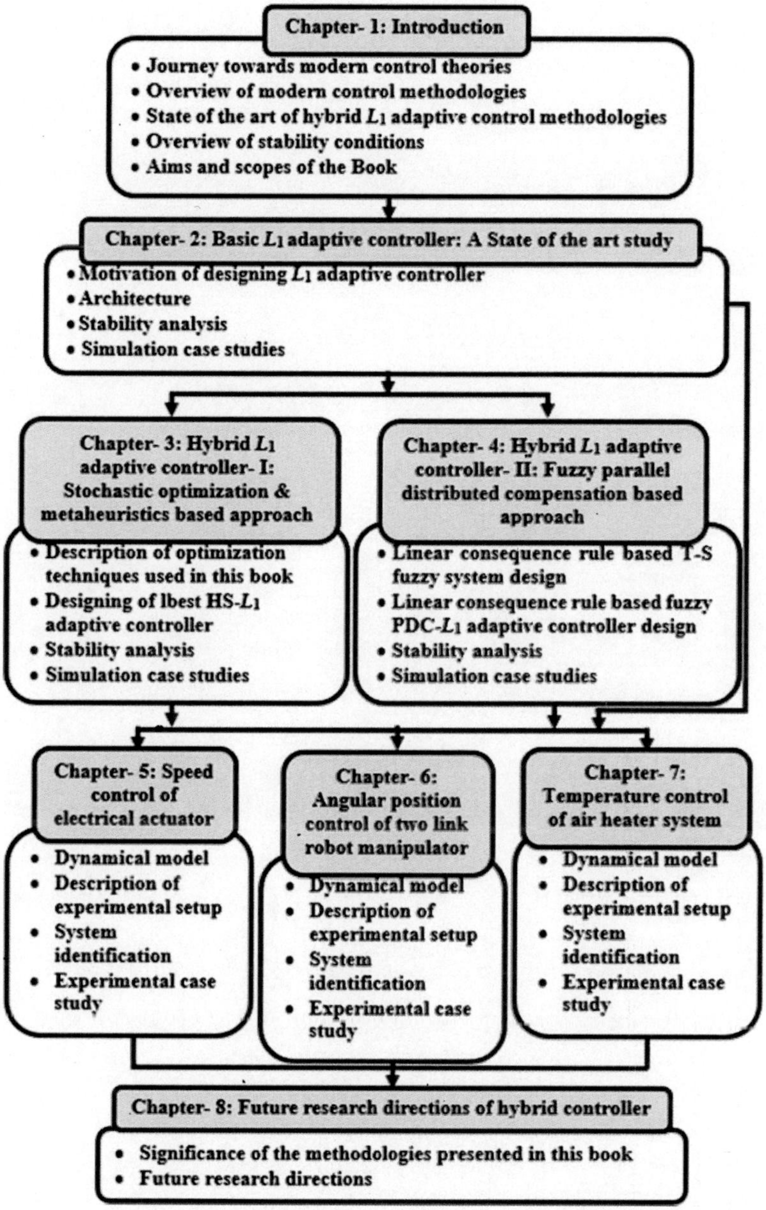

Kolkata, India

Roshni Maiti
Kaushik Das Sharma
Gautam Sarkar

Acknowledgements

Heartfelt gratitude of the authors go towards **Prof. Anjan Rakshit**, Retired Professor, Department of Electrical Engineering, Jadavpur University for fabricating air heater experimental setup which is utilized in this work. Author would like to express her gratefulness to **Prof. Jitendra Nath Bera**, Electrical Engineering Section, Department of Applied Physics, University of Calcutta, for building up the driver circuit of the DC motor experimental setup which is used in this work.

The authors would like to extend their special thanks to all the Professors and research scholars of the Department of Applied Physics for their moral support.

Last but not the least, authors would like to express heart-felt gratitude towards their family members for their selfless help, unflinching support and incessant encouragement during this work.

Kolkata, India
September 2021

Roshni Maiti
Kaushik Das Sharma
Gautam Sarkar

Contents

About the Authors

Roshni Maiti has completed her Ph.D. (Tech.) in Electrical Engineering from Department of Applied Physics, University of Calcutta, India in 2021 with University Research Fellowship. She received her B.Tech. and M.Tech. degrees in Electrical Engineering from the West Bengal University of Technology and University of Calcutta, India in 2012 and 2014 respectively.

Her research interests include adaptive controller design, controller design for non-linear system, fuzzy controller design, stochastic optimization techniques (GA, PSO), metaheuristics techniques (HS), Control applications in robotics, etc. She has authored 5 SCI journals including 2 IEEE Transactions journal and many conferences. She actively serves as reviewer of different journals like IEEE Transactions on Circuits and Systems I: Regular Papers, IEEE Transactions on Neural Networks and Learning Systems, ISA Transactions, Measurement and Control (SAGE journals) etc. She is a member of IEEE (USA).

Kaushik Das Sharma received his B.Tech and M.Tech degrees in Electrical Engineering from the University of Calcutta, India, in 2001 and 2004 respectively and Ph.D. (Engineering) degree from the Jadavpur University, India in 2012. Presently he is Professor in Electrical Engineering Section, Department of Applied Physics, University of Calcutta, India.

His research interests include fuzzy control system design, stochastic optimization applications, robotics etc. He has published more than 60 research articles in international and national journals or conferences. He has authored a book titled 'Intelligent Control: A Stochastic Optimization Based Adaptive Fuzzy Approach', Springer Nature, Singapore publisher, 2018. He is a senior member of IEEE (USA), and life member of The Indian Science Congress Association (Engineering Section).

Gautam Sarkar received his B.Tech, M.Tech and Ph.D. degrees from the University of Calcutta, India, in 1975, 1977 and 1991, respectively. He has retired as LD Chair Professor in Electrical Engineering Section, Department of Applied Physics, University of Calcutta, India.

His research interests include control system design, smart grid technologies etc. He has published more than 80 research articles in international and national journals or conferences.

Symbols

A_0	Open loop system matrix
A_m	Closed loop system matrix
b	Input matrix
c	Output matrix
$\underline{x}(t) = [x_1 \quad x_2 \quad \ldots x_n]^T \in \Re^n$	System state
$\underline{u}_1(t)$	L_1 adaptive control signal
$\underline{u}_2(t)$	State feedback control signal
$\underline{r}$	Input to the system
$\underline{y}$	Output of the system
$\underline{e} = \underline{r} - \underline{y}$	Tracking error
ω	Unknown constant
$\underline{\theta}$	Uncertainties
$\underline{\sigma}$	Disturbances
k	Pre-filter gain
k_g	Feed-forward gain
K	State feedback gain
Γ_C	Adaptation gain
$\hat{x}$	Predictor state
$\hat{y}$	Predictor output
$C(s)$	Low-pass filter
$D(s)$	Strictly proper stable transfer function
$\underline{Z}$	Candidate solution vector
$\underline{\xi}$	Fuzzy basis function
t	Time
h	Time step
V	Lyapunov function
Ψ	Barbalat's lemma operator

Acronyms

bw	Bandwidth
GMCR	Group memory considering rate
HM	Harmony memory
HMCR	Harmony memory considering rate
HS	Harmony search
IAE	Integral absolute error
lbest HS	Local best harmony search
MIMO	Multi-input multi-output
PAR	Pitch adjustment rate
PDC	Parallel distributed compensation
PI	Performance index
PSO	Particle swarm optimization
SIMO	Single-input multi-output
SISO	Single-input single-output
TLRM	Two link robot manipulator
TRMS	Twin rotor MIMO system
T-S	Takagi-Sugeno

List of Figures

List of Tables

Part I
Prologue

Chapter 1
Introduction

Abstract This chapter introduces the book with brief description of the journey towards modern control theories. Motivations of the book are portrayed owing to the drawbacks of present literatures. The designed methodologies are described briefly with an emphasis on contributions of the book. Then, the structure of the book is provided.

1.1 Journey Towards Modern Control Theories

Classical controllers, viz., proportional integral derivative (PID) controller [1–3], state feedback controller [4], observer based state feedback controller [5], output feedback controller [6, 7], linear quadratic regulator (LQR) [8], linear quadratic Gaussian (LQG) [8, 9], etc., are suitable for linear time invariant (LTI) systems. Though, they yield unsatisfactory control performances for systems with time varying quantities [10]. In various literatures, different modifications were furnished over classical controllers to overcome their drawbacks. LQG controller was augmented with vibration compensator to suppress the vibration of a piezoelectric tube actuator [11]. LQG was also combined with proportional integral (PI) controller to nullify steady state error [12]. Grid voltage disturbance was tackled with the help of frequency-adaptive multi-resonator augmented LQG current controller [13]. Two separately designed PID controllers were brought together with resonant controller to suppress the current harmonic components [14]. Data driven controller was augmented with d-step ahead prediction optimal controller and PID controller to control a pulp neutralization process by compensating unmodelled dynamics [15]. To tune the gains of the PID controller, different improved strategies were adopted rather than the Ziegler-Nichols method [10] in different literatures. Variable gains of PID controller were designed for a linear model of unmanned aerial vehicle (UAV) [16]. Relay feedback tuning method was also utilized to design the gains of the PID controller [17]. However, the performances of the classical controllers were unsatisfactory to tackle systems with uncertainties, disturbances, time varying parameters, nonlinearities and delays [18]. To overcome such drawbacks, modern control theories draw attention.

R. Maiti et al., *Hybrid L_1 Adaptive Control*, Studies in Systems, Decision and Control 422, https://doi.org/10.1007/978-3-030-97102-1_1

1.2 Overview of Modern Control Methodologies

Proper estimations of time varying uncertainties and disturbances are required to eliminate these quantities, which can be obtained through adaptive controllers [19, 20]. To eliminate the drawbacks of conventional adaptive controller, a newly invented adaptive controller, viz., L_1 adaptive controller comes into place and increases its applications in different fields. The parameters of different controllers are tuned employing stochastic optimization and metaheuristics techniques to obtain optimized performance of the system under control. In different literatures, fuzzy logic systems are utilized to approximate nonlinearities [21], to produce controller by approximating uncertainties [22], disturbances [23], unmodelled dynamics [24], etc. Literature surveys of these modern control methodologies are provided in the following sections.

1.2.1 Stochastic Optimization and Metaheuristics Techniques

Arbitrarily chosen parameter values of controllers, within the ranges obtained from the stability condition, do not guarantee optimal performance. Optimization techniques acquire great attention to choose the optimal parameter values of controller during last two decades. Different optimization techniques, viz., genetic algorithm (GA) [25, 26], particle swarm optimization (PSO) [27–29], ant colony optimization (ACO) [30, 31], firefly (FF) algorithm [32–34], cuckoo search (CS) [35, 36], artificial bee colony (ABC) [37, 38], harmony search algorithm (HS) [39, 40] were used in different literatures to obtain optimized performance of the controllers. It is observed from the literatures that, the computational cost reduces considerably in case of harmony search algorithm compared to other optimization techniques [39]. Involvement of lesser number of internal parameters and their easy selections increase the use of HS algorithm in different optimization problems, viz., pipe network designing [41], vehicle routing [42], structural designing [43], etc. Different parameter modifications of the HS algorithm provides the easy enhancement of exploration and exploitation capability of it. The advantages of HS algorithm over other optimization process increase its utilization gradually as elaborated follows.

1.2.1.1 Harmony Search Algorithm

HS resembles with the group of musicians playing pitches with the aim to develop an aesthetic harmony all together [40]. Being beneficial than other optimization techniques, the utilization of HS algorithm increases gradually. The HS algorithm was successfully utilized in low impact development and management of urban storm water systems to avoid flooding risk and assure environmental interventions [44]. The HS algorithm performed better than random sample consensus algorithm to detect vanishing point present at the extended lines of road lanes for self-driving autonomous vehicle [45]. In power system engineering, HS was employed in different literatures [46–50]. A distributed-generation (DG) system was controlled through proportional integral (PI) controller whose gains were tuned by means of HS algorithm [51]. The parameters of fuzzy controller were tuned by utilizing HS for air heater system with transport-delay [52]. HS was also applied in medical fields to RNA secondary structure prediction [53, 54], medical physics [55], hearing AIDS [56] etc.

It is of great importance to balance between exploration and exploitation capability of HS algorithm [57]. Good exploration phenomenon indicates finding more search spaces which avoids local minima trapping problem. Good exploitation signifies proper convergence of the optimization process [58]. Different modifications of basic HS algorithm were manifested by designing its parameter values in different ways to obtain optimal balance between diversification and intensification of the optimization technique [59]. Dynamic pitch adjustment rate (PAR) [60], dynamic bandwidth (bw) selection formulas [60, 61] were developed in different literatures. Gaussian distribution was utilized to formulate the PAR with the aim to improvise exploration at first and then exploitation at the end of the harmony improvisation process of HS [62]. The HS algorithm was hybridized with hill climbing and global best particle swarm optimization to improve local exploitation and global convergence respectively which was employed to design university course timetable [63]. Basic HS algorithm was also combined with wavelet mutation and new harmony was produced based on roulette wheel mechanism to improve convergence speed, accuracy and robustness for economic load dispatch (ELD) problem [64]. Differential mutation in pitch adjustment rate was incorporated to solve job shop scheduling problem [65]. Position update formula of PSO and mutation phenomenon of GA replaced the harmony memory considering rate (HMCR) and pitch adjustment rate (PAR) parameters of HS respectively in [66]. A detail review of harmony search algorithm, its developments and applications can be found in [67] and the references therein. To improvise the exploitation capability of basic HS algorithm, local search based phenomenon was incorporated in it [68–70].

Literatures show that, HS algorithm and its different modifications can be used efficiently to solve many problems. It can be used to tune the parameter values of the controller to provide optimize performance of the system. However, these modified HS may not find the proper system matrix values to model unmodelled dynamics, nonlinearities, delays, etc. This phenomenon can be addressed properly employing fuzzy logic. Literature reviews of fuzzy logic systems to model nonlinear systems, unmodelled dynamics, delays, etc., are elaborated in the next section.

1.2.2 Fuzzy Logic Systems

To tackle nonlinearities, delays, unmodelled dynamics with the uncertainties and disturbances, fuzzy logic based approaches gain much attention [71]. Uncertainties and disturbances can be nullified through adaptive controllers but they fail to provide satisfactory performance in case of nonlinearities, delays, unmodelled dynamics, etc., present in the system simultaneously. Universal approximator fuzzy logic system [72] was used extensively to tackle nonlinearities [73–77] during last few decades. Fuzzy logic system can be utilized to approximate system models with nonlinearities, unmodelled dynamics, uncertainties [78, 79] as well as to design different types of controllers [80–84]. Approximation of nonlinear system and developing controller for that system should be synchronized. With this aim, fuzzy parallel distributed compensation (PDC) type controller was introduced [85, 86] and utilized in many applications [87–90]. The nonlinear system can be approximated by fuzzy parallel distributed compensation logic [91–94]. The nonlinear system is considered as the fuzzy blending of finite number of linear systems. Then, for each linearized zone, linear state feedback controller is designed and fuzzy augmented to produce overall control action [95–97]. This is known as PDC control law which is nonlinear in nature and can handle nonlinear systems efficiently [98, 99]. At first, fuzzy PDC logic is employed to design the system and then same premise part is utilized to generate PDC control law. Fuzzy PDC control law was applied in different cases, viz., attitude control of spacecraft [99], speed control of surface-mounted permanent magnet synchronous motor [100], wind energy system control [101], maximum power point tracking control of solar photovoltaic electricity generation system [102], nonlinear vehicle dynamics control [103], etc. Yet, the performance of PDC controller degrades when uncertainties and disturbances appear in the system [101]. Fuzzy controller can be efficiently hybridized with other controllers to tackle different quantities consequently.

To tackle uncertainties and disturbances, a special type of adaptive controller, viz., L_1 adaptive controller gains much attention during few years. The literature reviews of the L_1 adaptive controller are provided in the following section.

1.2.3 L_1 Adaptive Control Methodologies

Systems with time varying uncertainties, disturbances can be controlled through adaptive controllers [104–106]. Conventional adaptive controller utilizes low adaptation gain to retain the robustness of the system [107]. However, utilization of low adaptation gain possesses sluggish transient performance [108]. To overcome this drawback in 2006, L_1 adaptive controller was proposed with high adaptation gain to incur quick transient performance and a low-pass filter is attached in the control channel to retain robust performance [109–111]. L_1 adaptive controller comprises of predictor, adaptation laws, state feedback control action, L_1 control law and low-pass filter. Predictor predicts the unknown constant, time varying uncertainties and time varying disturbances present in the system and their values are adapted following the adaptation laws. From these adaptive estimates, L_1 adaptive control law is formulated which contains high frequency component. These high frequencies, may yield instability of the system, are nullified by a low-pass filter attached after the control block. The filtered output is augmented with state feedback control signal and provided to the system under control. Designing of these components and selection of parameters from L_1 norm condition were elaborated in the seminal papers of L_1 adaptive controller, published in 2006 by Chengyu Cao and Naira Hovakimyan [107, 108]. Despite of utilizing high adaptation gain, the guaranteed transient performance was analyzed in [109–111]. The exponential decay of non-zero trajectory initialization error, transient and steady state system performances with respect to the reference system were also analyzed by Cao and Hovakimyan [112]. Stability analysis of L_1 adaptive controller was performed in digital mode [113], linear matrix inequality based approach [114, 115], etc. L_1 adaptive controller can efficiently handles systems with unknown time varying parameters [112], unmatched uncertainties [116, 117], unmatched disturbances [118], unmodelled dynamics [119], etc. Nonlinear systems with uncertainties and disturbances were also controlled through L_1 adaptive controller in different literatures [120–123]. Though, consideration of uncertainties, disturbances, nonlinearities, unmodelled dynamics, delays, etc., simultaneously is quite difficult and missing in the literatures, as far as author's belief and knowledge goes.

Advantages of L_1 adaptive controller made it so prevalent that, its applications grow very fast and spread to different fields. L_1 adaptive controller was successfully employed to control different types of flight systems, viz., aircraft [124, 125], wing-rock [126], X-wing tail-sitter micro aerial vehicle (MAV) [127], unmanned aerial vehicle (UAV) [128, 129], fighter aircraft [130], autopilot [131, 132]. Pitch break uncertainty as well as actuator failure of an unmanned military tailless unstable aircraft were controlled through L_1 adaptive controller and compared with model reference adaptive controller (MRAC) to show the effectiveness of L_1 adaptive controller [133]. L_1 adaptive controller was used to control magnetic torque coil in attitude control system of a Pico-scale satellite test bed [134]. Depth and pitch angle of a multi-input multi-output (MIMO) submarine system was controlled through this controller [135]. Wind turbine was controlled by regulating the speed of a generator for maximum power point tracking through the L_1 adaptive controller [136]. Precise controlling phenomenon of L_1 adaptive controller extended its application in medical field for delivering anaesthesia to the patient during surgery [137].

To enhance the performance of L_1 adaptive controller, it is augmented with other strategies which are elaborated in the subsequent section.

1.3 State of the Art of Hybrid L_1 Adaptive Control Methodologies

Practical systems contain nonlinearities, time varying structured and unstructured uncertainties, disturbances, unknown constants, cross-couplings, delays, etc. To tackle all of these quantities simultaneously, different controllers are hybridized together. L_1 adaptive controller provides fast transient performance as well as retains high robustness to tackle uncertainties and disturbances. Arbitrary parameter selection, within the ranges of L_1 norm condition, does not guarantee optimal performance of it, although ensures the stability of the closed loop system. Hence, proper balance between quick transient performance and high robustness should be maintained by designing the components and selecting the parameter values of L_1 adaptive controller appropriately. Different strategies were applied to design its different components. A greedy randomized optimization with multi-criteria was utilized to design the filter of L_1 adaptive controller [138]. Fuzzy logic was employed to design the pre-filter gain of L_1 adaptive controller where the filter structure was kept fixed [139]. Adaptive estimates of uncertainties, in designing L_1 adaptive controller, were also approximated utilizing fuzzy logic system for electropneumatic actuator [140] and wind energy conversion systems [141]. Being beneficial in controlling systems with time varying uncertainties and disturbances which present in practical system, L_1 adaptive controller was combined with other control strategies in many cases. Baseline dynamic inversion controller was augmented with L_1 adaptive controller to eliminate time-varying uncertainty of a combination of Novlit-3 micro aerial vehicle and a novel non-orthogonal X shaped wing layout with a single propeller system [142]. Dynamic inversion based L_1 adaptive controller was implemented for missile [143], generic transport model [144], etc. L_1 adaptive controller was augmented with differential proportional-integral (PI) baseline controller to tackle longitudinal dynamics of a F16 aircraft [145]. L_1 adaptive back-stepping controller was designed for unmanned aerial vehicles (UAVs) with position kinematics and dynamics present in strict feedback form [146]. PID controller was appended with L_1 adaptive controller to obtain proper tracking by eliminating the time lag for depth control of an under-actuated underwater vehicle [147].

Though in different literatures, different components of the L_1 adaptive controller were designed through optimization technique or utilizing fuzzy logic but the entire controller structure design was absent there, as far as author's knowledge goes. In case of designing L_1 adaptive controller for systems consisting of uncertainties, disturbances and nonlinearities, separate consideration of nonlinearities were also absent in the literatures.

After modelling the system and developing controller for that system, it is important to analyze the stability conditions of it. There are different ways to investigate the stability conditions of stochastic optimization and metaheuristics techniques, fuzzy logic controllers, etc., whose literature surveys are provided in the subsequent section.

1.4 Overview of Stability Conditions

It is essential to investigate the exploration (/intensification) and exploitation (/diversification) capabilities in case of any optimization process. Good balance between exploration and exploitation shows that, the optimization process searches huge spaces with the aim to converge at the end of the process without diverging from the stability region. Owing to the easy implementation and lower computational time of HS, in this book, the literatures of stability analysis of HS algorithm are provided. It is important to analyze mathematically the exploration and exploitation capabilities of the HS algorithm which was performed in [148]. The authors of [148] showed that, the population variance of the HS algorithm increased gradually with the iteration which exhibited its better exploration capability. They also proved the spectral radius convergence of the iterative matrix of the HS algorithm. This demonstrated the better exploitation capability of the HS algorithm [148]. Stability analysis of different modified HS algorithm was also provided in different literatures [149]. Mathematical analysis of the exploration and exploitation capabilities of the local search based HS (lbest HS) algorithm was absent in the literatures to the best of author's understanding. The stochastic optimization and metaheuristics techniques were utilized to provide the optimal parameter setting of the controllers. Therefore, it is important to analyze the stability condition of the optimization techniques along with the controller which was also not present in the literatures as per as author's knowledge goes.

In another case, for nonlinear system, a set of stability conditions have to be derived which are difficult from a single Lyapunov function. In case of fuzzy PDC logic, multiple number of fuzzy rules are formulated and combined to model the system and controller [88, 150]. In case of such large number of fuzzy rules, use of quadratic Lyapunov function [151, 152] introduces conservative conditions [153, 154]. To reduce conservatism, different approaches were made to formulate the Lyapunov functions in different literatures, viz., piecewise Lyapunov function [153, 155–158], polynomial Lyapunov function [159, 160], multiple Lyapunov function [161–165]. Based on this multiple Lyapunov function approach, fuzzy Lyapunov function was introduced [166]. Fuzzy Lyapunov function is smooth in nature which is advantageous over piecewise Lyapunov function. This fuzzy Lyapunov function is developed by augmenting finite number of zone-wise Lyapunov functions where in each zone, positive definite matrix is present [166, 167]. The same premise part utilized to design the nonlinear system and controller is employed to formulate the fuzzy Lyapunov function [168–171].

From these literature surveys of the existing controllers, optimization techniques, fuzzy logic systems, etc., and their stability analysis, the motivations of designing the control methodologies presented in this book arise which are as follows.

1.5 Aims and Scopes of the Book

1.5.1 Aims

System with uncertainties and disturbances can be tackled through adaptive controllers. Though, conventional adaptive controllers provide sluggish transient performance to retain robustness. To overcome such problem, in 2006, L_1 adaptive controller originates with high adaptation gain to achieve quick transient performance and a low-pass filter to retain the robust performance [109]. Literature shows that, the components of L_1 adaptive controller should be chosen sensibly within the L_1 norm bound to obtain fast transient performance with high robustness [111]. Different components of L_1 adaptive controller were designed utilizing different optimization techniques in different literatures [138, 139]. Nevertheless, designing the overall controller optimally is of great importance to achieve the accurate parameter setting and optimal control performance of L_1 adaptive controller. Again the literatures show that, HS and its different modifications can be efficiently utilized to choose the parameters of any controller. This motivates to design local best HS (lbest HS) based L_1 adaptive controller and to analyze overall stability condition [172].

On the other hand, though nonlinear systems with uncertainties and disturbances are controlled through L_1 adaptive controller but tackling nonlinearities separately was absent in the literatures [120–123]. Decentralized L_1 adaptive controller was designed for large scale nonlinear system with unknown interconnection and unmodelled dynamics with the help of decentralized passive identifiers [120]. Non-affine nonlinear systems, consist of unmeasured states, were controlled by implementing piece-wise continuous adaptive laws of L_1 adaptive controller [121]. Nonlinear time varying reference system was also considered to control nonlinear system where an upper bound had to be considered additionally with complicated calculation which increased the conservatism [173, 174]. To deal with nonlinearities, in particular, fuzzy parallel distributed compensation (PDC) logic gained much attention in last two decades [175]. Motivated with these ideas, the L_1 adaptive controller can be hybridized with fuzzy PDC controller to handle system with nonlinearities, unmodelled dynamics, delays, uncertainties and disturbances all in conjunction. Stability of the overall system can be analyzed utilizing fuzzy Lyapunov function due to its less conservatism.

1.5.2 Scopes

Motivated from the aforementioned concerns, in this book, the advantages of different methodologies are hybridized to tackle different quantities simultaneously with the aim to produce superior control performances for the practical systems. Owing to the fast transient performance as well as high robustness, in this book, a special type of adaptive controller, viz., L_1 adaptive controller is utilized. All the components of L_1 adaptive controller are designed by tuning the parameters utilizing stochastic optimization and metaheuristics techniques to obtain overall optimal performance. Considering the advantages of HS algorithm and to enhance the exploration and exploitation capabilities, a variant of HS algorithm, viz., metaheuristic lbest HS is utilized in this book. The overall stability analysis of the concurrent lbest HS based L_1 (lbest HS-L_1) adaptive controller is also performed. On the other hand, in many literatures L_1 adaptive controller was employed to tackle nonlinear system with uncertainties and disturbances where nonlinearity was not considered in a separate way. To tackle uncertainties, disturbances as well as nonlinearities, delays, etc., in this book, fuzzy logic is utilized to design PDC controller augmented L_1 adaptive controller. This method is termed as fuzzy parallel distributed compensation type L_1 (fuzzy PDC-L_1) adaptive controller. The stability of the system with controller is investigated by means of fuzzy Lyapunov function. The proposals and the contributions of this book are elaborated as follows.

The layout of the book proposals are depicted in Fig. 1.1. In this book, to obtain fast transient performance by eliminating uncertainties and disturbances present in the system, L_1 adaptive controller with fast adaptation and high robustness is employed. To obtain optimal performance from L_1 adaptive controller, its parameters are tuned utilizing optimization techniques. Being aware of easy implementation and good balance between exploration and exploitation capability, a novel variant of HS algorithm, viz., local best harmony search (lbest HS) algorithm is utilized. The values of unknown constant, time varying uncertainties and time varying disturbances present in the system are at first obtained utilizing lbest HS algorithm. Then, their values are adapted concurrently following L_1 adaptation laws to fine tune those values. In local search based HS, the total harmony memory is divided into some groups and from each group local best solution is found out. From those local best solutions, global best solution is obtained. In this local best search phenomenon, the exploitation capability of the HS algorithm increases enormously. The exploration and exploitation capabilities of the whole optimization technique augmented controller are investigated. It is proved analytically that, the population variance of the lbest HS-L_1 adaptive controller increases gradually with iteration which shows its better exploration capability [172]. With good exploration capability, high exploitation capability of the optimization process is also required. The exploitation capability of the designed method is assured by means of spectral radius convergence phenomenon [172].

Again to handle nonlinearities present in the system along with uncertainties and disturbances, in this book, fuzzy parallel distributed compensation (PDC) logic

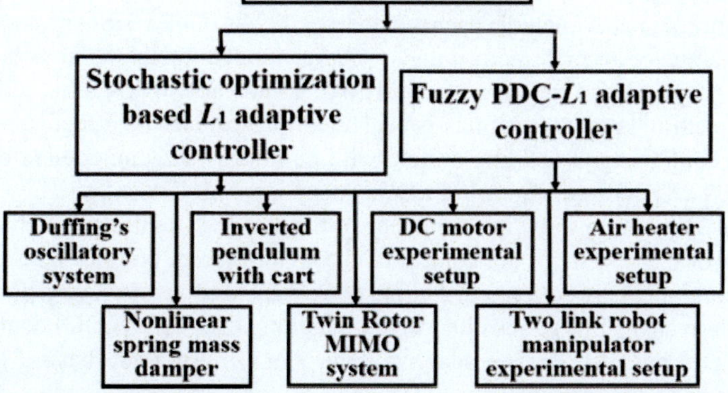

Fig. 1.1 Layout of the book proposal

is hybridized with L_1 adaptive controller [175]. In this fuzzy PDC-L_1 adaptive controller, the nonlinear system is considered as the combination of finite number of linear fuzzy systems. Then, the same fuzzy premise part, utilized to design the system, is employed to design different components of L_1 adaptive controller, viz., predictor, L_1 adaptation law, L_1 control law, PDC control law and low-pass filter. For each linearized zone, fuzzy predictor is developed and combined to produce overall predictor model. Uncertainties, disturbances present in these linearized zones are adapted following zone-wise L_1 adaptation laws and are accumulated to develop whole L_1 adaptation law. From zone-wise adaptation laws, L_1 control signals are developed keeping the zonal nature and then fuzzy blended to produce whole L_1 control law. Due to the use of high adaptation gain, the control law of each zone contains high frequency components. To nullify that high frequency from each of the linearized zone, the produced control signals are filtered through low-pass filter dedicated for each of those zones. The state feedback controller of basic L_1 adaptive controller is designed utilizing fuzzy PDC logic. The zone-wise developed fuzzy PDC controllers are accumulated to produce overall fuzzy PDC control law. Filtered L_1 adaptive control law and fuzzy PDC control signal are augmented and provided to the system under control. All of these components of the L_1 adaptive controller are designed separately from the fuzzy adaptation laws, developed from the stability criteria of fuzzy Lyapunov functions. The overall stability of the closed loop system is also investigated utilizing fuzzy Lyapunov function. The transient and steady state stable performances of the system are guaranteed analytically.

The designed controllers are employed on four simulation case studies and three experimental case studies to show their effectiveness. In simulation case studies, two single-input single-output (SISO) system, viz., Duffing's oscillatory system with disturbance and nonlinear spring mass damper system are considered to show the controller performance when disturbances are present in the system. Then, one single-input multi-output (SIMO) system, viz., unstable non-minimum phase 4th order inverted pendulum with cart is controlled and stabilized. A nonlinear, cross-coupled twin rotor multi-input multi-output (MIMO) system (TRMS) is investigated to show the controller performance in case of nonlinearities and cross-couplings. Furthermore, in experimental case studies, speed control of a SISO system, viz., an electrical actuator is performed owing to its large number of applications in practical systems. Angular position control of a MIMO, nonlinear two link robot manipulator (TLRM) with unknown dynamics is evaluated to examine the effectiveness of the designed controllers in tackling system with nonlinearities, unmodelled dynamics, etc. Finally, temperature control of a delay dependent air heater system with disturbance is performed to show the efficiency of the designed controllers to tackle large delay. All of these systems are controlled through different types of controllers and their performances are compared. At first, two classical controllers, viz., proportional integral derivative (PID) controller, linear quadratic Gaussian (LQG) controller are tested. Then, model reference adaptive controller (MRAC) and basic L_1 adaptive controller are imposed on them. After that, a stochastic optimization, viz., PSO based L_1 (PSO-L_1) adaptive controller; metaheuristic, viz., basic HS based L_1 (HS-L_1) adaptive controller and metaheuristic lbest HS based L_1 (lbest HS-L_1) adaptive controller are examined. Finally, the fuzzy PDC-L_1 adaptive controller is employed. Results show that, disturbance rejection phenomenon of basic L_1 adaptive controller is better than the classical controllers and MRAC. The lbest HS-L_1 adaptive controller provides better transient performance and tracking control than the basic L_1 adaptive controller. The fuzzy PDC-L_1 adaptive controller outperforms the basic L_1 adaptive controller as well as lbest HS-L_1 adaptive controller.

1.6 Summary

This chapter illustrates the introduction of the book. The journey of classical controllers towards modern control methods are portrayed with the current status of modern control methodologies. To deal with nonlinear system containing uncertainties and disturbances, modern controllers, viz., adaptive control, fuzzy control, optimization based control techniques come into place. Then, the motivations of this book are discussed owing to the shortcomings of basic L_1 adaptive controller as well as strengths of local best harmony search optimization technique and fuzzy parallel distributed compensation logic. Next the proposals of the book are elaborated with the architecture of the book proposal. In this book, two hybridized L_1 adaptive controllers are implemented with mathematical analysis and experimental demonstrations. lbest HS algorithm is hybridized with basic L_1 adaptive controller with the aim to obtain optimal parameter setting of L_1 adaptive controller. The lbest HS-L_1 adaptive controller provides satisfactory transient and steady state performance as well as high robustness. To tackle nonlinearities, delays along with uncertainties and disturbances, fuzzy PDC logic is hybridized with L_1 adaptive controller. The fuzzy PDC-L_1 adaptive controller provides fast transient performance by means of fuzzy L_1 adaptive controller as well as robust and satisfactory steady state tracking performance utilizing fuzzy PDC logic. The performances of the designed control methodologies are evaluated on four simulation case studies and three experimental case studies to show the effectiveness. After that, contributions of the book are summarized. Then, the structure and orientation of the book are elaborated followed by this summary of the present chapter.

Next chapter introduces the formulation of basic L_1 adaptive controller and its stability analysis. Four simulation case studies are performed to show the effectiveness of basic L_1 adaptive controller over PID, LQG and MRAC.

References

1. Liu, J., Wang, H., Zhang, Y.: New result on PID controller design of LTI systems via dominant eigenvalue assignment. Automatica **62**, 93–97 (2015)
2. Sujitjorn, S., Wiboonjaroen, W.: State-PID feedback for pole placement of LTI systems. Math. Probl. Eng. **2011**, 1–20 (2011)

3. Ramírez, A., Mondié, S., Garrido, R., Sipahi, R.: Design of proportional-integral-retarded (PIR) controllers for second-order LTI systems. IEEE Trans. Autom. Control **61**(6), 1688–1693 (2016)
4. Tarczewski, T., Grzesiak, L.M.: Constrained state feedback speed control of PMSM based on model predictive approach. IEEE Trans. Industr. Electron. **63**(6), 3867–3875 (2016)
5. Bhattarai, R., Gurung, N., Thakallapelli, A., Kamalasadan, S.: Reduced-order state observer-based feedback control methodologies for doubly fed induction machine. IEEE Trans. Ind. Appl. **54**(3), 2845–2856 (2018)
6. Zhang, D., Han, Q., Jia, X.: Network-based output tracking control for a class of T-S fuzzy systems that cannot be stabilized by nondelayed output feedback controllers. IEEE Trans. Cybern. **45**(8), 1511–1524 (2015)
7. Dong, X., Hu, G.: Time-varying output formation for linear multiagent systems via dynamic output feedback control. IEEE Trans. Control Netw. Syst. **4**(2), 236–245 (2017)
8. Silva, E.I., Vargas, F.J., Maass, A.I.: Optimal design of a class of controllers and data-dropout compensators for LTI plants controlled over erasure channels. In: 52nd IEEE Conference on Decision and Control, Florence, pp. 6741–6746 (2013)
9. Guaracy, F.H.D., Silva, D.L.F.D., Ferreira, L.H.C.: On the properties of augmented open-loop stable plants using LQG/LTR control. IEEE Trans. Autom. Control **60**(8), 2172–2176 (2015)
10. Astrom, K.J., Hagglund, T.: The future of PID control. Control. Eng. Pract. **9**(2001), 1163–1175 (2001)
11. Habibullah, H., Pota, H.R., Petersen, I.R., Rana, M.S.: Tracking of triangular reference signals using LQG controllers for lateral positioning of an AFM scanner stage. IEEE/ASME Trans. Mechatron. **19**(4), 1105–1114 (2014)
12. Angélico, B.A., Toriumi, F.Y., Barbosa, F.D.S., Neves, G.P.D.: On guaranteeing convergence of discrete LQG/LTR when augmenting it with forward PI controllers. IEEE Access **5**, 27203–27210 (2017)
13. Huerta, F., Pérez, J., Cóbreces, S., Rizo, M.: Frequency-adaptive multiresonant LQG state-feedback current controller LCL-Filtered VSCs under distorted grid voltages. IEEE Trans. Industr. Electron. **65**(11), 8433–8444 (2018)
14. Pan, Z., Dong, F., Zhao, J., Wang, L., Wang, H., Feng, Y.: Combined resonant controller and two-degree-of-freedom PID controller for PMSLM current harmonics suppression. IEEE Trans. Industr. Electron. **65**(9), 7558–7568 (2018)
15. Zhang, Y., Jia, Y., Chai, T., Wang, D., Dai, W., Fu, J.: Data-driven PID controller and its application to pulp neutralization process. IEEE Trans. Control Syst. Technol. **26**(3), 828–841 (2018)
16. Lee, J.Y., Jin, M., Chang, P.H.: Variable PID gain tuning method using backstepping control with time-delay estimation and nonlinear damping. IEEE Trans. Industr. Electron. **61**(12), 6975–6985 (2014)
17. Bazanella, A.S., Pereira, L.F.A., Parraga, A.: A new method for PID tuning including plants without ultimate frequency. IEEE Trans. Control Syst. Technol. **25**(2), 637–644 (2017)
18. Fukamachi, S., Kanno, M., Hara, S.: Sum of imaginary parts and performance limitation of the LQG problem for SIMO systems. In: 53rd IEEE Conference on Decision and Control, Los Angeles, CA, pp. 6665–6670 (2014)
19. Vajedi, M., Azad, N.L.: Ecological adaptive cruise controller for plug-in hybrid electric vehicles using nonlinear model predictive control. IEEE Trans. Intell. Transp. Syst. **17**(1), 113–122 (2016)
20. Cai, J., Wen, C., Su, H., Liu, Z., Xing, L.: Adaptive backstepping control for a class of nonlinear systems with non-triangular structural uncertainties. IEEE Trans. Autom. Control **62**(10), 5220–5226 (2017)
21. Tong, S., Liu, C., Li, Y.: Fuzzy-adaptive decentralized output-feedback control for large-scale nonlinear systems with dynamical uncertainties. IEEE Trans. Fuzzy Syst. **18**(5), 845–861 (2010)
22. Boulkroune, A., Tadjine, M., M'Saad, M., Farza, M.: Design of a unified adaptive fuzzy observer for uncertain nonlinear systems. Inf. Sci. **265**, 139–153 (2014)

23. Wang, Y., Xia, Y., Zhou, P.: Fuzzy-model-based sampled-data control of chaotic systems: a fuzzy time-dependent Lyapunov-Krasovskii functional approach. IEEE Trans. Fuzzy Syst. **25**(6), 1672–1684 (2017)
24. Tong, S., Sui, S., Li, Y.: Adaptive fuzzy decentralized control for stochastic large-scale nonlinear systems with unknown dead-zone and unmodeled dynamics. Neurocomputing **135**, 367–377 (2014)
25. Huang, S., Jiau, M., Lin, C.: Optimization of the carpool service problem via a fuzzy-controlled genetic algorithm. IEEE Trans. Fuzzy Syst. **23**(5), 1698–1712 (2015)
26. Loop, B.P., Sudhoff, S.D., Zak, S.H., Zivi, E.L.: Estimating regions of asymptotic stability of power electronics systems using genetic algorithms. IEEE Trans. Control Syst. Technol. **18**(5), 1011–1022 (2010)
27. Das Sharma, K., Chatterjee, A., Rakshit, A.: A PSO–Lyapunov hybrid stable adaptive fuzzy tracking control approach for vision-based robot navigation. IEEE Trans. Instrum. Measurem. **61**(7), 1908–1914 (2012)
28. Xu, F., Chen, H., Gong, X., Mei, Q.: Fast nonlinear model predictive control on FPGA using particle swarm optimization. IEEE Trans. Industr. Electron. **63**(1), 310–321 (2016)
29. Yuan, Q., Yin, G.: Analyzing convergence and rates of convergence of particle swarm optimization algorithms using stochastic approximation methods. IEEE Trans. Autom. Control **60**(7), 1760–1773 (2015)
30. Liao, T., Socha, K., Montes de Oca, M.A., Stützle, T., Dorigo, M.: Ant colony optimization for mixed-variable optimization problems. IEEE Trans. Evol. Comput. **18**(4), 503–518 (2014)
31. Zheng, F., Zecchin, A.C., Newman, J.P., Maier, H.R., Dandy, G.C.: An adaptive convergence-trajectory controlled ant colony optimization algorithm with application to water distribution system design problems. IEEE Trans. Evol. Comput. **21**(5), 773–791 (2017)
32. Dekhici, L., Redjem, R., Belkadi, K., Mhamedi, A.E.: Discretization of the firefly algorithm for home care. Can. J. Electr. Comput. Eng. **42**(1), 20–26 (2019)
33. Alb, M., Alotto, P., Magele, C., Renhart, W., Preis, K., Trapp, B.: Firefly algorithm for finding optimal shapes of electromagnetic devices. IEEE Trans. Magn. **52**(3), 1–4 (2016)
34. Gao, M., He, X., Luo, D., Jiang, J., Teng, Q.: Object tracking using firefly algorithm. IET Comput. Vision **7**(4), 227–237 (2013)
35. Chrouta, J., Chakchouk, W., Zaafouri, A., Jemli, M.: Modeling and control of an irrigation station process using heterogeneous cuckoo search algorithm and fuzzy logic controller. IEEE Trans. Ind. Appl. **55**(1), 976–990 (2019)
36. Chitara, D., Niazi, K.R., Swarnkar, A., Gupta, N.: Cuckoo search optimization algorithm for designing of a multimachine power system stabilizer. IEEE Trans. Ind. Appl. **54**(4), 3056–3065 (2018)
37. Huang, L., Ai, J.: Automatic software fault localization based on artificial bee colony. J. Syst. Eng. Electron. **26**(6), 1325–1332 (2015)
38. Li, T.S., Kuo, P., Ho, Y., Kao, M., Tai, L.: A biped gait learning algorithm for humanoid robots based on environmental impact assessed artificial bee colony. IEEE Access **3**, 13–26 (2015)
39. Geem, Z.W., Kim, J.H., Loganathan, G.V.: A new heuristic optimization algorithm: harmony search. Simulation **76**(2), 60–68 (2001)
40. Geem, Z.W. (ed.): Music-Inspired Harmony Search Algorithm: Theory and Applications. Springer, Berlin (2009)
41. Geem, Z.W., Kim, J.H., Loganathan, G.V.: Harmony search optimization: application to pipe network design. Int. J. Model. Simul. **22**(2), 125–133 (2002)
42. Geem, Z.W., Lee, K.S., Park, Y.: Application of harmony search to vehicle routing. Amer. J. Appl. Sci. **2**(12), 1552–1557 (2005)
43. Geem, Z.W.: Harmony Search Algorithms for Structural Design Optimization, vol. 239. Springer, Berlin, Germany (2009)
44. Paola, F.D., Giugni, M., Pugliese, F., Romano, P.: Optimal design of LIDs in urban stormwater systems using a harmony-search decision support system. Water Resour. Manage. **32**(3), 4933–4951 (2018)

45. Moon, Y.Y., Geem, Z.W., Han, G.-T.: Vanishing point detection for self-driving car using harmony search algorithm. Swarm Evol. Comput. **41**, 111–119 (2018)
46. Geem, Z.W., Yoon, Y.: Harmony search optimization of renewable energy charging with energy storage system. Int. J. Elect. Power Energy Syst. **86**, 120–126 (2017)
47. Vasebi, A., Fesanghary, M., Bathaee, S.M.T.: Combined heat and power economic dispatch by harmony search algorithm. Int. J. Elect. Power Energy Syst. **29**(3), 713–719 (2007)
48. Khorram, E., Jaberipour, M.: Harmony search algorithm for solving combined heat and power economic dispatch problems. Energy Convers. Manag. **52**(2), 1550–1554 (2011)
49. Khazali, A.H., Kalantar, M.: Optimal reactive power dispatch based on harmony search algorithm. Int. J. Elect. Power Energy Syst. **33**(3), 684–692 (2011)
50. Ngonkham, S., Buasri, P.: Harmony search algorithm to improve cost reduction in power generation system integrating large scale wind energy conversion system. In: Proceedings of World Non-Grid-Connected Wind Power Energy Conference (WNWEC), September 2009, pp. 1–5 (2009)
51. Ambia, M.N., Hasanien, H.M., Al-Durra, A., Muyeen, S.M.: Harmony search algorithm-based controller parameters optimization for a distributed-generation system. IEEE Trans. Power Deliv. **30**(1), 246–255 (2015)
52. Das Sharma, K., Chatterjee, A., Rakshit, A.: Harmony search algorithm and Lyapunov theory based hybrid adaptive fuzzy controller for temperature control of air heater system with transport-delay. Appl. Soft Comput. **25**, 40–50 (2014)
53. Mohsen, A.M., Khader, A.T., Ghallab, A.: The effect of using different thermodynamic models with harmony search algorithm in the accuracy of RNA secondary structure prediction. In: Proceedings of International Conference on Soft Computing and Pattern Recognition (SOCPAR), December 2009, pp. 505–510 (2009)
54. Mohsen, A.M., Khader, A.T., Ramachandram, D.: Hsrnafold: a harmony search algorithm for RNA secondary structure prediction based on minimum free energy. In: Proceedings of International Conference on Innovations in Information Technology (IIT), December 2008, pp. 11–15 (2008)
55. Panchal, A.: Harmony search in therapeutic medical physics. In: Music—Inspired Harmony Search Algorithm, pp. 189–203. Springer, Berlin, Germany (2009)
56. Amor, J., Alexandre, E., Gil, R., Álvarez, L., Huerta, E.: Music inspired harmony search algorithm applied to feature selection for sound classification in hearing AIDS. In: Proceedings of Audio Engineering Society Convention, pp. 124 (2008)
57. Chen, J., Xin, B., Peng, Z., Dou, L., Zhang, J.: Optimal contraction theorem for exploration–exploitation tradeoff in search and optimization. IEEE Trans. Syst. Man Cybern. Part A Syst. Humans **39**(3), 680–691 (2009)
58. Tayarani, N.M., Yao, X., Xu, H.: Meta-heuristic algorithms in car engine design: a literature survey. IEEE Trans. Evol. Comput. **19**(5), 609–629 (2015)
59. Das Sharma, K., Chatterjee, A., Rakshit, A.: Design of a hybrid stable adaptive fuzzy controller employing Lyapunov theory and harmony search algorithm. IEEE Trans. Control Syst. Technol. **18**(6), 1440–1447 (2010)
60. Mahdavi, M., Fesanghary, M., Damangir, E.: An improved harmony search algorithm for solving optimization problems. Appl. Math. Comput. **188**(2), 1567–1579 (2007)
61. Pan, Q.-K., Suganthan, P.N., Tasgetiren, M.F., Liang, J.J.: A self-adaptive global best harmony search algorithm for continuous optimization problems. Appl. Math. Comput. **216**(3), 830–848 (2010)
62. El-Abd, M.: An improved global-best harmony search algorithm. Appl. Math. Comput. **222**, 94–106 (2013)
63. Al-Betar, M.A., Khader, A.T., Zaman, M.: University course timetabling using a hybrid harmony search metaheuristic algorithm. IEEE Trans. Syst. Man Cybern. Part C (Applications and Reviews) **42**(5), 664–681 (2012)
64. Jeddi, B., Vahidinasab, V.: A modified harmony search method for environmental/economic load dispatch of real-world power systems. Energy Convers. Manag. **78**, 661–675 (2014)

65. Zhao, F., Qin, S., Yang, G., Ma, W., Zhang, C., Song, H.: A differential-based harmony search algorithm with variable neighborhood search for job shop scheduling problem and its runtime analysis. IEEE Access **6**, 76313–76330 (2018)
66. Zou, D., Gao, L., Wu, J., Li, S.: Novel global harmony search algorithm for unconstrained problems. Neurocomputing **73**(3), 3308–3318 (2010)
67. Al-Omoush, A.A., Alsewari, A.A., Alamri, H.S., Zamli, K.Z.: Comprehensive review of the development of the harmony search algorithm and its applications. IEEE Access **7**, 14233–14245 (2019)
68. Fesanghary, M., Mahdavi, M., Minary-Jolandan, M., Alizadeh, Y.: Hybridizing harmony search algorithm with sequential quadratic programming for engineering optimization problems. Comput. Methods Appl. Mech. Eng. **197**(3), 3080–3091 (2008)
69. Wang, L., Pan, Q.-K., Tasgetiren, M.F.: Minimizing the total flow time in a flow shop with blocking by using hybrid harmony search algorithms. Expert Syst. Appl. **37**(12), 7929–7936 (2010)
70. Lin, C., Deng, D., Wang, S.: Extending the lifetime of dynamic underwater acoustic sensor networks using multi-population harmony search algorithm. IEEE Sens. J. **16**(11), 4034–4042 (2016)
71. Das Sharma, K., Chatterjee, A., Rakshit, A.: A hybrid approach for design of stable adaptive fuzzy controllers employing Lyapunov theory and particle swarm optimization. IEEE Trans. Fuzzy Syst. **17**(2), 329–342 (2009)
72. Wang, L.X.: Fuzzy systems are universal approximators. In: 1992 Proceedings of IEEE International Conference on Fuzzy Systems, San Diego, CA, USA pp. 1163–1170 (1992)
73. Er, M.J., Mandal, S.: A survey of adaptive fuzzy controllers: nonlinearities and classifications. IEEE Trans. Fuzzy Syst. **24**(5), 1095–1107 (2016)
74. Chen, B., Liu, X., Lin, C.: Observer and adaptive fuzzy control design for nonlinear strict-feedback systems with unknown virtual control coefficients. IEEE Trans. Fuzzy Syst. **26**(3), 1732–1743 (2018)
75. Yu, W., Karkoub, M., Wu, T., Her, M.: Delayed output feedback control for nonlinear systems with two-layer interval fuzzy observers. IEEE Trans. Fuzzy Syst. **22**(3), 611–630 (2014)
76. Tong, S., Li, Y., Sui, S.: Adaptive fuzzy output feedback control for switched nonstrict-feedback nonlinear systems with input nonlinearities. IEEE Trans. Fuzzy Syst. **24**(6), 1426–1440 (2016)
77. Gao, Q., Feng, G., Dong, D., Liu, L.: Universal fuzzy models and universal fuzzy controllers for discrete-time nonlinear systems. IEEE Trans. Cybern. **45**(5), 880–887 (2015)
78. Teng, L., Wang, Y., Cai, W., Li, H.: Robust fuzzy model predictive control of discrete-time Takagi-Sugeno systems with nonlinear local models. IEEE Trans. Fuzzy Syst. **26**(5), 2915–2925 (2018)
79. Zhang, J., Deng, Z., Choi, K., Wang, S.: Data-driven elastic fuzzy logic system modeling: constructing a concise system with human-like inference mechanism. IEEE Trans. Fuzzy Syst. **26**(4), 2160–2173 (2018)
80. Nasiri, A., Nguang, S.K., Swain, A., Almakhles, D.J.: Robust output feedback controller design of discrete-time Takagi–Sugeno fuzzy systems: a non-monotonic Lyapunov approach. IET Control Theory Appl. **10**(5), 545–553 (2016)
81. Chen, B., Liu, X., Liu, K., Lin, C.: Fuzzy approximation-based adaptive control of nonlinear delayed systems with unknown dead zone. IEEE Trans. Fuzzy Syst. **22**(2), 237–248 (2014)
82. Liu, X., Zhai, D., Li, T., Zhang, Q.: Fuzzy-approximation adaptive fault-tolerant control for nonlinear pure-feedback systems with unknown control directions and sensor failures. Fuzzy Sets Syst. **356**, 28–43 (2019)
83. Xiong, W., Yu, W., Lü, J., Yu, X.: Fuzzy modelling and consensus of nonlinear multiagent systems with variable structure. IEEE Trans. Circuits Syst. I Regul. Pap. **61**(4), 1183–1191 (2014)
84. Zhao, X., Yin, Y., Zhang, L., Yang, H.: Control of switched nonlinear systems via T-S fuzzy modeling. IEEE Trans. Fuzzy Syst. **24**(1), 235–241 (2016)

85. Wang, H.O., Tanaka, K., Griffin, M.: Parallel distributed compensation of nonlinear systems by Takagi-Sugeno fuzzy model. In: Proceedings of FUZZ-IEEE/IFES'95, pp. 531–538 (1995)
86. Wang, H., Tanaka, K., Griffin, M.F.: An approach to fuzzy control of nonlinear systems: stability and design issues. IEEE Trans. Fuzzy Syst. **4**(1), 14–23 (1996)
87. Derakhshan, S.F., Fatehi, A., Sharabiany, M.G.: Nonmonotonic observer-based fuzzy controller designs for discrete time T-S fuzzy systems via LMI. IEEE Trans. Cybern. **44**(12), 2557–2567 (2014)
88. Han, C., Wu, L., Lam, H.K., Zeng, Q.: Nonfragile control with guaranteed cost of T-S fuzzy singular systems based on parallel distributed compensation. IEEE Trans. Fuzzy Syst. **22**(5), 1183–1196 (2014)
89. Leite, D., Palhares, R.M., Campos, V.C.S., Gomide, F.: Evolving granular fuzzy model-based control of nonlinear dynamic systems. IEEE Trans. Fuzzy Syst. **23**(4), 923–938 (2015)
90. Li, H., Liu, H., Gao, H., Shi, P.: Reliable fuzzy control for active suspension systems with actuator delay and fault. IEEE Trans. Fuzzy Syst. **20**(2), 342–357 (2012)
91. Tanaka, K., Sugeno, M.: Stability analysis and design of fuzzy control systems. Fuzzy Sets Syst. **45**(2), 135–156 (1992)
92. Tanaka, K., Ikeda, T., Wang, H.: Fuzzy regulators and fuzzy observers: relaxed stability conditions and LMI-based designs. IEEE Trans. Fuzzy Syst. **6**(2), 250–265 (1998)
93. Tanaka, K., Wang, H.: Fuzzy Control Systems Design and Analysis: A Linear Matrix Inequality Approach. Wiley Interscience, New York, NY, USA (2001)
94. Wang, T., Tong, S.: Observer-based output-feedback asynchronous control for switched fuzzy systems. IEEE Trans. Cybern. **47**(9), 2579–2591 (2017)
95. Lee, S.: Novel stabilization criteria for T-S fuzzy systems with affine matched membership functions. IEEE Trans. Fuzzy Syst. **27**(3), 540–548 (2019)
96. Su, X., Zhou, H., Song, Y.: An optimal divisioning technique to stabilization synthesis of T-S fuzzy delayed systems. IEEE Trans. Cybern. **47**(5), 1147–1156 (2017)
97. Choi, H.D., Ahn, C.K., Shi, P., Wu, L., Lim, M.T.: Dynamic output-feedback dissipative control for T-S fuzzy systems with time-varying input delay and output constraints. IEEE Trans. Fuzzy Syst. **25**(3), 511–526 (2017)
98. Baranyi, P., Korondi, P., Tanaka, K.: Parallel distributed compensation based stabilization of a 3-DOF RC helicopter: a tensor product transformation based approach. J. Adv. Comput. Intell. Intell. Informat. **13**, 25–34 (2009)
99. Mazmanyan, L., Ayoubi, M.A.: Fuzzy attitude control of spacecraft with fuel sloshing via linear matrix inequalities. IEEE Trans. Aerosp. Electron. Syst. **54**(5), 2526–2536 (2018)
100. Chang, Y., Chen, C., Zhu, Z., Huang, Y.: Speed control of the surface-mounted permanent-magnet synchronous motor based on Takagi–Sugeno fuzzy models. IEEE Trans. Power Electron. **31**(9), 6504–6510 (2016)
101. Kamal, E., Oueidat, M., Aitouche, A., Ghorbani, R.: Robust scheduler fuzzy controller of DFIG wind energy systems. IEEE Trans. Sustain. Energy **4**(3), 706–715 (2013)
102. Rakhshan, M., Vafamand, N., Khooban, M., Blaabjerg, F.: Maximum power point tracking control of photovoltaic systems: a polynomial fuzzy model-based approach. IEEE J. Emerg. Sel. Top. Power Electron. **6**(1), 292–299 (2018)
103. Jin, X., Yu, Z., Yin, G., Wang, J.: Improving vehicle handling stability based on combined AFS and DYC system via robust Takagi-Sugeno fuzzy control. IEEE Trans. Intell. Transp. Syst. **19**(8), 2696–2707 (2018)
104. Wei, M., Liu, T.: Design and implementation of an online tuning adaptive controller for synchronous reluctance motor drives. IEEE Trans. Industr. Electron. **60**(9), 3644–3657 (2013)
105. Zou, Y., Cheung, N.C., Pan, J.F.: An adaptive high-precision tracking controller for the coupled switched reluctance two-finger gripper. IEEE Trans. Magn. **51**(11), 1–4 (2015)
106. Wang, C., Quan, L., Jiao, Z., Zhang, S.: Nonlinear adaptive control of hydraulic system with observing and compensating mismatching uncertainties. IEEE Trans. Control Syst. Technol. **26**(3), 927–938 (2018)
107. Cao, C., Hovakimyan, N.: Design and analysis of a novel L_1 adaptive controller, part I: control signal and asymptotic stability. In: Proceedings of American Control Conference, USA, pp. 3397–3402 (2006)

108. Cao, C., Hovakimyan, N.: Design and analysis of a novel L_1 adaptive controller, part II: guaranteed transient performance. In: Proceedings of American Control Conference, USA, pp. 3403–3408 (2006)
109. Cao, C., Hovakimyan, N.: Guaranteed transient performance with L_1 adaptive controller for systems with unknown time-varying parameters and bounded disturbances: part I. In: Proceedings of the 2007 American Control Conference, Marriott Marquis Hotel at Times Square, New York City, USA, pp. 3925–3930 (2007)
110. Cao, C., Hovakimyan, N.: Design and analysis of a novel L_1 adaptive control architecture with guaranteed transient performance. IEEE Trans. Autom. Control **53**(2), 586–591 (2008)
111. Hovakimyan, N., Cao, C.: L_1 Adaptive Control Theory Guaranteed Robustness with Fast Adaptation. SIAM, Philadelphia (2010)
112. Cao, C., Hovakimyan, N.: L_1 adaptive controller for systems with unknown time-varying parameters and disturbances in the presence of non-zero trajectory initialization error. Int. J. Control **81**(7), 1147–1161 (2008)
113. Elnaggar, M., Saad, M.S., Fattah, H.A., Elshafei, A.L.: Discrete time L_1 adaptive control for systems with time-varying parameters and disturbances. In: Proceedings of IEEE Conference on Decision and Control (CDC), USA, December 12–14, pp. 2115–2120 (2016)
114. Li, D., Hovakimyan, N.: Filter design for feedback-loop trade-off of L_1 adaptive controller: a linear matrix inequality approach. In: AIAA Guidance, Navigation and Control Conference and Exhibit, Honolulu, Hawaii, August 2008 (2008)
115. Jalali, A.A., Golmohammad, H.: Digital L_1 controller based on Linear Matrix Inequality (LMI) for Line-of-sight stabilization. In: Proceedings of the 7th IFAC Symposium on Robust Control Design: The International Federation of Automatic Control, Aalborg, Denmark, June 2012 (2012)
116. Xargay, E., Hovakimyan, N., Cao, C.: L_1 adaptive controller for multi-input multi-output systems in the presence of nonlinear unmatched uncertainties. In: American Control Conference, MD, USA, pp. 874–879 (2010)
117. Maiti, R., Das Sharma, K., Sarkar, G.: On-line adaptation of parameter uncertainties of a practical plant employing L_1 adaptive controller. In: International Conference on Modelling and Simulation (MS-17), Kolkata, India, 4–5 Nov. 2017 (2017)
118. Cao, C., Hovakimyan, N.: L_1 adaptive controller for Multi-Input Multi-Output systems in the presence of unmatched disturbances. In: Proceedings of American Control Conference, Seattle, WA, pp. 4105–4110 (2008)
119. Cao, C., Hovakimyan, N.: L_1 adaptive controller for nonlinear systems in the presence of unmodelled dynamics: Part II. In: Proceedings of American Control Conference, Seattle, Washingtion, pp. 4099–4104 (2008)
120. Yoo, S.J., Hovakimyan, N., CAO, C.: Decentralised L_1 adaptive control for large-scale non-linear systems with interconnected unmodelled dynamics. IET Control Theory Appl. **4**(10), 1972–1988 (2010)
121. Luo, J., Cao, C., Yang, Q.-M.: L_1 adaptive controller for a class of non-affine multi-input multi-output nonlinear systems. Int. J. Control **86**(2), 348–359 (2013)
122. Luo, J., Cao, C.: L_1 Adaptive Controller for a Class of Nonlinear Systems. ASME J Dyn. Sys. Meas. Control **136**(3), 031023–031030 (2014)
123. Choe, R., Xargay, E., Hovakimya, N.: L_1 adaptive control for a class of uncertain nonaffine-in-control nonlinear systems. IEEE Trans. Autom. Contr. **61**, 840–846 (2016)
124. Leman, T., Xargay, E., Dullerud, G., Hovakimyan, N., Wendel, T.: L_1 adaptive control augmentation system for the X-48B aircraft. In: AIAA Guidance, Navigation and Control Conference, Chicago, IL (2009)
125. Yan, Y., Dong, W., Zou, Q., Xie, W., Yin, H.: Longitudinal inner loop flight controller design by using L_1 adaptive control theory. In: Proceeding of the 2015 IEEE International Conference on Information and Automation, Lijiang, China, August 2015, pp. 2965–2970 (2015)
126. Evgeny, K., Hovakimyan, N.: Application of L_1 adaptive controller to wing-rock. In: AIAA Infotech at Aerospace, Atlanta, Georgia, April 2010, pp. 1–14 (2010)

127. Jin, W., Bifeng, S., Liguang, W., Wei, T.: L_1 adaptive dynamic inversion controller for an X-wing tail-sitter MAV in hover flight. Procedia Eng. **99**, 969–974 (2010)
128. Mallikarjunan, S., Bill, N., Kharisov, E., Xargay, E., Hovakimyan, N., Cao, C.: L_1 adaptive controller for attitude control of multirotors. In: AIAA Guidance, Navigation, and Control Conference, Minneapolis, Minnesota, AIAA (2012)
129. Gao, L., Wu, W., Zhou, S.: Application of L_1 adaptive control augmentation to the flying wing unmanned aerial vehicle. In: Proceeding of 2011 Fourth International Symposium on Knowledge Acquisition and Modeling, Sanya, pp. 222–225 (2011)
130. Peterson, A., Åstrom, K., Robertsson, A., Johansson, R.: Augmenting L_1 adaptive control of piecewise constant type to a fighter aircraft. Performance and robustness evaluation for rapid maneuvering. In: AIAA Guidance, Navigation, and Control Conference, Minneapolis, Minnesota (2012)
131. Wang, J., Cao, C., Hovakimyan, N.: L_1 Adaptive Controller for a Missile Longitudinal Autopilot Design. In: Proceedings of AIAA Guidance, Navigation and Control Conference and Exhibit, Honolulu, Hawaii, 18–21 August 2008, pp. 1–21 (2008)
132. Erdos, D., Shima, T., Kharisov, E., Hovakimyan, N.: L_1 adaptive control integrated missile autopilot and guidance. In: AIAA Guidance, Navigation, and Control Conference, Minneapolis Minnesota (2012)
133. Patel, V.V., Cao, C., Hovakimyan, N., Wise, K.A., Lavretsky, E.: L_1 adaptive controller for tailless unstable aircraft in the presence of unknown actuator failures. Int. J. Control **82**(4), 705–720 (2009)
134. Sease, B., Yang, Q., Xu, Y., Che, J., Cao, C.: L_1 adaptive attitude control for a picoscale satellite test bed. IEEE Trans. Aerosp. Electron. Syst. **51**, 1147–1154 (2015)
135. Lee, K.W., Singh, S.N.: Multi-input submarine control via L_1 adaptive feedback despite uncertainties. Proc. Inst. Mech. Eng. Part I J. Syst. Control Eng. **228**(5), 330–347 (2014)
136. Zhao, H., Wu, Q., Rasmussen, C.N., Blanke, M.: L_1 adaptive speed control of a small wind energy conversion system for maximum powerpoint tracking. IEEE Trans. Energy Convers. **29**(3), 576–584 (2014)
137. Kharisov, E., Beck, C.L., Bloom, M.: Design of L_1 adaptive controllers for human patient anesthesia. Control. Eng. Pract. **44**, 65–77 (2015)
138. Kim, K.-K.K., Hovakimyan, N.: Multi-criteria optimization for filter design of L_1 adaptive control. J. Optim. Theory Appl. **161**(2), 557–581 (2014)
139. Hashim, H.A., Ferik, S.E., Abido, M.A.: A fuzzy logic feedback filter design tuned with PSO for L_1 adaptive controller. Expert Syst. Appl. **42**(23), 9077–9085 (2015)
140. Boubakir, A., Souanef, T., Labiod, S., Plestan, F., Boudjema, F.: An L_1 fuzzy adaptive controller for a class of SISO nonaffine nonlinear systems: application to the control of an electropneumatic actuator. Proc. Inst. Mech. Eng. Part I J. Syst. Control Eng. **230**(8), 736–748 (2016)
141. Elnaggar, M., Saad, M.S., Fattah, H.A.A., Elshafei, A.L.: L_1 adaptive fuzzy control of wind energy conversion systems via variable structure adaptation for all wind speed regions. IET Renew. Power Gener. **12**(1), 18–27 (2018)
142. Jin, W., Bifeng, S., Liguang, W., Wei, T.: L_1 adaptive dynamic inversion controller for an x-wing tail-sitter MAV in hover flight. Procedia Eng. **99**(2015), 969–974 (2015)
143. Kutluay, K.T., Yavrucuk, Ð.: Dynamic inversion based control of a missile with L_1 adaptive control augmentation. In: Proceedings of 2010 IEEE International Symposium on Intelligent Control, Part of 2010 IEEE Multi-Conference on Systems and Control, Yokohama, Japan, September 8–10, 2010, pp. 2374–2379 (2010)
144. Campbell, S.F., Kaneshige, J.T.: A nonlinear dynamic inversion L_1 adaptive controller for a generic transport model. In: Proceedings of the 2010 American Control Conference, Baltimore, MD, pp. 862–867 (2010)
145. Hellmundt, F., Wildschek, A., Maier, R., Osterhube, R., Holzapfel, F.: Comparison of L_1 adaptive augmentation strategies for a differential PI baseline controller on a longitudinal F16 aircraft model. In: Advances in Aerospace Guidance, Navigation and Control, April 2015, pp. 99–118 (2015)

146. Zuo, Z., Mallikarjunan, S.: L_1 adaptive backstepping for robust trajectory tracking of UAVs. IEEE Trans. Ind. Electron. **64**(4), 2944–2954 (2017)
147. Maalouf, D., Chemori, A., Creuze, V.: Stability analysis of a new extended L_1 controller with experimental validation on an underwater vehicle. In: Proceedings of 52nd IEEE Conference on Decision and Control, Florence, Italy, pp. 6149–6155 (2013)
148. Das, S., Mukhopadhyay, A., Roy, A., Abraham, A., Panigrahi, B.K.: "Exploratory power of the harmony search algorithm: analysis and improvements for global numerical optimization. IEEE Trans. Syst. Man Cybern. Part B (Cybernetics) **41**(1), 89–106 (2011)
149. Yang, H.-B.O., Gao, L.-Q., Li, S., Kong, X., Zou, D.-X.: On the iterative convergence of harmony search algorithm and a proposed modification. Appl. Math. Comput. **247**, 1064–1095 (2014)
150. Zeiaee, A., Kharrati, H., Khanmohammadi, S.: Optimized fuzzy PDC controller for nonlinear systems with T-S model mismatch. In: The 2011 International Conference on Advanced Mechatronic Systems, Zhengzhou, pp. 61–66 (2011)
151. Lu, L., Lin, Z.: Design of a nonlinear anti-windup gain by using a composite quadratic Lyapunov function. IEEE Trans. Autom. Control **56**(12), 2997–3001 (2011)
152. Lipkovich, M., Fradkov, A.: Equivalence of MIMO circle criterion to existence of quadratic Lyapunov function. IEEE Trans. Autom. Control **61**(7), 1895–1899 (2016)
153. Iervolino, R., Vasca, F., Iannelli, L.: Cone-copositive piecewise quadratic Lyapunov functions for conewise linear systems. IEEE Trans. Autom. Control **60**(11), 3077–3082 (2015)
154. Gering, S., Eciolaza, L., Adamy, J., Sugeno, M.: A piecewise approximation approach to nonlinear systems: stability and region of attraction. IEEE Trans. Fuzzy Syst. **23**(6), 2231–2244 (2015)
155. Chen, Y., Ohtake, H., Tanaka, K., Wang, W., Wang, H.O.: Relaxed stabilization criterion for T-S fuzzy systems by minimum-type piecewise-Lyapunov-function-based switching fuzzy controller. IEEE Trans. Fuzzy Syst. **20**(6), 1166–1173 (2012)
156. Li, L., Ding, S.X., Qiu, J., Yang, Y., Zhang, Y.: Weighted fuzzy observer-based fault detection approach for discrete-time nonlinear systems via piecewise-fuzzy Lyapunov functions. IEEE Trans. Fuzzy Syst. **24**(6), 1320–1333 (2016)
157. Zhang, K., Jiang, B., Shi, P.: Fault estimation observer design for discrete-time Takagi-Sugeno fuzzy systems based on piecewise Lyapunov functions. IEEE Trans. Fuzzy Syst. **20**(1), 192–200 (2012)
158. Zhang, C., Hu, J., Qiu, J., Chen, Q.: Event-triggered nonsynchronized H∞ filtering for discrete-time T-S fuzzy systems based on piecewise Lyapunov functions. IEEE Trans. Syst. Man Cybern. Syst. **47**(8), 2330–2341 (2017)
159. Sassi, M.A.B., Sankaranarayanan, S., Chen, X., Ábrahám, E.: Linear relaxations of polynomial positivity for polynomial Lyapunov function synthesis. MA Jo. Math. Control Inf. **33**(3), 723–756 (2016)
160. Han, D., Chesi, G., Luk, C.K.: Homogeneous polynomial Lyapunov functions for robust local synchronisation with time-varying uncertainties. IET Control Theory Appl. **8**(10), 855–862 (2014)
161. Long, L.: Integral ISS for switched nonlinear time-varying systems using indefinite multiple Lyapunov functions. IEEE Trans. Autom. Control **64**(1), 404–411 (2019)
162. Long, L., Zhao, J.: An integral-type multiple Lyapunov functions approach for switched nonlinear systems. IEEE Trans. Autom. Control **61**(7), 1979–1986 (2016)
163. Long, L., Zhao, J.: H∞ control of switched nonlinear systems in p-normal form using multiple Lyapunov functions. IEEE Trans. Autom. Control **57**(5), 1285–1291 (2012)
164. Su, S., Lin, Z.: A multiple Lyapunov function approach to distributed synchronization control of multi-agent systems with switching directed communication topologies and unknown nonlinearities. IEEE Trans. Control Netw. Syst. **5**(1), 23–33 (2018)
165. Long, L.: Multiple Lyapunov functions-based small-gain theorems for switched interconnected nonlinear systems. IEEE Trans. Autom. Control **62**(8), 3943–3958 (2017)
166. Tanaka, K., Hori, T., Wang, H.O.: A multiple Lyapunov function approach to stabilization of fuzzy control systems. IEEE Trans. Fuzzy Syst. **11**(4), 582–589 (2003)

167. Tanaka, K., Ohtake, H., Wang, H.O.: A descriptor system approach to fuzzy control system design via fuzzy Lyapunov functions. IEEE Trans. Fuzzy Syst. **15**(3), 333–341 (2007)
168. Liu, Y., Wu, F., Ban, X.: Dynamic output feedback control for continuous-time T-S fuzzy systems using fuzzy Lyapunov functions. IEEE Trans. Fuzzy Syst. **25**(5), 1155–1167 (2017)
169. Faria, F., Silva, G., Oliveira, V.: Reducing the conservatism of LMI-based stabilization conditions for TS fuzzy systems using fuzzy Lyapunov functions. Int. J. Syst. Sci. **44**, 1956–1969 (2012)
170. Faria, F.A., Valentino, M.C., Oliveira, V.A.: A fuzzy Lyapunov function approach for stabilization and H1 control of switched TS fuzzy systems. Appl. Math. Model. **38**(19–20), 4817–4834 (2014)
171. Lam, H.K., Lauber, J.: Membership-function-dependent stability analysis of fuzzy-model-based control systems using fuzzy Lyapunov functions. Inf. Sci. **232**, 253–266 (2013)
172. Maiti, R., Das Sharma, K., Sarkar, G.: lBest-HS algorithm based concurrent L_1 adaptive control for nonlinear systems. ISA Trans. **96**, 390–414 (2020)
173. Wang, X.-F., Hovakimyan, N.: L_1 adaptive controller for nonlinear time-varying reference systems. Syst. Control Lett. **61**(4), 455–463 (2012)
174. Song, H.-T., Zhang, T., Zhang, G.-L.: L_1 adaptive controller of nonlinear reference system in presence of unmatched uncertainties. J. Central South Univ. **23**, 834–840 (2016)
175. Maiti, R., Das Sharma, K., Sarkar, G.: Linear consequence based fuzzy parallel distributed compensation type L_1 adaptive controller for two link robot manipulator. IEEE Trans. Circ. Syst. I Regul. Pap. **66**(10), 3978–3990 (2019)

Part II
Preliminaries

Chapter 2
Basic L_1 Adaptive Controller: A State of the Art Study

Abstract This chapter briefly describes the phenomenon of basic L_1 adaptive controller. The slow transient performance of conventional adaptive controller is eliminated in L_1 adaptive controller by utilization of high adaptation gain. At the same time robustness is also guaranteed with the aid of a low-pass filter augmentation into the control channel. The L_1 adaptive controller quickly nullifies uncertainties and disturbances present in the system and provides fast transient performance. The performance of basic L_1 adaptive controller is evaluated on four simulation case studies and compared with PID, LQG and MRAC to show its effectiveness.

2.1 Introduction

Classical controllers are incapable of controlling systems with time varying uncertainties, disturbances, and nonlinearities [1, 2]. Model reference adaptive controllers (MRAC) can handle systems with time varying uncertainties and time varying disturbances but they suffer from sluggish transient performance to retain the stability of the system [3, 4]. Their performances are very much dependent upon initial condition of the system, variation of input to the system, etc., [5]. To expedite fast transient performance possessing high robustness, L_1 adaptive controller was introduced [6–10]. Utilization of high adaptation gain facilitates quick transient performance and inclusion of a low-pass filter in the control channel keeps high robustness. Advantages of L_1 adaptive controller enhance its applications for systems with unknown time-varying parameters [11], disturbances [12], unknown nonlinearities [13], unmodelled dynamics [14], etc. In this chapter, the design of L_1 adaptive controller and its stability analysis are described. The control law of L_1 adaptive controller is composed of L_1 adaptive control signal and state feedback control action. The L_1 adaptive control signal is formulated from the adaptive estimates of unknown constant, time varying uncertainties and disturbances. Adaptation laws of unknown constant, time varying uncertainties and disturbances are formulated from the Lyapunov stability theory where high adaptation gain is utilized to obtain fast transient performance. Thus, L_1 adaptive control law contains high frequency component due to the high adaptation gain. Direct occurrence of this high frequency control signal may incur instability

R. Maiti et al., *Hybrid L_1 Adaptive Control*, Studies in Systems, Decision and Control 422, https://doi.org/10.1007/978-3-030-97102-1_2

of the system. Hence, the L_1 adaptive control signal is low-pass filtered and then provided to the system to guarantee stable performance with quick transient performance. L_1 norm condition provides the ranges of the parameter values to design components of L_1 adaptive controller.

To demonstrate the effectiveness of the L_1 adaptive controller over proportional integral derivative (PID), linear quadratic Gaussian (LQG) and MRAC, four simulation case studies are performed. The controllers are imposed on the systems with unknown constant, time varying uncertainties and time varying disturbances, viz., Duffing's oscillatory system, nonlinear spring mass damper system, inverted pendulum with cart and twin rotor MIMO system. The designed controllers are then subjected to the unknown time varying disturbance without further tuning and their performances are compared.

2.2 Motivation of Designing L_1 Adaptive Controller

Two classical controllers, viz., PID and LQG are tested in this work. Classical controllers are incapable of tackling systems with time varying uncertainties, disturbances, etc. MRAC comes into place to tackle uncertainties though, it possesses sluggish transient performance to keep robustness. Their brief descriptions are provided in the subsequent sections. To overcome their drawbacks, L_1 adaptive controller arises, whose formulation is elaborated in Sect. 2.3.

2.2.1 *Proportional Integral Derivative Controller*

PID controller is capable of controlling linear time invariant (LTI) system. The mathematical formulation of PID controller is given in this section.

The PID controller is considered as [15]:

$$u(t) = K_P\left(e(t) + \frac{1}{T_i}\int_0^t e(t)dt + T_d\frac{de(t)}{dt}\right)$$

or

$$u(t) = a_0e(t) + a_1e(t-1) + a_2e(t-2) + u(t-1) \tag{2.1}$$

where $e(t) = r(t) - y(t)$ represents the error between reference and system output. T_i and T_d are the integral time constant and derivative time constant respectively. K_p is the proportional gain. a_0, a_1 and a_2 are given as respectively [15]:

$$a_0 = K_P\left(1 + \frac{h}{T_i} + \frac{T_d}{h}\right) = \left(K_P + hK_i + \frac{K_d}{h}\right) \tag{2.2}$$

$$a_1 = -K_P\left(\frac{2T_d}{h} + 1\right) = -\left(\frac{2K_d}{h} + K_P\right) \tag{2.3}$$

$$a_2 = \left(\frac{K_P T_d}{h}\right) = \left(\frac{K_d}{h}\right) \tag{2.4}$$

where $K_i = \frac{K_p}{T_i}$, $K_d = K_p T_d$ are the integral gain and derivative gain respectively. h is the time step.

In spite of very simple construction and operation of PID controller, the problem is to select the proper gains for keeping the error minimum. As, the gains of PID controller are fixed therefore, it is incapable of tackling time varying uncertainties and time varying disturbances present in the system. To eliminate disturbances, linear quadratic Gaussian (LQG) comes into place whose description is provided as follows.

2.2.2 *Linear Quadratic Gaussian*

Linear quadratic regulator (LQR) with Kalman filter as observer is known as linear quadratic Gaussian (LQG) [16]. Using observer with state feedback controller introduces process disturbance $d(t)$ and measurement noise z into the system model. Noise and disturbance inclusion into the system degrade its performance which can be nullified by utilizing Kalman filter.

Let, the system model is considered as:

$$\dot{\underline{x}} = A_0\underline{x} + b\underline{u} + Fd(t) \tag{2.5}$$

$$\underline{y} = c\underline{x} + z \tag{2.6}$$

where A_0 is the open loop system matrix, b and c are the input matrix and output matrix respectively. F is a constant. $\underline{x}$ and $\underline{y}$ are the state and output of the system respectively. $\underline{u}$ is the input to the system.

The LQG can be considered as [16]:

$$\dot{\underline{x}}_e = A_0\underline{x}_e + b\underline{u} + L(\underline{y} - \underline{y}_e) + Fd(t) \tag{2.7}$$

$$\underline{y}_e = c\underline{x}_e + z \tag{2.8}$$

where $\underline{x}_e$ and $\underline{y}_e$ are the state and output of the observer respectively. L is the Kalman gain matrix.

The input to the system is given as:

$$\underline{u} = -K\underline{x}_e \tag{2.9}$$

where K is the state feedback gain matrix.

The error between system state $\underline{x}$ and observer state $\underline{x}_e$ is given as $\underline{e} = \underline{x} - \underline{x}_e$. The performance index is considered as $PI = \int_{t_0}^{t_f} (\underline{e}^T Q_e \underline{e} + \underline{u}^T R_e \underline{u})dt$, where t_0 is the initial time and t_f is the final time. Q_e and R_e are two weight matrices.

LQG is capable of estimating and rejecting fixed disturbances for LTI systems. It cannot cope up with time varying uncertainties and time varying disturbances with its fixed gains. To handle these time varying quantities, adaptive controllers arise.

2.2.3 Model Reference Adaptive Controller

The description of model reference adaptive controller (MRAC) is provided in this section. Let, the system model is considered as:

$$\dot{\underline{x}}(t) = A_0\underline{x}(t) + b(\underline{u}_1(t) + \underline{\theta}\left\|\underline{x}\right\|_{L_\infty}) + b\underline{u}_2(t) \tag{2.10}$$

where $A_0 \in \Re^{n \times n}$ is the open loop system matrix, $b \in \Re^{n \times p}$ is the input matrix. $\underline{x}(t) = [x_1 \;\; x_2 \;\; \ldots x_n]^T \in \Re^n$ with $\underline{x}(0) = \underline{x}_0$ is the state vector. $\underline{\theta}(t) \in \Re^p$ is the unknown parameter uncertainties vector. $\underline{u}_1(t) \in \Re^p$, $\underline{u}_2(t) \in \Re^p$ are the MRAC control signal and state feedback control law respectively.

Putting the value of $\underline{u}_2(t) = -K^T\underline{x}(t)$ in (2.10) provides:

$$\dot{\underline{x}}(t) = A_m\underline{x}(t) + b(\underline{u}_1(t) + \underline{\theta}\left\|\underline{x}\right\|_{L_\infty}) \tag{2.11}$$

where $A_m = A_0 - bK^T \in \Re^{n \times n}$ is the closed loop system matrix, K is the state feedback control gain.

Now, the objective is to find the nominal MRAC controller:

$$u_1(t) = -\underline{\theta}\left\|\underline{x}\right\|_{L_\infty} + k_g r(t) \tag{2.12}$$

This control law (2.12) provides perfect cancellation of unknown parameter uncertainties present in (2.11). This leads to the ideal reference system as:

$$\dot{\underline{x}}_m(t) = A_m\underline{x}_m(t) + bk_g r(t) \tag{2.13}$$

where $\underline{x}_m(t) \in \Re^n$ is the state of the reference model. k_g is the feed-forward gain, $r(t)$ is the reference to the system.

The model reference adaptive controller is considered as [17]:

$$u_1(t) = -\underline{\hat{\theta}}(t) \left\|\underline{x}\right\|_{L_\infty} + k_g r(t) \tag{2.14}$$

where $\underline{\hat{\theta}}$ is the estimate of $\underline{\theta}$.

Putting the value of (2.14) in (2.11) yields:

$$\underline{\dot{x}}(t) = A_m \underline{x}(t) - b\underline{\tilde{\theta}}(t) \left\|\underline{x}\right\|_{L_\infty} + bk_g r(t) \tag{2.15}$$

where $\underline{\tilde{\theta}}(t) = \underline{\hat{\theta}}(t) - \underline{\theta}(t)$ is the parametric uncertainties estimation error.

Let, the Lyapunov candidate vector is considered as:

$$V_{MRAC} = \frac{1}{2}\underline{e}^T_{MRAC}(t) P \underline{e}_{MRAC}(t) + \frac{1}{2}\underline{\tilde{\theta}}^T(t)\Gamma^{-1}_{MRAC}\underline{\tilde{\theta}}(t) \tag{2.16}$$

where $\underline{e}_{MRAC}(t) = \underline{x}_m(t) - \underline{x}(t)$ is the error between reference system state and system state. Γ_{MRAC} is the adaptation gain of MRAC. P is the solution of the algebraic Lyapunov equation $\frac{1}{2}A_m^T P + \frac{1}{2}PA_m = -Q$, where $Q = Q^T > 0$ is a positive definite symmetric matrix.

The derivative of (2.16) can be provided as:

$$\begin{aligned} \dot{V}_{MRAC} = & - \underline{e}^T_{MRAC}(t) Q \underline{e}_{MRAC}(t) \\ & + \left[\underline{e}^T_{MRAC}(t) P b \underline{\tilde{\theta}}^T(t) \left\|\underline{x}\right\|_{L_\infty} + \underline{\tilde{\theta}}^T(t)\Gamma^{-1}_{MRAC}\dot{\underline{\hat{\theta}}}(t)\right] \end{aligned} \tag{2.17}$$

Now, the system will be stable if the following condition holds:

$$\dot{V}_{MRAC} = -\underline{e}^T_{MRAC}(t) Q \underline{e}_{MRAC}(t) \leq 0 \tag{2.18}$$

This yields the unknown parameter adaptation law as:

$$\dot{\underline{\hat{\theta}}}(t) = -\Gamma_{MRAC}\underline{e}^T_{MRAC}(t) P b \left\|\underline{x}\right\|_{L_\infty} \tag{2.19}$$

Now, $\dot{V}_{MRAC}$ and $\ddot{V}_{MRAC}$ are bounded. Therefore, from Lyapunov stability theory and Barbalat's lemma it is proved that, $\lim_{t\to\infty} \dot{V}_{MRAC}(t) = 0$ and consequently $\underline{e}_{MRAC}(t) \to 0$ as $t \to \infty$.

Remarks 2.1 In this case, the asymptotic convergence of $\underline{x}(t)$ to $\underline{x}_m(t)$ is guaranteed through the asymptotic convergence of parametric uncertainties estimation error. It does not guarantee that the parametric uncertainties estimation error $\underline{\tilde{\theta}}(t)$ will tend to zero as $t \to \infty$. In MRAC, it is proved through the adaptation rule that, the parametric uncertainties estimation error $\tilde{\underline{\theta}}(t)$ is bounded. To provide the guaranteed asymptotic convergence of parametric uncertainties estimation error $\tilde{\underline{\theta}}(t)$ to zero ($\tilde{\underline{\theta}}(t) \rightarrow 0$) as $t \rightarrow \infty$, a predictor model has to be considered. This predictor model estimates the value of parametric uncertainties and provides the condition that the value of parametric uncertainties error ($\hat{\underline{\theta}}(t)$) tend to zero as $t \rightarrow \infty$.

To derive the predictor based MRAC model a predictor model is considered as:

$$\dot{\hat{\underline{x}}}(t) = A_m \hat{\underline{x}}(t) + b(\underline{u}_1(t) + \hat{\underline{\theta}}(t) \left\|\underline{x}\right\|_{L_\infty}) \tag{2.20}$$

where $\hat{\underline{x}}(t)$ is the predictor state, $\hat{\underline{\theta}}(t)$ is the estimate of unknown uncertainties.

Now, the Lyapunov candidate vector is formulated as:

$$V_{MRAC} = \frac{1}{2}\tilde{\underline{x}}^T(t) P \tilde{\underline{x}}(t) + \frac{1}{2}\tilde{\underline{\theta}}^T(t) \Gamma_{MRAC}^{-1} \tilde{\underline{\theta}}(t) \tag{2.21}$$

where the prediction error dynamics is considered as $\dot{\tilde{\underline{x}}}(t) = A_m \tilde{\underline{x}}(t) + b\, \tilde{\underline{\theta}}^T(t) \left\|\underline{x}\right\|_{L_\infty}$.

Lyapunov stability criterion yields: $\dot{V}_{MRAC} = -\tilde{\underline{x}}^T(t) Q \tilde{\underline{x}}(t) \le 0$ which leads to the adaptation rule of parameter uncertainties as:

$$\dot{\hat{\underline{\theta}}}(t) = -\Gamma_{MRAC} \tilde{\underline{x}}^T(t) P b \left\|\underline{x}\right\|_{L_\infty} \tag{2.22}$$

From Lyapunov stability criterion and Barbalat's lemma it is proved that, $\tilde{\underline{x}}(t) \rightarrow 0$ as $t \rightarrow \infty$. Therefore, state error between predictor and system tends to zero as well as the parametric uncertainties estimation error also tends to zero ($\tilde{\underline{\theta}}(t) \rightarrow 0$) as $t \rightarrow \infty$.

From Lyapunov stability criterion, the following bound holds [10]:

$$\frac{1}{2}\lambda_{\min}(P) \left\|\tilde{\underline{x}}(t)\right\|_{L_\infty}^2 \le \frac{1}{2}\tilde{\underline{x}}^T(t) P \tilde{\underline{x}}(t) \le V_{MRAC}(t) \le V_{MRAC}(0), \forall t \ge 0 \tag{2.23}$$

or

$$\left\|\tilde{\underline{x}}(t)\right\|_{L_\infty} \le \frac{\left\|\tilde{\underline{\theta}}\right\|}{\sqrt{\lambda_{\min}(P)\Gamma_{MRAC}}} \tag{2.24}$$

From this bound, it is evident that the state error between the system and predictor can be reduced arbitrarily by increasing the adaptation gain Γ_{MRAC}. This introduces high gain feedback control signal which exhibits high frequency oscillation into the control channel and reduces the robustness. Therefore, MRAC controller cannot increase the adaptation gain Γ_{MRAC} arbitrarily to retain the robustness of the system [17, 18].

In the subsequent sections, the architecture and formulation of the L_1 adaptive controller are portrayed and it is demonstrated that, the high adaptation gain facilitates quick reduction of error between system states and predictor states as well as increases robustness.

2.3 Architecture of Basic L_1 Adaptive Controller

L_1 adaptive controller is comprised of predictor, adaptation laws to adapt unknown constant, time varying uncertainties, disturbances present in the system, control block and low-pass filter. The unknown constant, time varying uncertainties and time varying disturbances, present in the system, are predicted by employing the predictor and their values are adapted following the adaptation laws formulated from the Lyapunov stability criteria. High adaptation gain is utilized to achieve quick transient performance. L_1 adaptive control signal is generated from the adaptive estimates of unknown constant, time varying uncertainties and time varying disturbances. Hence, high frequency component includes into the control channel which may incurs instability of the system. To obtain robust controller, the L_1 adaptive control signal is filtered through a low-pass filter. State feedback control signal is generated and augmented with the filtered L_1 adaptive control signal which are given to the system under control. The architecture of basic L_1 adaptive controller is provided in Fig. 2.1.

Let, the open loop system dynamics containing unknown constant, time varying uncertainties and time varying disturbances is considered as:

$$\dot{\underline{x}}(t) = A_0\underline{x}(t) + b(\omega\underline{u}_1(t) + \underline{\theta}(t)\left\|\underline{x}\right\|_{L_\infty} + \underline{\sigma}(t)) + b\underline{u}_2(t) \tag{2.25}$$

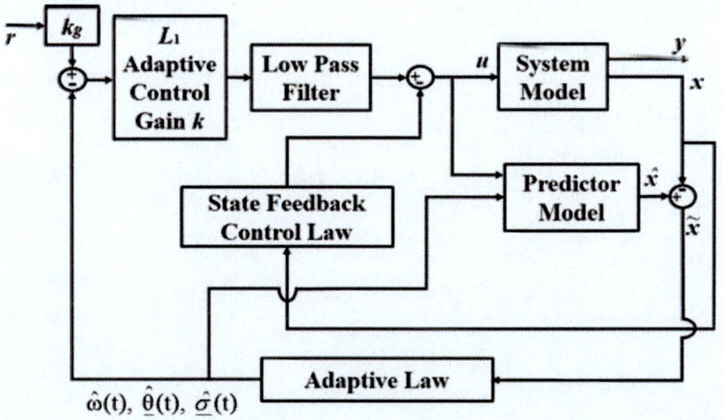

Fig. 2.1 Architecture of basic L_1 adaptive controller

$$\underline{y}(t) = c^T \underline{x}(t) \tag{2.26}$$

where $A_0 \in \Re^{n\times n}$ is the open loop system matrix, $b \in \Re^{n\times p}$ is the input matrix and $c \in \Re^{m\times n}$ is the output matrix. $\underline{x}(t) = [x_1 \ \ x_2 \ \ \ldots x_n]^T \in \Re^n$ with $\underline{x}(0) = \underline{x}_0$ is the state vector, $\underline{y}(t) \in \Re^m$ is the output of the system. Unknown constant $\omega \in \Re^{p\times p}$, time varying uncertainties $\underline{\theta}(t) \in \Re^p$ and time varying disturbances $\underline{\sigma}(t) \in \Re^p$ are bounded to compact set $\left[\omega\middle|\underline{\theta}(t)\middle|\underline{\sigma}(t)\right] \in \left[\Omega\middle|\underline{\Theta}\middle|\underline{\Sigma}\right]$ for $t \geq 0$. $\underline{u}_1(t) \in \Re^p$, $\underline{u}_2(t) \in \Re^p$ are the L_1 adaptive control input and state feedback control law respectively.

The L_1 adaptive control signal, formulated from the adaptive estimates of unknown constant, time varying uncertainties and disturbances, is considered as [10]:

$$\underline{u}_1(s) = -kD(s)(\hat{\underline{r}}_{ud}(s) - k_g\underline{r}(s)) \tag{2.27}$$

where $\hat{\underline{r}}_{ud}(s)$ is the Laplace transform of:

$$\hat{\underline{r}}_{ud}(t) = \hat{\omega}(t)\underline{u}_1(t) + \hat{\underline{\theta}}(t)\left\|\underline{x}\right\|_{L\infty} + \hat{\underline{\sigma}}(t) \tag{2.28}$$

k and k_g are the pre-filter gain and feed-forward gain respectively. $\underline{r}(s)$ is the input to the system. $D(s)$ is a strictly proper stable transfer function.

The $D(s)$ is to be designed in such a way that it incurs a low-pass filter of strictly proper stable transfer function as [10]:

$$C(s) = \frac{\omega k\, D(s)}{(1 + \omega k\, D(s))} \tag{2.29}$$

The DC gain of $C(s)$ is $C(0) = 1$.

The state feedback control law is considered as:

$$\underline{u}_2(t) = -K^T \underline{x}(t) \tag{2.30}$$

where $K \in \Re^{n\times m}$ is the state feedback gain matrix. This K has to ensure the closed loop system matrix

$$A_m = A_0 - bK^T \tag{2.31}$$

is Hurwitz in nature.

The overall control law is given as:

$$\underline{u}(t) = \underline{u}_1(t) + \underline{u}_2(t), \ \in \Re^p \tag{2.32}$$

Putting the values from (2.27) and (2.30) in (2.32) yield:

$$\underline{u}(s) = -kD(s)(\hat{\underline{r}}_{ud}(s) - k_g\underline{r}(s)) - K^T\underline{x}(s) \tag{2.33}$$

Now, the system model (2.25) in closed loop configuration with L_1 adaptive control law becomes:

$$\dot{\underline{x}}(t) = A_m\underline{x}(t) + b(\omega\underline{u}_1(t) + \underline{\theta}(t)\left\|\underline{x}\right\|_{L_\infty} + \underline{\sigma}(t)) \tag{2.34}$$

Then, the predictor model is considered as:

$$\dot{\hat{\underline{x}}}(t) = A_m\hat{\underline{x}}(t) + b(\hat{\omega}\underline{u}_1(t) + \hat{\underline{\theta}}(t)\left\|\underline{x}\right\|_{L_\infty} + \hat{\underline{\sigma}}(t)) \tag{2.35}$$

$$\hat{y}(t) = c^T\hat{\underline{x}}(t) \tag{2.36}$$

where $\hat{\underline{x}}(t)$ is the state vector of predictor model, $\hat{\underline{y}}(t)$ is the output of the predictor. $\hat{\omega}(t), \hat{\underline{\theta}}(t)$ and $\hat{\underline{\sigma}}(t)$ are the estimates of unknown constant, time varying uncertainties and disturbances respectively. Their values are adapted following the adaptation laws formulated from the Lyapunov stability criteria.

Let, the Lyapunov candidate vector is considered as:

$$V = \frac{1}{2}\tilde{\underline{x}}^T(t)P\tilde{\underline{x}}(t) + \frac{1}{2}\tilde{\omega}^T(t)\Gamma_C^{-1}\tilde{\omega}(t) + \frac{1}{2}\tilde{\underline{\theta}}^T(t)\Gamma_C^{-1}\tilde{\underline{\theta}}(t) + \frac{1}{2}\tilde{\underline{\sigma}}^T(t)\Gamma_C^{-1}\tilde{\underline{\sigma}}(t) \tag{2.37}$$

where $\tilde{\omega}(t) = \hat{\omega}(t) - \omega$ is the error of unknown constant present in the system. $\tilde{\underline{\theta}}(t) = \hat{\underline{\theta}}(t) - \underline{\theta}(t)$ is the error of uncertainties and $\tilde{\underline{\sigma}}(t) = \hat{\underline{\sigma}}(t) - \underline{\sigma}(t)$ is the error of disturbances present in the system. $\tilde{\underline{x}}(t) = \hat{\underline{x}}(t) - \underline{x}(t)$ is the state error between predictor and system. $\Gamma_C > 0$ is the high adaption gain.

P is the solution of the algebraic Lyapunov equation given as:

$$\frac{1}{2}A_m^TP + \frac{1}{2}PA_m = -Q$$

or

$$A_m^TP + PA_m = -2Q \tag{2.38}$$

where $Q = Q^T > 0$ is a positive definite symmetric matrix.

Differentiating (2.37) yields:

$$\dot{V} = \frac{1}{2}\dot{\tilde{\underline{x}}}^T(t)P\tilde{\underline{x}}(t) + \frac{1}{2}\tilde{\underline{x}}^T(t)P\dot{\tilde{\underline{x}}}(t) + \frac{1}{2}\dot{\tilde{\omega}}^T(t)\Gamma_C^{-1}\tilde{\omega}(t) + \frac{1}{2}\tilde{\omega}^T(t)\Gamma_C^{-1}\dot{\tilde{\omega}}(t)$$
$$+\frac{1}{2}\dot{\tilde{\underline{\theta}}}^T(t)\Gamma_C^{-1}\tilde{\underline{\theta}}(t) + \frac{1}{2}\tilde{\underline{\theta}}^T(t)\Gamma_C^{-1}\dot{\tilde{\underline{\theta}}}(t) + \frac{1}{2}\dot{\tilde{\underline{\sigma}}}^T(t)\Gamma_C^{-1}\tilde{\underline{\sigma}}(t) + \frac{1}{2}\tilde{\underline{\sigma}}^T(t)\Gamma_C^{-1}\dot{\tilde{\underline{\sigma}}}(t) \tag{2.39}$$

Putting the value of $\dot{\tilde{\underline{x}}}(t) = A_m\tilde{\underline{x}}(t) + b\left[\tilde{\omega}(t)\underline{u}_1(t) + \tilde{\underline{\theta}}^T(t)\|\underline{x}\|_{L_\infty} + \tilde{\underline{\sigma}}(t)\right]$ in (2.39) provides:

$$\begin{aligned}\dot{V} =& \frac{1}{2}\tilde{\underline{x}}^T(t)A_m^T P\tilde{\underline{x}}(t) + \frac{1}{2}\left[\tilde{\omega}(t)\underline{u}_1(t) + \tilde{\underline{\theta}}^T(t)\|\underline{x}\|_{L_\infty} + \tilde{\underline{\sigma}}(t)\right]^T b^T P\tilde{\underline{x}}(t)\\ &+ \frac{1}{2}\tilde{\underline{x}}^T(t)PA_m\tilde{\underline{x}}(t) + \frac{1}{2}\tilde{\underline{x}}^T(t)Pb\left[\tilde{\omega}(t)\underline{u}_1(t) + \tilde{\underline{\theta}}^T(t)\|\underline{x}\|_{L_\infty} + \tilde{\underline{\sigma}}(t)\right]\\ &+ \tilde{\omega}^T(t)\Gamma_C^{-1}\dot{\hat{\omega}}(t) + \tilde{\underline{\theta}}^T(t)\Gamma_C^{-1}\dot{\hat{\underline{\theta}}}(t) - \tilde{\underline{\theta}}^T(t)\Gamma_C^{-1}\dot{\underline{\theta}}(t)\\ &+ \tilde{\underline{\sigma}}^T(t)\Gamma_C^{-1}\dot{\hat{\underline{\sigma}}}(t) - \tilde{\underline{\sigma}}^T(t)\Gamma_C^{-1}\dot{\underline{\sigma}}(t)\end{aligned}$$

or

$$\begin{aligned}\dot{V} =& \frac{1}{2}\tilde{\underline{x}}^T(t)\left[A_m^T P + PA_m\right]\tilde{\underline{x}}(t) + \tilde{\underline{x}}^T(t)Pb\left[\tilde{\omega}(t)u_1(t) + \tilde{\underline{\theta}}^T(t)\|\underline{x}\|_{L_\infty} + \tilde{\underline{\sigma}}(t)\right]\\ &+ \tilde{\omega}^T(t)\Gamma_C^{-1}\dot{\hat{\omega}}(t) + \tilde{\underline{\theta}}^T(t)\Gamma_C^{-1}\dot{\hat{\underline{\theta}}}(t) - \tilde{\underline{\theta}}^T(t)\Gamma_C^{-1}\dot{\underline{\theta}}(t)\\ &+ \tilde{\underline{\sigma}}^T(t)\Gamma_C^{-1}\dot{\hat{\underline{\sigma}}}(t) - \tilde{\underline{\sigma}}^T(t)\Gamma_C^{-1}\dot{\underline{\sigma}}(t)\end{aligned}$$

or

$$\begin{aligned}\dot{V} =& -\tilde{\underline{x}}^T(t)Q\tilde{\underline{x}}(t) + \left[\tilde{\underline{x}}^T(t)Pb\tilde{\omega}(t)\underline{u}_1(t) + \tilde{\omega}^T(t)\Gamma_C^{-1}\dot{\hat{\omega}}(t)\right]\\ &+ \left[\tilde{\underline{x}}^T(t)Pb\tilde{\underline{\theta}}^T(t)\|\underline{x}\|_{L_\infty} + \tilde{\underline{\theta}}^T(t)\Gamma_C^{-1}\dot{\hat{\underline{\theta}}}(t)\right]\\ &+ \left[\tilde{\underline{x}}^T(t)Pb\tilde{\underline{\sigma}} + \tilde{\underline{\sigma}}^T(t)\Gamma_C^{-1}\dot{\hat{\underline{\sigma}}}(t)\right]\\ &+ \left|\tilde{\underline{\theta}}^T(t)\Gamma_C^{-1}\dot{\underline{\theta}}(t)\right| + \left|\tilde{\underline{\sigma}}^T(t)\Gamma_C^{-1}\dot{\underline{\sigma}}(t)\right|\end{aligned} \tag{2.40}$$

Now, the system will be stable in the sense of Lyapunov if the following criterion takes place:

$$\dot{V} = \left(-\tilde{\underline{x}}^T(t)Q\tilde{\underline{x}}(t) + \frac{1}{\Gamma_C}\left(\left|\tilde{\underline{\theta}}^T(t)\dot{\underline{\theta}}(t)\right| + \left|\tilde{\underline{\sigma}}^T(t)\dot{\underline{\sigma}}(t)\right|\right)\right) \le 0 \tag{2.41}$$

This Lyapunov stability criterion incurs that:

$$\left[\tilde{\underline{x}}^T(t)Pb\tilde{\omega}(t)\underline{u}_1(t) + \tilde{\omega}^T(t)\Gamma_C^{-1}\dot{\hat{\omega}}(t)\right] = 0$$

$$\left[\underline{\tilde{x}}^T(t)Pb\underline{\tilde{\theta}}^T(t)\,\|\underline{x}\|_{L_\infty}+\underline{\tilde{\theta}}^T(t)\Gamma_C^{-1}\dot{\underline{\hat{\theta}}}(t)\right]=0$$

$$\left[\underline{\tilde{x}}^T(t)Pb\underline{\tilde{\sigma}}+\underline{\tilde{\sigma}}^T(t)\Gamma_C^{-1}\dot{\underline{\hat{\sigma}}}(t)\right]=0$$

These lead the following adaptation laws for unknown constant, time varying uncertainties and time varying disturbances respectively [10]:

$$\dot{\hat{\omega}}(t)=\Gamma_C\,\mathrm{Proj}(\hat{\omega}(t),-\underline{\tilde{x}}^T(t)Pb\,u_1(t)) \tag{2.42}$$

$$\dot{\underline{\hat{\theta}}}(t)=\Gamma_C\,\mathrm{Proj}(\underline{\hat{\theta}}(t),-\underline{\tilde{x}}^T(t)Pb\|\underline{x}\|_{L_\infty}) \tag{2.43}$$

$$\dot{\underline{\hat{\sigma}}}(t)=\Gamma_C\,\mathrm{Proj}(\underline{\hat{\sigma}}(t),-\underline{\tilde{x}}^T(t)Pb) \tag{2.44}$$

Proj (· , ·) denotes the Pomet-Praly projection operator [19, 20].

2.4 Stability Analysis of Basic L_1 Adaptive Controller

Let, minimum eigenvalue of P is $\lambda_{\min}(P)$. Then, following upper bound holds:

$$\frac{1}{2}\lambda_{\min}(P)\|\underline{\tilde{x}}(t)\|^2\le\frac{1}{2}\underline{\tilde{x}}^T(t)P\underline{\tilde{x}}(t)\le V(t)\le V(0),\forall t\ge 0 \tag{2.45}$$

As the projection operator ensures the bounds of $[\omega|\underline{\theta}(t)|\underline{\sigma}(t)]\in[\Omega|\underline{\Theta}|\underline{\Sigma}]$, $\forall t\ge 0$, therefore, following upper bounds take place:

$$\begin{aligned}&\max\left(\frac{1}{\Gamma_C}\left(\tilde{\omega}^T(t)\omega+\underline{\tilde{\theta}}^T(t)\,\underline{\theta}(t)+\underline{\tilde{\sigma}}^T(t)\,\underline{\sigma}(t)\right)\right)\\&\le\frac{4}{\Gamma_C}\left(\max_{\omega\in\Omega}tr(\omega^T\omega)+\max_{\theta\in\Theta}(\underline{\theta}^T\underline{\theta})+\max_{\sigma\in\Sigma}(\underline{\sigma}^T\underline{\sigma})\right)\end{aligned} \tag{2.46}$$

and

$$\left(\underline{\tilde{\theta}}^T(t)\,\dot{\underline{\theta}}(t)+\underline{\tilde{\sigma}}^T(t)\,\dot{\underline{\sigma}}(t)\right)\le 2\left(\max_{\theta\in\Theta}\|\underline{\theta}\|d\theta+\max_{\sigma\in\Sigma}\|\underline{\sigma}\|d\sigma\right) \tag{2.47}$$

Here, $\dot{\underline{\theta}}(t)$ and $\dot{\underline{\sigma}}(t)$ are completely differentiable and bounded as $\dot{\underline{\theta}}(t)\in d\theta$ and $\dot{\underline{\sigma}}(t)\in d\sigma$ respectively.

Now, it can be written that:

$$\underline{\tilde{x}}^T(t)Q\underline{\tilde{x}}(t) \leq \frac{1}{2}\frac{\lambda_{\min}(Q)}{\lambda_{\max}(P)}\underline{\tilde{x}}^T(t)P\underline{\tilde{x}}(t) \leq 2\Gamma_C^{-1}\left(\max_{\theta\in\Theta}\|\underline{\theta}\|d\theta + \max_{\sigma\in\Sigma}\|\underline{\sigma}\|d\sigma\right) \tag{2.48}$$

Again, it can be shown that [10]:

$$V(t) \leq \theta_m/\Gamma_C \tag{2.49}$$

where

$$\begin{aligned}\theta_m &= 2\left(\max_{\omega\in\Omega} tr(\omega^T\omega) + \max_{\theta\in\Theta}(\underline{\theta}^T\underline{\theta}) + \max_{\sigma\in\Sigma}(\underline{\sigma}^T\underline{\sigma})\right)\\ &+4\frac{\lambda_{\min}(Q)}{\lambda_{\max}(P)}\left(\max_{\theta\in\Theta}\|\underline{\theta}\|d\theta + \max_{\sigma\in\Sigma}\|\underline{\sigma}\|d\sigma\right)\end{aligned} \tag{2.50}$$

This incurs:

$$\left\|\underline{\tilde{x}}(t)\right\|_{L\infty} \leq \sqrt{\frac{2\theta_m}{\lambda_{\min}(P)\ \Gamma_C}}$$

or

$$\left|\left\|\underline{\hat{x}}(t)\right\|_{L\infty} - \left\|\underline{x}(t)\right\|_{L\infty}\right| \leq \sqrt{\frac{2\theta_m}{\lambda_{\min}(P)\ \Gamma_C}} \tag{2.51}$$

Therefore, the state error between the predictor and system is bounded.
The predictor model (2.35) in frequency domain follows the form given as:

$$\left\|\underline{\hat{x}}(t)\right\|_{L_\infty} \leq \left(\|G(s)\|_{L_1}\|r_{ud}(t)\|_{L_\infty} + \|H(s)C(s)\|_{L_1}\left\|k_g r(t)\right\|_{L_\infty}\right) \tag{2.52}$$

where

$$H(s) = (SI - A_m)^{-1}b \tag{2.53}$$

and

$$G(s) = H(s)(1 - C(s)) \tag{2.54}$$

Now, putting the values of $r_{ud}(t)$ and then $\left\|\underline{x}(t)\right\|_{L\infty}$ in (2.52) incur:

$$\left\|\underline{\hat{x}}(t)\right\|_{L_\infty} \le \left(\frac{\begin{array}{c}\|G(s)\|_{L_1}\omega_{\max}\left\|\underline{u}_1(t)\right\|_{L_\infty} + \|G(s)\|_{L_1}\theta_{\max}\sqrt{\dfrac{2\theta_m}{\lambda_{\min}(P)\,\Gamma_C}} \\ +\|G(s)\|_{L_1}\underline{\sigma}_{\max} + \|H(s)\,C(s)\|_{L_1}\|r(t)\|_{L_\infty}\end{array}}{(1-\|G(s)\|_{L_1}\theta_{\max})} \right) \tag{2.55}$$

where $\omega_{\max} = \max\limits_{\omega\in\Omega}(\omega)$, $\underline{\theta}_{\max} = \max\limits_{\underline{\theta}\in\Theta}(\underline{\theta})$ and $\underline{\sigma}_{\max} = \max\limits_{\underline{\sigma}\in\Sigma}(\underline{\sigma})$.

Therefore, to stabilize the system, L_1 norm condition becomes:

$$\|G(s)\|_{L_1}\theta_{\max} < 1 \tag{2.56}$$

Now, $\|G(s)\|$, $\|H(s)\|$, $\omega_{\max}$, $\underline{\theta}_{\max}$, $\underline{\sigma}_{\max}$ all are bounded, therefore, $\left\|\underline{\hat{x}}(t)\right\|_{L\infty}$ is finite for any $t > 0$, hence $\underline{\hat{x}}(t)$ is bounded. Now, (2.51) yields $\left\|\underline{x}(t)\right\|_{L\infty}$ is also finite $\forall t > 0$, and hence $\underline{x}(t)$ is bounded.

Again, Q is positive definite, which incurs $\dot{V}$ is negative semi-definite, viz., $V\left(\underline{\tilde{x}}(t), \hat{\omega}(t), \underline{\hat{\theta}}(t), \underline{\hat{\sigma}}(t)\right) \le V\left(\underline{\tilde{x}}(0), \hat{\omega}(0), \underline{\hat{\theta}}(0), \underline{\hat{\sigma}}(0)\right)$. It in turn provides $\underline{\tilde{x}}^T(t)P\underline{\tilde{x}}$ is bounded as $\hat{\omega}(t), \underline{\hat{\theta}}(t), \underline{\hat{\sigma}}(t)$ are bounded. Let, $\Psi = \underline{\tilde{x}}^T(t)Q\underline{\tilde{x}}(t) \le -\dot{V}$. Integration of this is given as $\int\limits_0^t \Psi(\tau)\,d\tau \le V\left(\underline{\tilde{x}}(0), \hat{\omega}(0), \underline{\hat{\theta}}(0), \underline{\hat{\sigma}}(0)\right) - V\left(\underline{\tilde{x}}(t), \hat{\omega}(t), \underline{\hat{\theta}}(t), \underline{\hat{\sigma}}(t)\right)$. The first term $V\left(\underline{\tilde{x}}(0), \hat{\omega}(0), \underline{\hat{\theta}}(0), \underline{\hat{\sigma}}(0)\right)$ is bounded and the second term $V\left(\underline{\tilde{x}}(t), \hat{\omega}(t), \underline{\hat{\theta}}(t), \underline{\hat{\sigma}}(t)\right)$ is non-increasing and bounded which incurs $\lim\limits_{t\to\infty} \int\limits_0^t \Psi(\tau)\,d\tau < \infty$ and hence $\dot{\Psi}(t)$ is bounded. Therefore, the Barbalat's lemma provides $\lim\limits_{t\to\infty} \Psi(t) = 0$ which implies that $\lim\limits_{t\to\infty}\left|\underline{\tilde{x}}\right| = 0$. Therefore, the state error between predictor and system diminishes and the system follows the predictor model as well as become stable.

2.5 Transient and Steady State Performance Analysis of Basic L_1 Adaptive Controller

A reference system model is considered to analyze the transient and steady state performance of the system. The closed loop system has to follow the transient and steady state responses of closed loop reference system.

The system model and control input in frequency domain are expressed as respectively:

$$\underline{x}(s) = H(s)((1 - C(s))\underline{r}_2(s) + C(s)k_g\underline{r}(s) - C(s)\underline{\tilde{r}}_{ud}(s)) \tag{2.57}$$

$$\underline{u}(s) = \frac{C(s)}{\omega}\left[k_g\underline{r}(s) - \underline{r}_2(s) - \underline{\tilde{r}}_{ud}(s)\right] - K\underline{x}(s) \tag{2.58}$$

where $\underline{r}_2(s)$ is the Laplace transform of $\underline{r}_2(t) = \theta \left\|\underline{x}\right\|_{L_\infty} + \sigma$.

Let, the closed loop LTI reference system model with control input are considered as respectively:

$$\underline{x}_{ref}(s) = H(s)((1 - C(s))\underline{r}_{2ref}(s) + C(s)k_g\underline{r}(s)) \tag{2.59}$$

$$\underline{u}_{ref}(s) = \frac{C(s)}{\omega}\left[k_g\underline{r}(s) - \underline{r}_{2ref}(s)\right] - K\underline{x}_{ref}(s) \tag{2.60}$$

where $\underline{r}_{2ref}(s)$ is the Laplace transform of $\underline{r}_{2ref}(t) = \theta \left\|\underline{x}_{ref}\right\|_{L_\infty} + \sigma$. $\underline{r}_{2ref}(t)$ will satisfy the boundary condition $\left\|\underline{r}_{2ref}\right\|_{L_\infty} \leq \left(\|\theta\|_{L_1}\left\|\underline{x}_{ref}\right\|_{L_\infty} + \|\sigma\|_{L_\infty}\right)$.

The state error and control input error between the system and reference system model hold the following bounds respectively [10]:

$$\left\|\underline{e}\right\|_{L_\infty} = \left\|\underline{x}_{ref} - \underline{x}\right\|_{L_\infty} \leq val_1 \tag{2.60}$$

$$\left\|\underline{u}_{ref} - \underline{u}\right\|_{L_\infty} \leq val_2 \tag{2.61}$$

where

$$val_1 = \frac{\|C(s)\|_{L_1}}{1 - \|H(s)(1 - C(s))\|_{L_1}\theta_{\max}}\sqrt{\frac{2\theta_m}{\lambda_{\min}(P)\Gamma_C}}$$

and

$$val_2 = \left(-\left\|\frac{C(s)\theta}{\omega} + K\right\|_{L_1} val_1 + \left\|\frac{C(s)}{\omega}\cdot\frac{1}{c_0^T H(s)}\cdot c_0^T\right\|_{L_1}\sqrt{\frac{2\theta_m}{\lambda_{\min}(P)\Gamma_C}}\right)$$

Remarks 2.2 From these findings, it can be stated that, as the adaptation gain Γ_C increases, the error between the system state and reference system state reduces arbitrarily. At the same time, the control signal error between the system and reference model can be arbitrarily reduced by increasing the adaptation gain. Therefore, the use of low-pass filter in the architecture of L_1 adaptive controller provides bounded control signal and hence the increase in adaptation gain increases the time delay margin, viz., robustness.

2.6 Simulation Case Studies

In this book, four systems are evaluated in simulation case studies. Two single-input single-output (SISO) systems, viz., chaotic Duffing's oscillatory system, nonlinear spring mass damper system are examined. One single-input multi-output (SIMO), unstable, non-minimum phase 4th order inverted pendulum with cart system is investigated to show the performances of controllers in tackling inherent unstable system. One cross-coupled, nonlinear twin rotor multi-input multi-output (MIMO) system (TRMS) is also considered to examine the performances of the controllers in tackling such complicated system. These systems are nonlinear and contain time varying uncertainties and disturbances. The performance of basic L_1 adaptive controller is compared with classical controllers, viz., PID controller, LQG and conventional adaptive controller, viz., MRAC. The system dynamics are simulated by employing Runge–Kutta (R-K) 4th order method and integral absolute errors (IAE) are calculated to analysis the performances of the controllers [21]. The four control methodologies, employed on the systems, are briefly described next.

PID controller: The proportional gain (K_P), integral gain (K_i) and derivative gain (K_d) of the PID controller have to be chosen sensibly to obtain better performance of this controller. The Ziegler-Nichols (Z-N) method of designing PID controller is not able to tackle the system with nonlinearity and significant cross-coupling [15]. Therefore, in this book, the gains of the PID controller are tuned employing a well-known stochastic optimization technique, viz., particle swarm optimization (PSO) [22]. PSO is employed in this work as the representative stochastic global optimization algorithm to design the PID controller. The candidate solution vector of PSO is formulated as $\underline{Z} = [K_p|K_i|K_d\ ]$. 30 number of particles of PSO are iterated for 200 times to generate the optimal parameter setting of controller that the system response can track reference trajectory providing minimum integral absolute error, $IAE = \int_0^{PST} |r(t) - y(t)|dt$, where PST is the plant simulation time. This process runs 10 times and from that minimum, viz., best, average and standard deviation (std. dev.) of IAEs are calculated and tabulated. The response of the system with best IAE is shown graphically.

LQG controller: The state feedback controller gain K and observer gain L are tuned utilizing a representative stochastic optimization technique, viz., PSO [16]. In this case, the performance index is considered as $PI = \int_{t_0}^{t_f} (\underline{e}^T Q_e \underline{e} + \underline{u}^T R_e \underline{u})dt$. The candidate solution vector is considered as $\underline{Z} = [K|L|Q_e|R_e\ ]$. The optimal LQG controller is designed employing 30 number of particles and 200 iterations of PSO. This controller provides optimal state feedback control signal with noise rejection.

Integral absolute error (IAE) of this controller is also calculated for the comparison purpose with the other controllers. This process runs 10 times to calculate and tabulate minimum, viz., best, average and standard deviation of IAEs. The response of the system with best IAE is portrayed graphically.

Model reference adaptive controller (MRAC): In case of MRAC, the parameter values k_g,Γ_{MRAC} and K are obtained from mathematical calculation (2.24) keeping the stability of the system. During the design of this controller, the adaptation gain Γ_{MRAC} is set at a low value which provides slow transient performance. The adaptive parameters are adapted following the adaptation laws (2.19) formulated from the Lyapunov stability criteria. The tracking performance of the system with this controller is evaluated for comparison purpose.

Basic L_1 adaptive controller: The parameter values of L_1 adaptive controller, viz., k , k_g, Γ_C , $\hat{\omega}$, $\underline{\hat{\theta}}$, $\underline{\hat{\sigma}}$ are selected within the mathematical ranges obtained from L_1 norm condition (2.56). State feedback controller gain K is also selected from the mathematical calculation of stability condition. Then, the unknown constant, time varying uncertainties and time varying disturbances are adapted following the adaptation laws (2.42)–(2.44). Proper estimation of these quantities helps to nullify them properly from the system. High adaptation gain facilitates quick transient performance and the low-pass filter provides robust performance by eliminating high frequency inclusion into the control channel.

It is to be noted that, in case of the controllers, designed utilizing any stochastic optimization and metaheuristics techniques, the statistical measures of performances, viz., best, average and standard deviation (std. dev.) of IAEs are taken. Whereas, for conventional adaptive controllers, these statistical measures are not required to be calculated. Thus, keeping that in mind, the table for comparative studies of different controllers in each case study is prepared.

The systems are at first tuned with small amount of disturbance or without disturbance and then, the tuned controllers are subjected to disturbances with higher magnitude and frequency to verify the effectiveness of basic L_1 adaptive controller over other controllers. The disturbance used for Duffing's oscillatory system and nonlinear spring mass damper system at the time of tuning is shown in Fig. 2.2. This disturbance has random amplitude varying between [-1 1] with time period 0.5 s. Then the tuned controllers, without further tuning, are subjected to the 10 dB white Gaussian noise as depicted in Fig. 2.3. This noise as depicted in Fig. 2.3 is of higher magnitude and frequency than the disturbance as depicted in Fig. 2.2.

2.6.1 Case Study-I: Duffing's Oscillatory System

Duffing's oscillatory system with disturbance is considered as [23]:

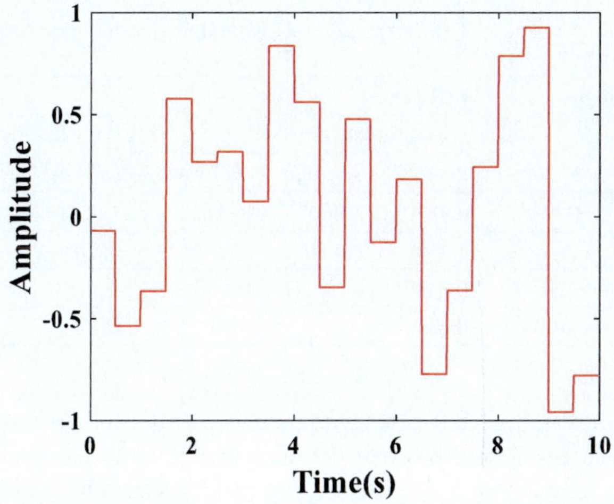

Fig. 2.2 Nature of the disturbance

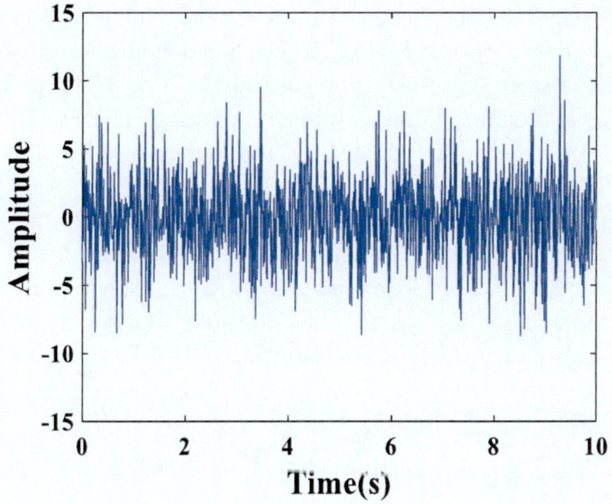

Fig. 2.3 Nature of 10 dB white Gaussian noise

$$\left.\begin{aligned} \dot{x}_1 &= x_2 \\ \dot{x}_2 &= -0.1x_2 - x_1^3 + 12\cos(t) + u(t) + d(t) \\ y &= x_1 \end{aligned}\right\} \tag{2.63}$$

where $d(t)$ represents the disturbance. $u(t)$ is the force applied to the system, represented as the control input, and $y(t)$ is the displacement output of the system. All units are in S.I. The controller produces desired control input for the Duffing's oscillatory system with disturbance that it can track a desired trajectory $r = 2 * \sin t$. The

Table 2.1 Comparative study of different control methodologies for case study-I (Duffing's oscillatory system)

Control methodologies	IAE value			
	For disturbance			For white Gaussian noise
	Best	Average	Std. Dev	
PID controller	0.9820	1.0602	0.1552	1.0581
LQG	0.4506	0.4646	0.0225	0.4580
MRAC	0.3307	–	–	0.5999
Basic L_1 adaptive controller	0.2347	–	–	0.4559

system dynamics is simulated for 10 s with time step 0.01 s and the responses, evaluated in terms of integral absolute error, are tabulated in Table 2.1. System responses with disturbance employing four different controllers, viz., PID, LQG, MRAC and basic L_1 adaptive controller are shown in Fig. 2.4a, Fig. 2.5a, Fig. 2.6a and Fig. 2.7a respectively. Corresponding control input required and tracking error are depicted in Fig. 2.4b, Fig. 2.5b, Fig. 2.6b and Fig. 2.7b respectively. From tabulated results and figures, it can be stated that, performance of the basic L_1 adaptive controller is better than the classical controllers and MRAC. These tuned controllers are then subjected to 10 dB white Gaussian noise. These results are also provided in Table 2.1. The system responses and control signal with error signal plots for PID, LQG, MRAC and basic L_1 adaptive controller when subjected to the noise are shown in Fig. 2.8, Fig. 2.9, Fig. 2.10 and Fig. 2.11 respectively.

From results it is evident that, the basic L_1 adaptive controller can eliminate time varying uncertainties and disturbances better than the classical controllers. The gains of the classical controllers are fixed hence, do not cope with time varying uncertainties and disturbances. Basic L_1 adaptive controller adapts the uncertainties and disturbances quickly than the MRAC and hence, provides fast transient performance.

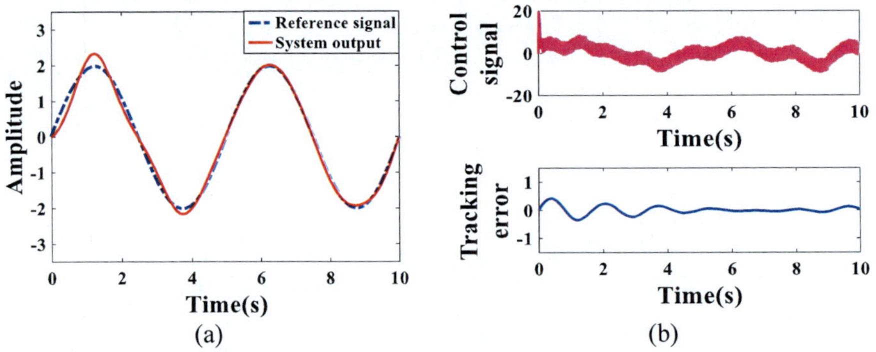

Fig. 2.4 Evaluation period **a** system response and corresponding **b** control signal with tracking error of case study-I (Duffing's oscillatory system) with disturbance for PID controller

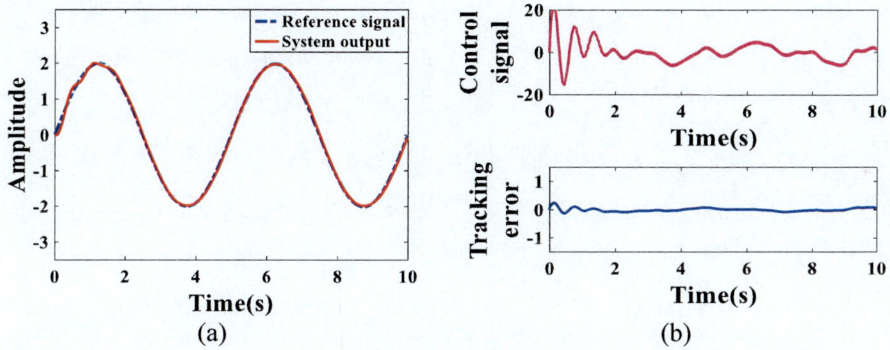

Fig. 2.5 Evaluation period **a** system response and corresponding **b** control signal with tracking error of case study-I (Duffing's oscillatory system) with disturbance for LQG

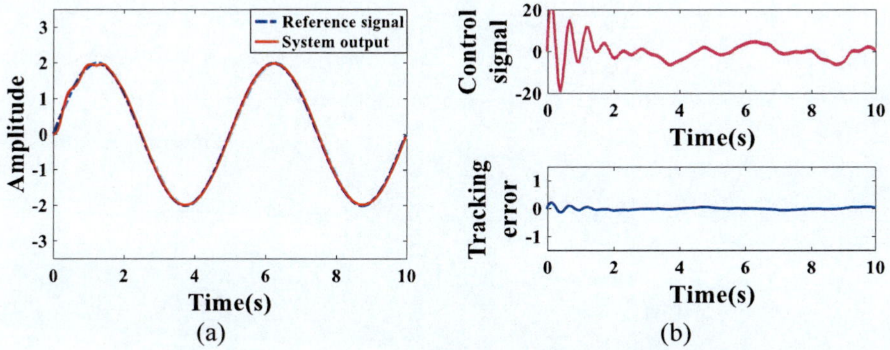

Fig. 2.6 Evaluation period **a** system response and corresponding **b** control signal with tracking error of case study-I (Duffing's oscillatory system) with disturbance for MRAC

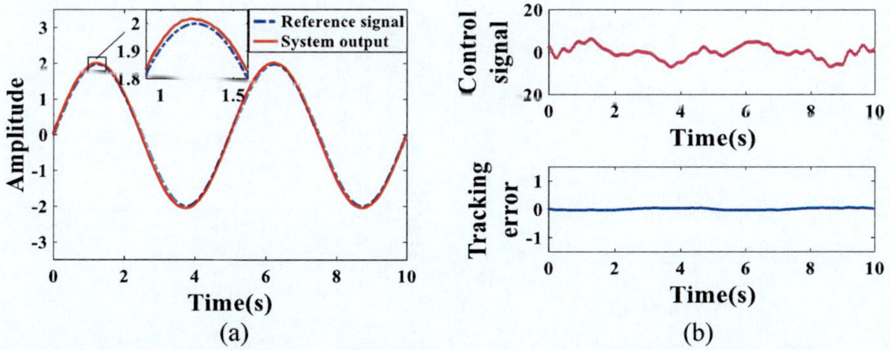

Fig. 2.7 Evaluation period **a** system response and corresponding **b** control signal with tracking error of case study-I (Duffing's oscillatory system) with disturbance for basic L_1 adaptive controller

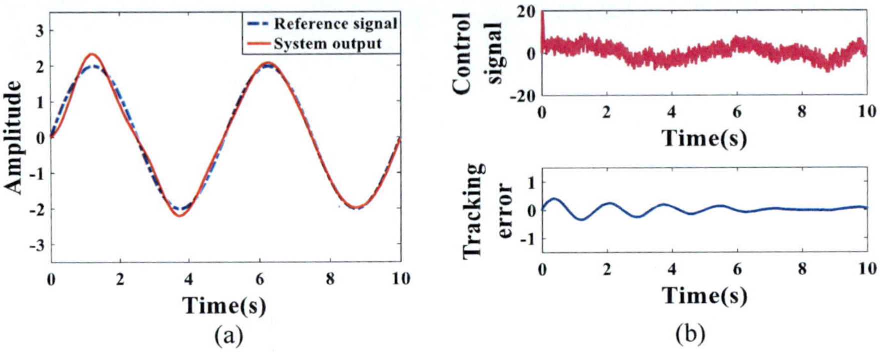

Fig. 2.8 Evaluation period **a** system response and corresponding **b** control signal with tracking error of case study-I (Duffing's oscillatory system) with white Gaussian noise for PID controller

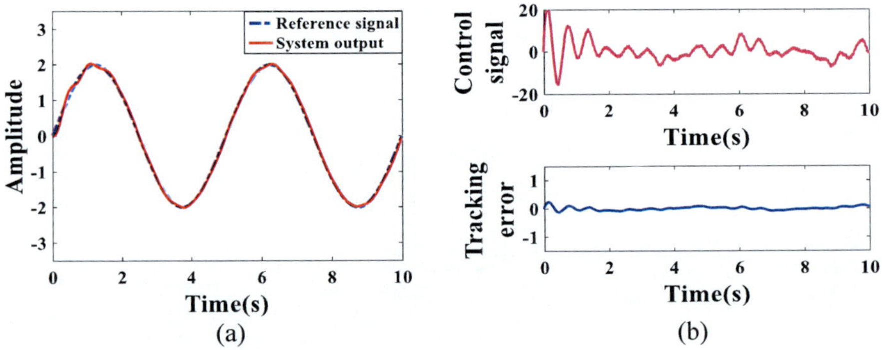

Fig. 2.9 Evaluation period **a** system response and corresponding **b** control signal with tracking error of case study-I (Duffing's oscillatory system) with white Gaussian noise for LQG

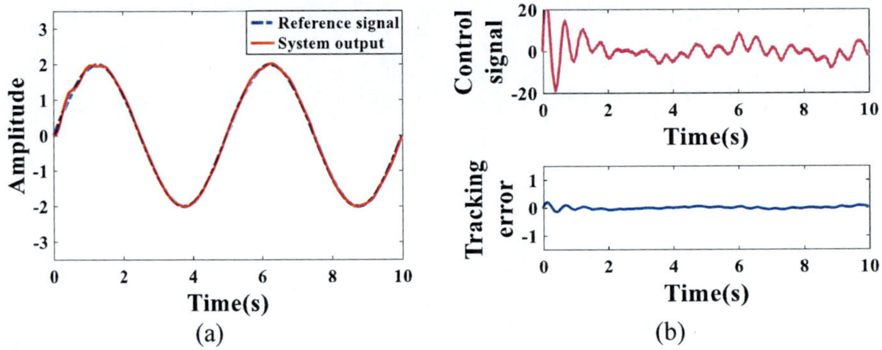

Fig. 2.10 Evaluation period **a** system response and corresponding **b** control signal with tracking error of case study-I (Duffing's oscillatory system) with white Gaussian noise for MRAC

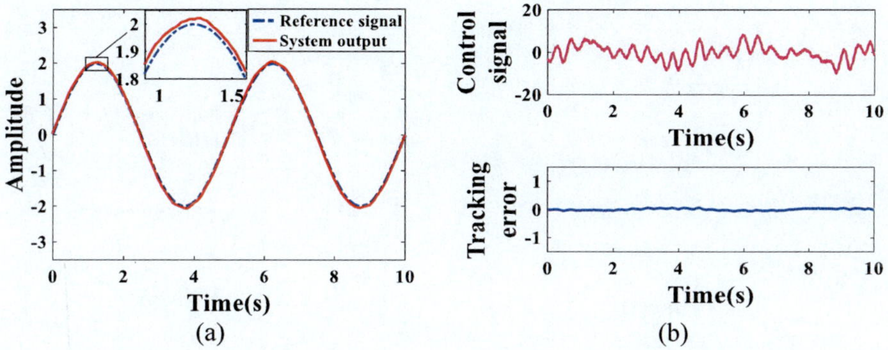

Fig. 2.11 Evaluation period **a** system response and corresponding **b** control signal with tracking error of case study-I (Duffing's oscillatory system) with white Gaussian noise for basic L_1 adaptive controller

2.6.2 Case Study-II: Nonlinear Spring Mass Damper System

The nonlinear spring mass damper system with disturbance is given as [23]:

$$\left.\begin{aligned} \dot{x}_1 &= x_2 \\ \dot{x}_2 &= -0.02x_1 - 0.67x_1^3 - 0.1x_2^3 + u(t) + d(t) \\ y &= x_1 \end{aligned}\right\} \tag{2.64}$$

where $d(t)$ represents time varying disturbance present in the system. $u(t)$ represents input force and $y(t)$ represents output displacement of the system. All parameters are in S.I. unit. The system is evaluated with the aim to track desired trajectory $r = 2 * \sin t$ for 10 s with time step 0.01 s. The controllers are tuned for disturbance as shown in Fig. 2.2. The tracking performances are tabulated in Table 2.2. The system responses and required control effort with error signal for PID, LQG, MRAC and L_1 adaptive controllers are shown in Fig. 2.12, Fig. 2.13, Fig. 2.14 and Fig. 2.15 respectively. Then the controllers are subjected to the 10 dB white Gaussian noise as

Table 2.2 Comparative study of different control methodologies for case study-II (nonlinear spring mass damper system)

Control methodologies	IAE value			
	For disturbance			For white Gaussian noise
	Best	Average	Std. Dev	
PID controller	0.1588	0.1651	0.0132	0.4007
LQG	0.3075	0.5783	0.2480	0.3285
MRAC	0.2549	–	–	0.2853
Basic L_1 adaptive controller	0.0813	–	–	0.2694

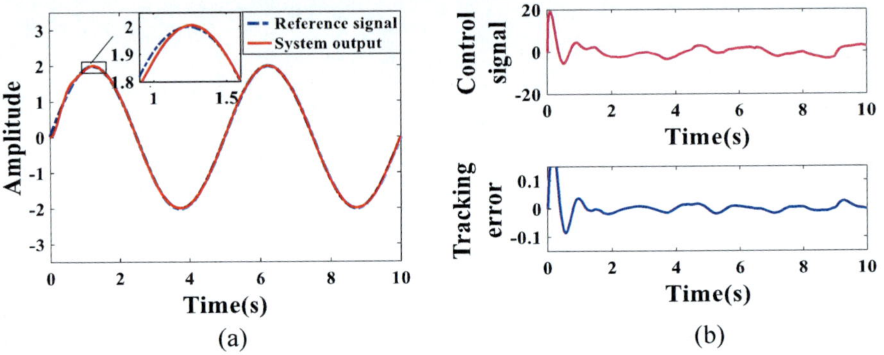

Fig. 2.12 Evaluation period **a** system response and corresponding **b** control signal with tracking error of case study-II (nonlinear spring mass damper) with disturbance for PID controller

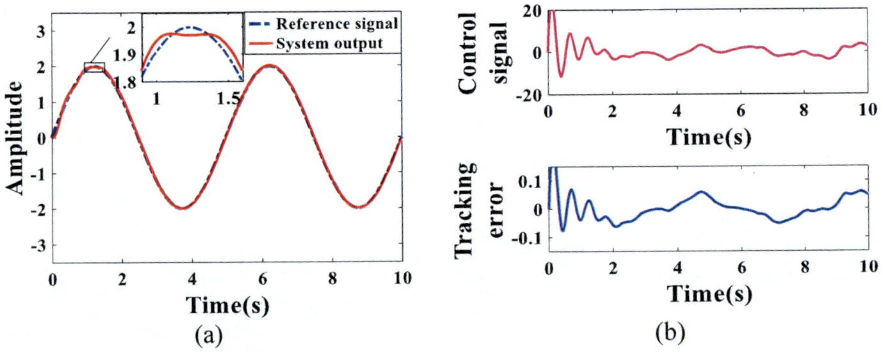

Fig. 2.13 Evaluation period **a** system response and corresponding **b** control signal with tracking error of case study-II (nonlinear spring mass damper) with disturbance for LQG

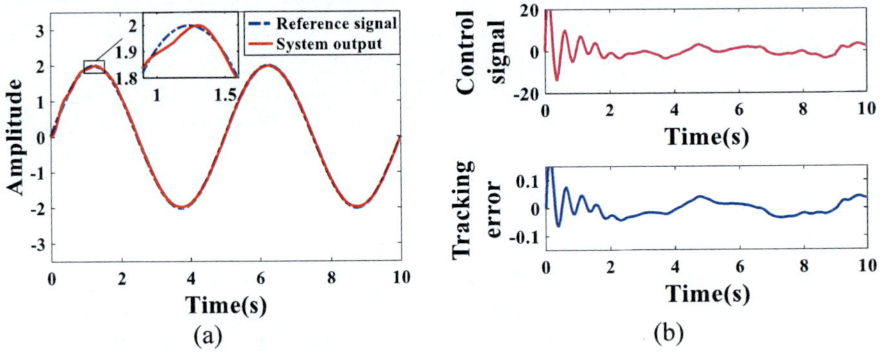

Fig. 2.14 Evaluation period **a** system response and corresponding **b** control signal with tracking error of case study-II (nonlinear spring mass damper) with disturbance for MRAC

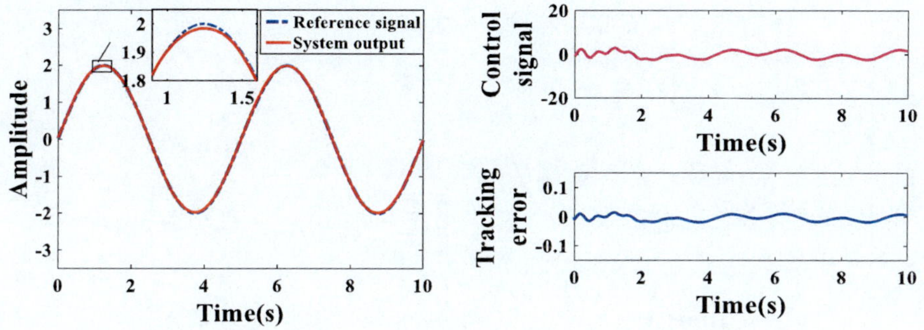

(a) (b)

Fig. 2.15 Evaluation period **a** system response and corresponding **b** control signal with tracking error of case study-II (nonlinear spring mass damper) with disturbance for basic L_1 adaptive controller

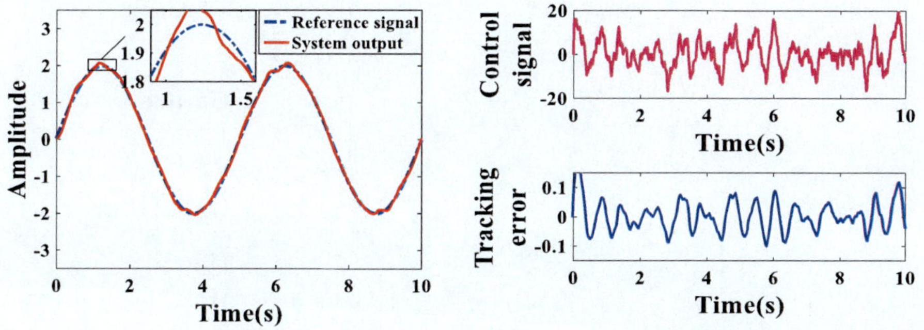

(a) (b)

Fig. 2.16 Evaluation period **a** system response and corresponding **b** control signal with tracking error of case study-II (nonlinear spring mass damper) with white Gaussian noise for PID controller

shown in Fig. 2.3 without further tuning. The performances of four above mentioned controllers with noise are portrayed in Fig. 2.16, Fig. 2.17, Fig. 2.18 and Fig. 2.19 respectively.

From results, it is evident that, the basic L_1 adaptive controller can eliminate time varying uncertainties and disturbances quickly and well enough than the classical controllers and MRAC.

2.6.3 Case Study III: Inverted Pendulum with Cart

In this case study, an unstable, non-minimum phase inverted pendulum with cart system [24–26] is examined. This system is an under-actuated system having two degree of freedoms with only one control input [27]. The schematic diagram of an

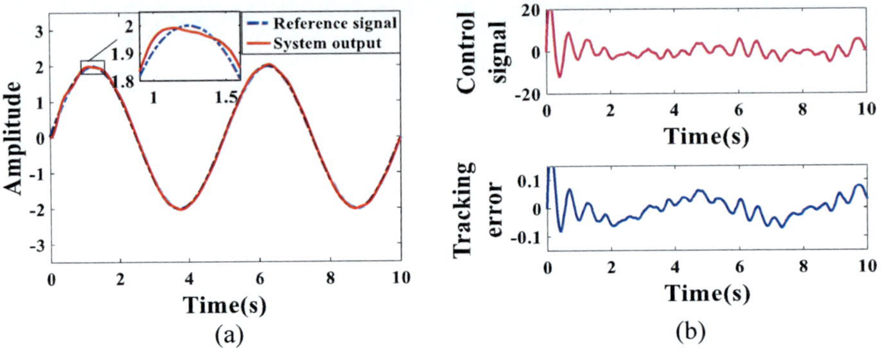

Fig. 2.17 Evaluation period **a** system response and corresponding **b** control signal with tracking error of case study-II (nonlinear spring mass damper) with white Gaussian noise for LQG

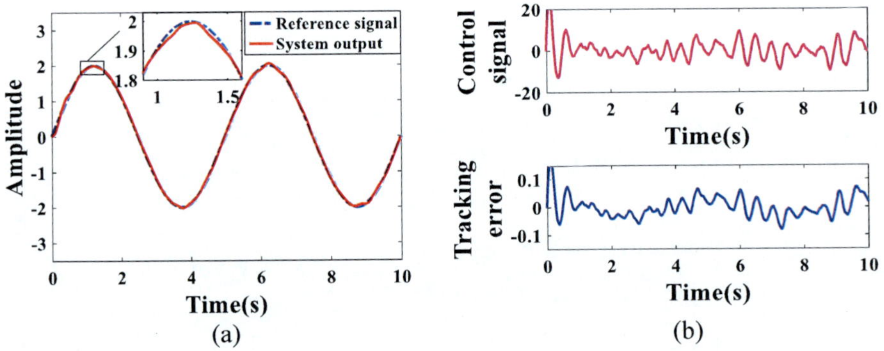

Fig. 2.18 Evaluation period **a** system response and corresponding **b** control signal with tracking error of case study-II (nonlinear spring mass damper) with white Gaussian noise for MRAC

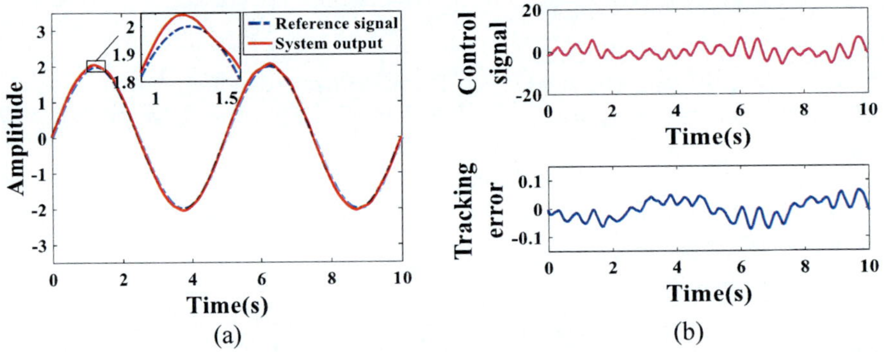

Fig. 2.19 Evaluation period **a** system response and corresponding **b** control signal with tracking error of case study-II (nonlinear spring mass damper) with white Gaussian noise for basic L_1 adaptive controller

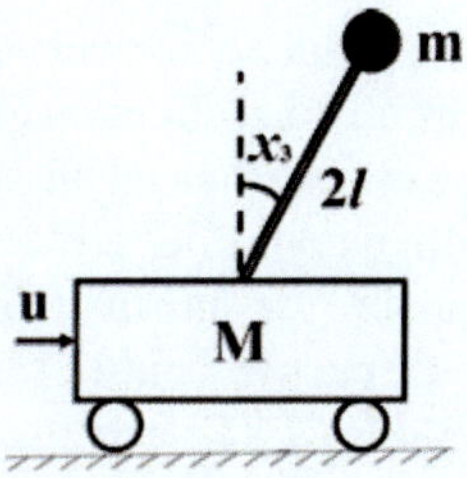

Fig. 2.20 Schematic diagram of inverted pendulum with cart

inverted pendulum with cart system [28] is provided in Fig. 2.20. Different types of classical controllers were designed to tackle this system in different literatures [29–31] though their performances were not satisfactory [32]. Fractional order controller [33], model predictive adaptive controllers [34] were also implemented for such system. L_1 adaptive controller was designed to tackle unmodelled dynamics of two inverted pendulum with cart system [24]. In this case study, the comparison of classical controllers, MRAC and L_1 adaptive controller are performed.

The dynamics of the system is given as follows [35]:

$$\left.\begin{aligned}
\dot{x}_1 &= x_2 \\
\dot{x}_2 &= \frac{-mg\,\cos x_3\,\sin x_3 + ml\sin x_3\, x_4^2 + u(t)}{M + m\sin^2 x_3} + d_1(t) \\
\dot{x}_3 &= x_4 \\
\dot{x}_4 &= \frac{-ml\,\cos x_3\,\sin x_3\, x_4^2 + (M+m)g\sin x_3 \,-\, u(t)\cos x_3}{Ml + ml\sin^2 x_3} + d_2(t)
\end{aligned}\right\} \tag{2.65}$$

where $x_1,\ x_2$ represent cart position and cart velocity respectively. $x_3,\ x_4$ are the pendulum angle and pendulum angular velocity respectively. $u(t)$ is the input force and the output displacement is $y = x_1$. All units are in S.I. $d_1(t)$ is the time varying disturbance of amplitude varying between $[-1, 1]$ with time period 0.5 s as shown in Fig. 2.2. $d_2(t)$ is the disturbance with magnitude four times that of the $d_1(t)$, viz., its magnitude varies between $[-4, 4]$ with time period 0.5 s. The parameter values of the system [35] are provided in Table 2.3.

Table 2.3 Parameter values of inverted pendulum with cart

Parameter	Description	Value
M	Mass of pivot	1 kg
m	Mass of pendulum rod	0.1 kg
l	Length of pendulum rod	0.3 m
g	Gravitational acceleration	9.81 m/s^2

In this case, the objective is to produce a control signal that the cart can track the desired reference $r = 0.1$ unit with all other states reach at zero. It is of most important for unstable, non-minimum phase system that, all the states should be stabilized. PID, LQG, MRAC and basic L_1 adaptive controller are designed for the system without disturbances and the system responses are shown in Fig. 2.21a, Fig. 2.22a, Fig. 2.23a and Fig. 2.24a respectively. The respective control signal with error signal plots are shown in Fig. 2.21b, Fig. 2.22b, Fig. 2.23b and Fig. 2.24b. Tuned controllers are then subjected to the disturbances as described above and the system responses and control signal with error signal plots for PID, LQG, MRAC and basic L_1 adaptive controller are provided in Fig. 2.25, Fig. 2.26, Fig. 2.27 and Fig. 2.28 respectively. The tracking performances for 10 s with 0.01 s time step are tabulated in Table 2.4 in terms of IAE.

As it is an unstable non-minimum phase system therefore, the difference of performances for classical controllers, MRAC and basic L_1 adaptive controller are distinct

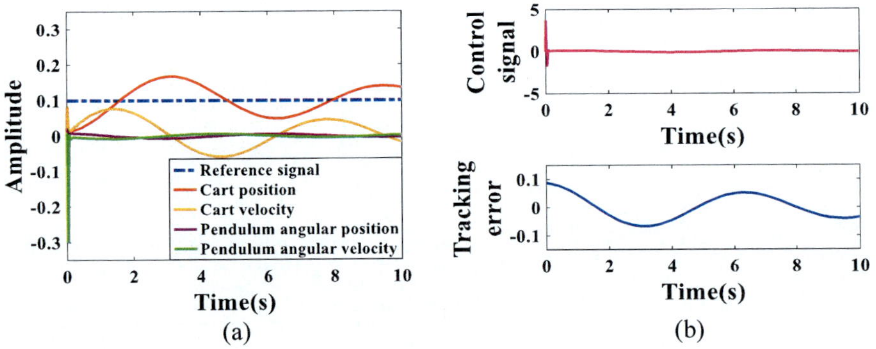

Fig. 2.21 Evaluation period **a** system response and corresponding **b** control signal with tracking error of case study-III (inverted pendulum with cart) without disturbance for PID controller

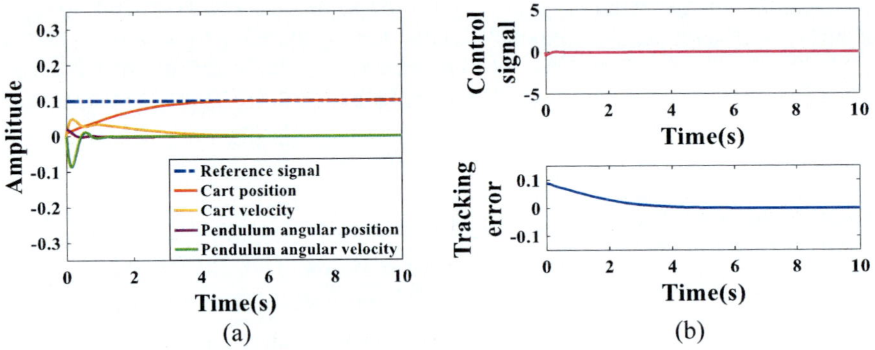

Fig. 2.22 Evaluation period **a** system response and corresponding **b** control signal with tracking error of case study-III (inverted pendulum with cart) without disturbance for LQG

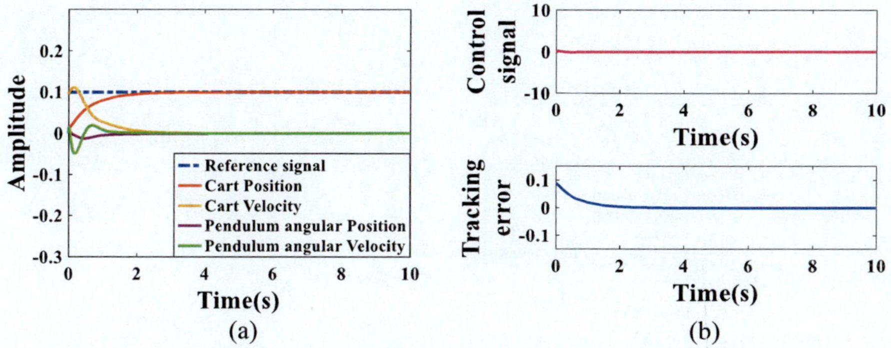

Fig. 2.23 Evaluation period **a** system response and corresponding **b** control signal with tracking error of case study-III (inverted pendulum with cart) without disturbance for MRAC

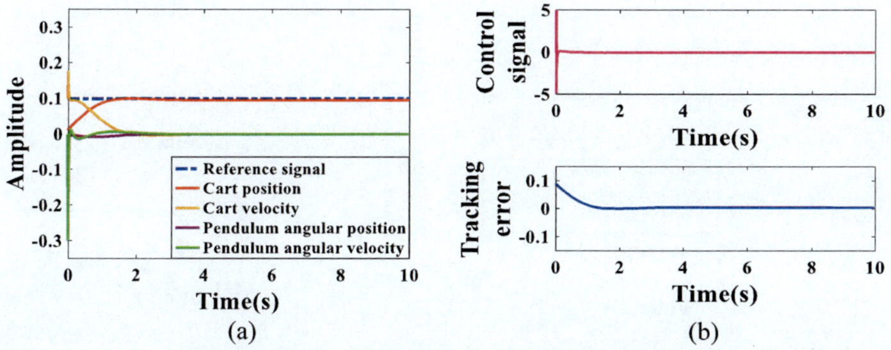

Fig. 2.24 Evaluation period **a** system response and corresponding **b** control signal with tracking error of case study-III (inverted pendulum with cart) without disturbance for basic L_1 adaptive controller

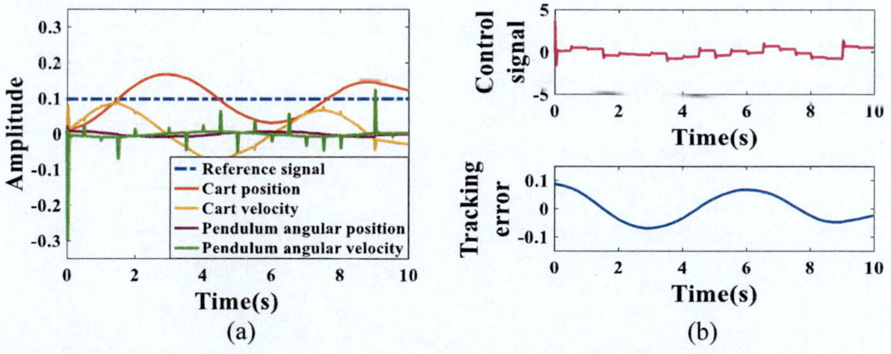

Fig. 2.25 Evaluation period **a** system response and corresponding **b** control signal with tracking error of III (inverted pendulum with cart) with disturbance for PID controller

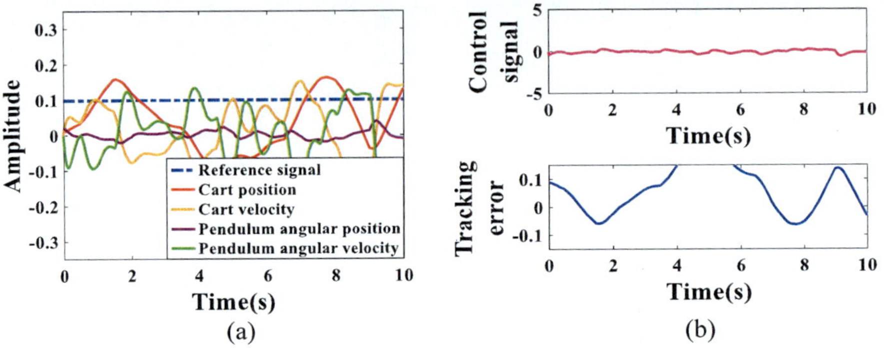

Fig. 2.26 Evaluation period **a** system response and corresponding **b** control signal with tracking error of case study-III (inverted pendulum with cart) with disturbance for LQG

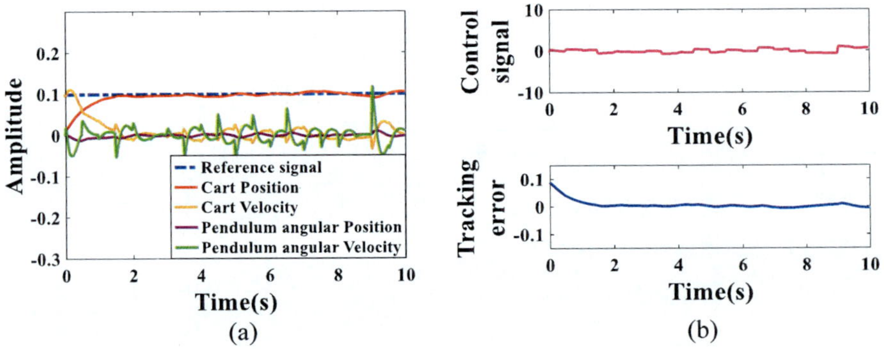

Fig. 2.27 Evaluation period **a** system response and corresponding **b** control signal with tracking error of case study-III (inverted pendulum with cart) with disturbance for MRAC

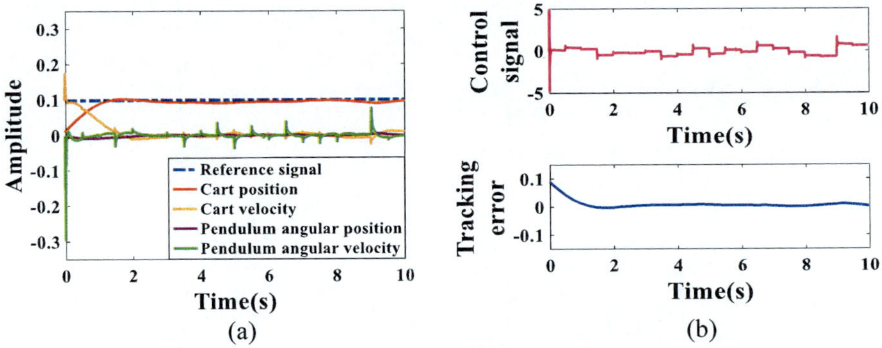

Fig. 2.28 Evaluation period **a** system response and corresponding **b** control signal with tracking error of case study-III (inverted pendulum with cart) with disturbance for basic L_1 adaptive controller

Table 2.4 Comparative study of different control methodologies for case study-III (inverted pendulum with cart)

Control methodologies	IAE value			
	Best	Average	Std. Dev	With disturbance
PID controller	0.5505	0.7266	0.1726	0.5696
LQG	0.2708	0.3326	0.0463	22.2317
MRAC	0.0612	–	–	0.0835
Basic L_1 adaptive controller	0.0465	–	–	0.0502

in this case. From Figs. 2.21 and 2.22, it can be stated that, the transient and tracking performances are very poor for classical controllers. The performance of MRAC as shown in Fig. 2.23 is better than the classical controllers though it produces sluggish transient response. In case of basic L_1 adaptive controller, as portrayed in Fig. 2.24, the transient response is fast and tracking performance is also satisfactory. When the tuned controllers are subjected to the disturbances then, the performances of classical controllers as well as MRAC degrade very much. They cannot eliminate the disturbances from the system. Whereas, in case of basic L_1 adaptive controller, elimination of disturbance is well enough and both transient and steady state performances are satisfactory.

2.6.4 *Case Study-IV: Twin Rotor MIMO System*

Twin rotor MIMO system (TRMS) exemplifies a highly nonlinear system with uncertainties, disturbances and significant cross-coupling between its two rotors, viz., pitch rotor and yaw rotor [36–40]. It resembles with helicopter system mounted on a tower where it can freely rotate [41, 42] as shown in Fig. 2.29. It has cross-coupled pitch

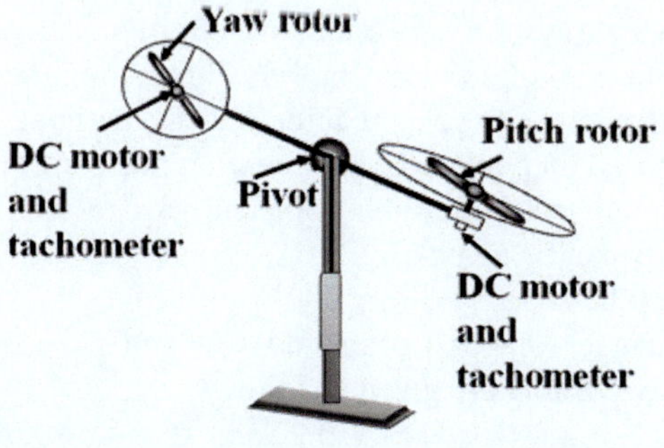

Fig. 2.29 Schematic diagram of twin rotor MIMO system

rotor and yaw rotor which produce thrust in vertical and horizontal direction respectively [43]. Two DC motors are attached to these two rotors and this whole arrangement is mounted on a beam [44]. These rotor arrangements and a counter balance is attached at the pivot on the tower [45]. Classical controllers provide unsatisfactory control performance for such system [46–48]. In this case study, performances of classical controllers and MRAC are compared with L_1 adaptive controller to control a TRMS.

Dynamics of the TRMS is given as follows [49]:

$$\left.\begin{aligned}
\dot{x}_1 &= x_2 \\
\dot{x}_2 &= \frac{1}{I_1}\begin{pmatrix} a_1x_5^2 + b_1x_5 - M_g \sin x_1 \\ -K_{gy}b_1 \cos x_1x_4x_5 + 0.016 \sin 2x_1x_4^2 \\ -K_{gy}a_1 \cos x_1x_4x_5^2 - B_{1\phi}x_2 \end{pmatrix} \\
\dot{x}_3 &= x_4 \\
\dot{x}_4 &= \frac{1}{I_2}\begin{pmatrix} a_2x_6^2 + b_2x_6 - 1.75(K_c a_1x_5^2 + K_c b_1x_5) \\ +0.375(K_c a_1x_5^2 + K_c b_1x_5)\exp^{-0.5t} - B_{1\varphi}x_4 \end{pmatrix} \\
\dot{x}_5 &= -\frac{T_{10}x_5}{T_{11}} + \frac{K_{11}u_1}{T_{11}} \\
\dot{x}_6 &= -\frac{T_{20}x_6}{T_{21}} + \frac{K_{22}u_2}{T_{21}}
\end{aligned}\right\} \tag{2.66}$$

where x_1 is the pitch angle, x_2 is the pitch angular speed, x_3 represents the yaw angle, x_4 is the yaw angular speed, x_5 is the pitch motor speed and x_6 is the yaw motor speed. u_1 and u_2 are the input to the pitch rotor and yaw rotor respectively.

The parameter values of TRMS are provided in Table 2.5. All units are considered in S.I. system.

The system dynamics are evaluated for 100 s with 0.01 s time step by employing PID, LQG, MRAC and basic L_1 adaptive controller. Pitch angle x_1 and yaw angle x_3 are taken as outputs of the system in radian. The control objective is to track variable step trajectories for pitch rotor and yaw rotor by eliminating the disturbances.

At first, the controllers are designed for TRMS without disturbance. Then, Disturbance-I is imposed on pitch rotor and the designed controllers without further tuning are utilized to control both pitch rotor and yaw rotor. The disturbance that used in this case study can be given as:

(i) Disturbance-I: $0.02\cos(2\pi t)$.

Tracking performances of pitch rotor and yaw rotor are evaluated where the references for these two rotors are given as follows.

$$\text{(i) Reference signal for pitch angle:} \begin{cases} 0.2 \text{ rad}, & 0 \text{ s} \le t \le 25 \text{ s} \\ 0.5 \text{ rad}, & 25 \text{ s} < t \le 50 \text{ s} \\ 0.1 \text{ rad}, & 50 \text{ s} < t \le 75 \text{ s} \\ 0.3 \text{ rad}, & 75 \text{ s} < t \le 100 \text{ s} \end{cases}$$

Table 2.5 Parameter values of twin rotor MIMO system

Parameter	Description	Value
K_{11}	Gain of pitch motor	1.1
T_{11}	Denominator parameter of pitch motor	1.1
T_{10}	Denominator parameter of pitch motor	1
a_1	Static characteristics parameter	0.0077
b_1	Static characteristics parameter	0.0949
M_g	Gravity momentum	0.3174 N m
$B_{1\psi}$	Friction momentum function parameter	0.0089 N m s/rad
K_{Gy}	Gyroscopic momentum parameter	0.0450 s/rad
I_1	Moment of inertia of vertical rotor	0.0800 kg m^2
K_{22}	Gain of tail motor	0.8
T_{21}	Denominator parameter of tail motor	1
T_{20}	Denominator parameter of tail motor	1
a_2	Static characteristics parameter	0.0289
b_2	Static characteristics parameter	0.0500
T_0	Cross reaction momentum parameter	3.5
T_P	Cross reaction momentum parameter	2
K_C	Cross reaction momentum gain	– 0.2
$B_{1\varphi}$	Friction momentum function parameter	0.1355 N m s/rad
I_2	Moment of inertia of horizontal rotor	0.0360 kg m^2

$$\text{(ii) Reference signal for yaw angle: } \begin{cases} 0.3 \text{ rad}, & 0 \text{ s} \le t \le 25 \text{ s} \\ -0.1 \text{ rad}, & 25 \text{ s} < t \le 50 \text{ s} \\ 0.2 \text{ rad}, & 50 \text{ s} < t \le 75 \text{ s} \\ 0.6 \text{ rad}, & 75 \text{ s} < t \le 100 \text{ s} \end{cases}$$

Without disturbance, the pitch angle and yaw angle tracking errors in terms of IAE are given in Table 2.6. The performances of these two rotor angles without disturbance for PID, LQG, MRAC and basic L_1 adaptive controller are shown in Fig 2.30,

Table 2.6 Comparative study of different control methodologies for case study-IV (twin rotor MIMO system) without disturbance

Control methodologies	IAE value without disturbance					
	Pitch angle			Yaw angle		
	Best	Average	Std. Dev	Best	Average	Std. Dev
PID controller	3.9087	3.9090	0.0005	5.5232	5.8195	0.1449
LQG	1.5600	2.0539	0.5457	5.7616	6.3882	0.4372
MRAC	1.5232	–	–	4.2526	–	–
Basic L_1 adaptive controller	1.4914	–	–	4.0749	–	–

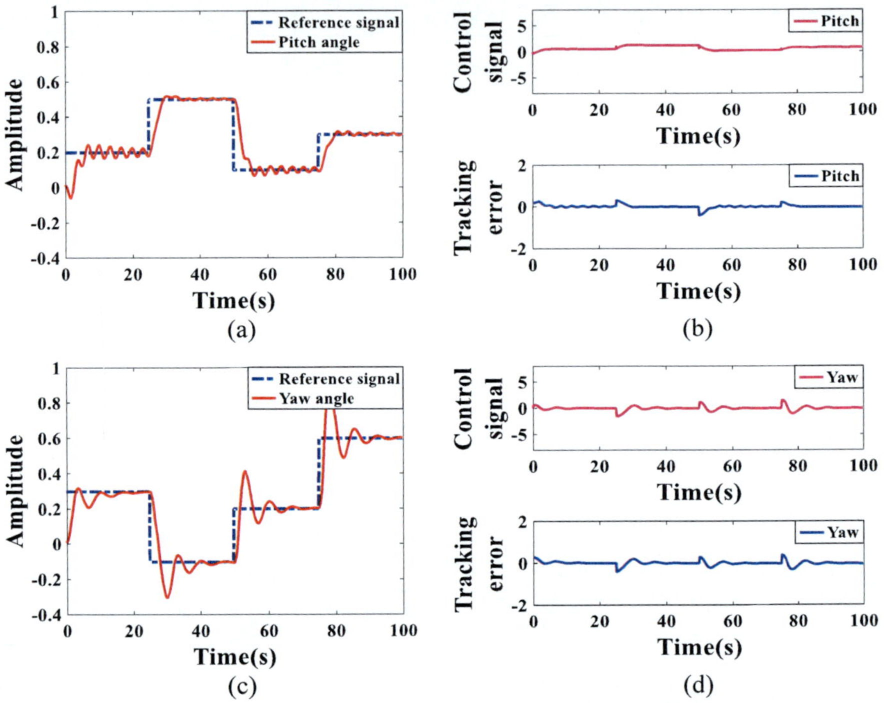

Fig. 2.30 Evaluation period **a** pitch angle and corresponding **b** control signal with tracking error, **c** yaw angle and corresponding **d** control signal with tracking error of case study-IV (twin rotor MIMO system) without disturbance for PID controller

Fig. 2.31, Fig. 2.32 and Fig. 2.33 respectively. When Disturbance-I is imposed on the pitch rotor then the system performances for these respective controllers without further tuning are provided in Figs. 2.34, 2.35, 2.36 and 2.37. IAEs with Disturbance-I for pitch and yaw rotor are given in Table 2.7. From results, it can be stated that, the transient and steady state tracking performances of basic L_1 adaptive controller is better than the classical controllers and MRAC. Basic L_1 adaptive controller is also capable of rejecting unknown time varying disturbances very well.

2.7 Summary

In this chapter, the motivations of designing a newly developed adaptive controller, viz., L_1 adaptive controller is portrayed owing to the drawbacks of classical controllers and MRAC. The architecture of basic L_1 adaptive controller is described and its performance is compared with classical controllers and MRAC by employing them on four systems, viz., Duffing's oscillatory system, nonlinear spring mass damper system, inverted pendulum with cart system and twin rotor MIMO system

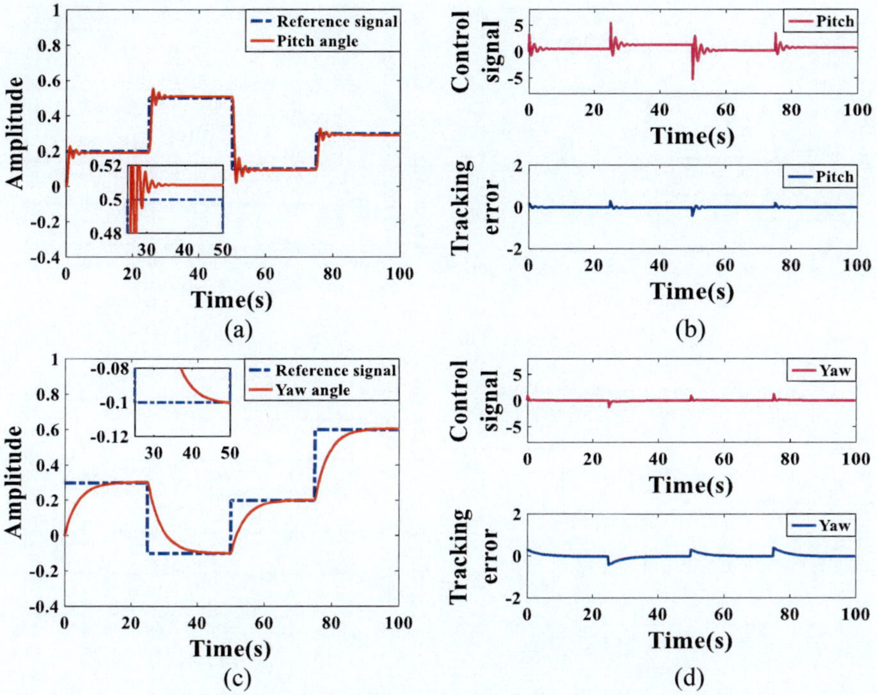

Fig. 2.31 Evaluation period **a** pitch angle and corresponding **b** control signal with tracking error, **c** yaw angle and corresponding **d** control signal with tracking error of case study-IV (twin rotor MIMO system) without disturbance for LQG

in simulation environment. The system performances with occurrence of sudden unknown disturbances are tested in terms of integral absolute error. Results show that, the basic L_1 adaptive controller provides fast transient performance and can afford satisfactory tracking performance by eliminating unknown time varying disturbances and uncertainties.

In this chapter, the parameter values of L_1 adaptive controller are selected arbitrarily within the ranges obtained from the mathematical calculation of L_1 norm condition, which do not guarantee optimal performance. To achieve optimal performance of L_1 adaptive controller, in the next chapter stochastic optimization and metaheuristics technique are utilized to tune the parameter values of L_1 adaptive controller.

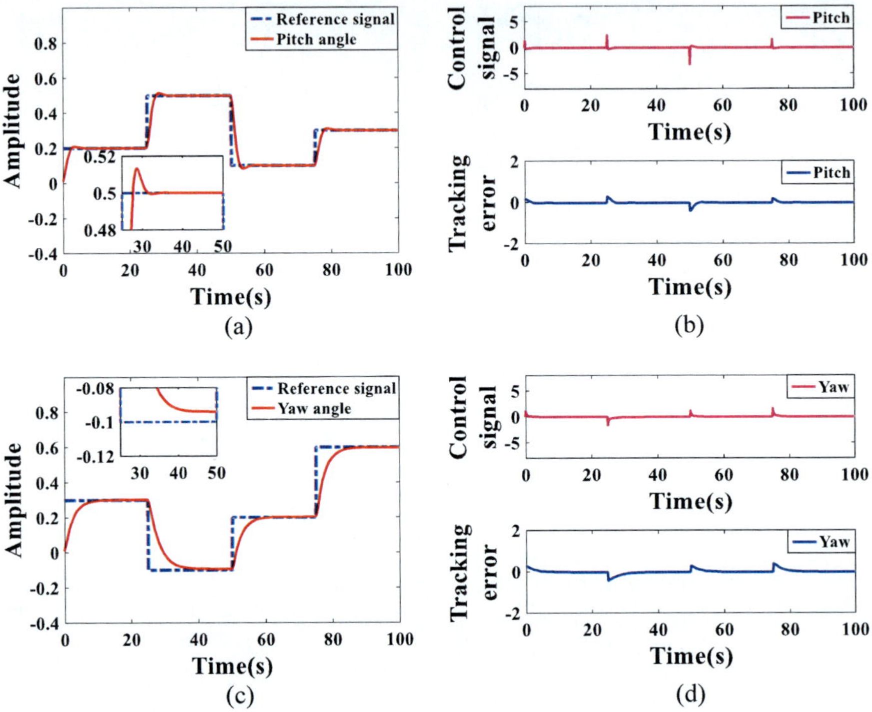

Fig. 2.32 Evaluation period **a** pitch angle and corresponding **b** control signal with tracking error, **c** yaw angle and corresponding **d** control signal with tracking error of case study-IV (twin rotor MIMO system) without disturbance for MRAC

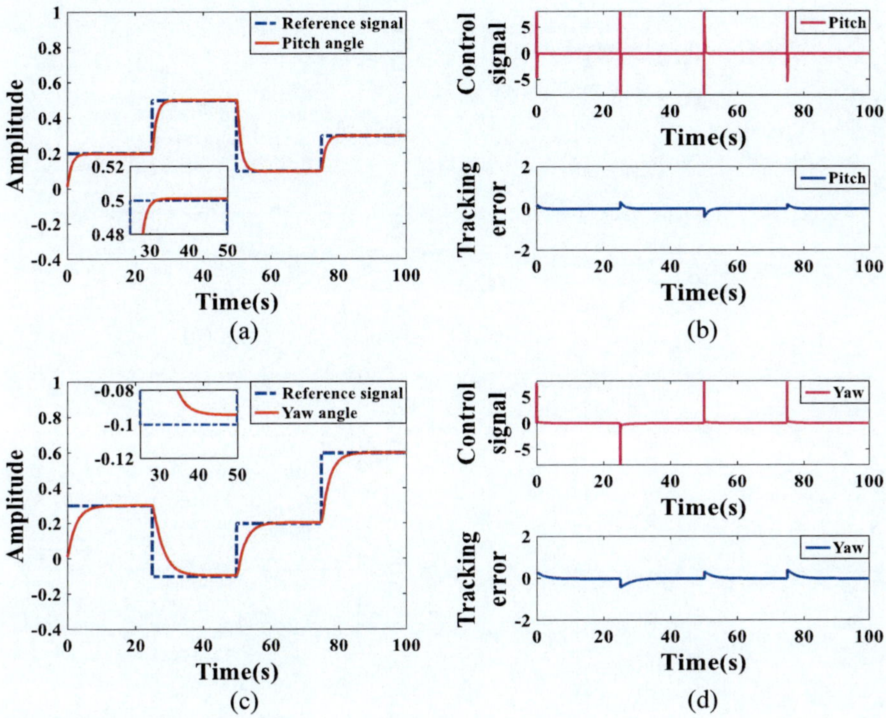

Fig. 2.33 Evaluation period **a** pitch angle and corresponding **b** control signal with tracking error, **c** yaw angle and corresponding **d** control signal with tracking error of case study-IV (twin rotor MIMO system) without disturbance for basic L_1 adaptive controller

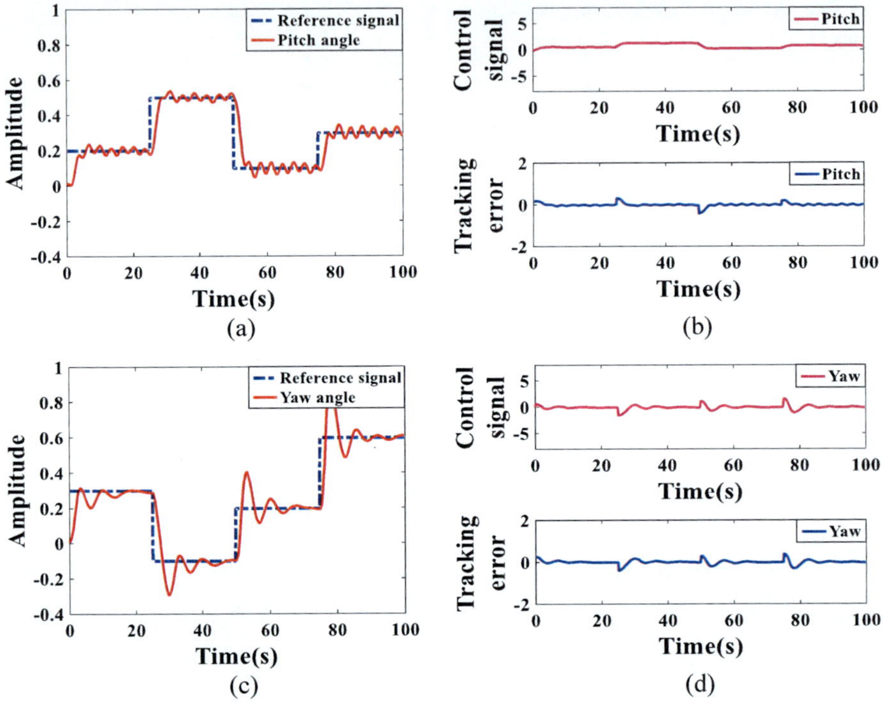

Fig. 2.34 Evaluation period **a** pitch angle and corresponding **b** control signal with tracking error, **c** yaw angle and corresponding **d** control signal with tracking error of case study-IV (twin rotor MIMO system) with Disturbance-I for PID controller

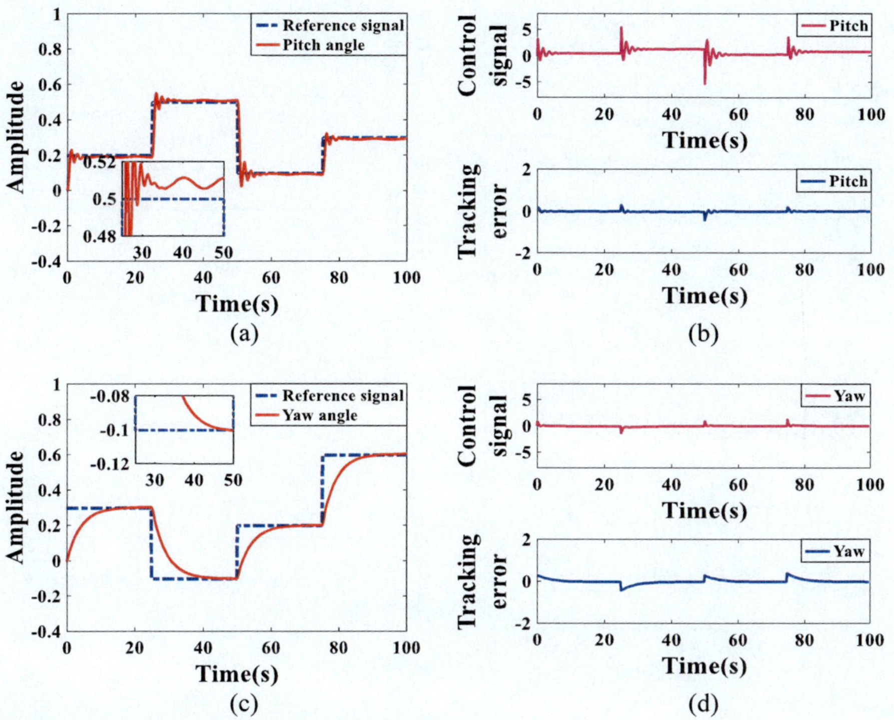

Fig. 2.35 Evaluation period **a** pitch angle and corresponding **b** control signal with tracking error, **c** yaw angle and corresponding **d** control signal with tracking error of case study-IV (twin rotor MIMO system) with Disturbance-I for LQG

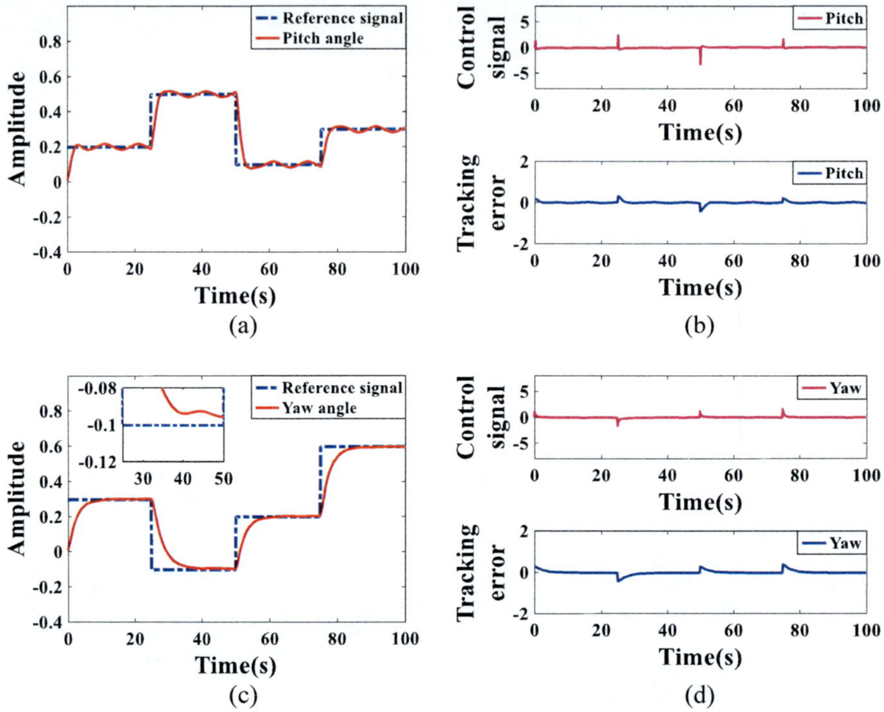

Fig. 2.36 Evaluation period **a** pitch angle and corresponding **b** control signal with tracking error, **c** yaw angle and corresponding **d** control signal with tracking error of case study-IV (twin rotor MIMO system) with Disturbance-I for MRAC

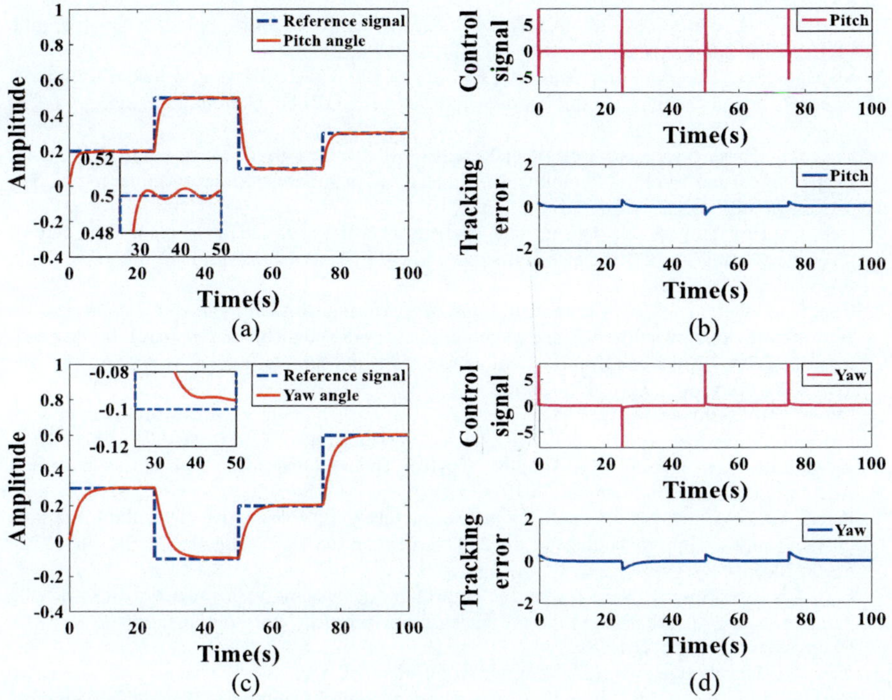

Fig. 2.37 Evaluation period **a** pitch angle and corresponding **b** control signal with tracking error, **c** yaw angle and corresponding **d** control signal with tracking error of case study-IV (twin rotor MIMO system) with Disturbance-I for basic L_1 adaptive controller

Table 2.7 Comparative study of different control methodologies for case study-IV (twin rotor MIMO system) with Disturbance-I

Control methodologies	IAE value with Disturbance-I	
	Pitch angle	Yaw angle
PID controller	3.9272	5.5484
LQG	1.5677	5.7616
MRAC	2.4760	4.2602
Basic L_1 adaptive controller	1.5655	4.0783

References

1. Dai, A., Zhou, X., Liu, X.: Design and simulation of a genetically optimized fuzzy immune PID controller for a novel grain dryer. IEEE Access **5**, 14981–14990 (2017)
2. Rahman, Y., Xie, A., Bernstein, D.S.: Retrospective cost adaptive control: pole placement, frequency response, and connections with LQG control. IEEE Control Syst. Mag. **37**(5), 28–69 (2017)
3. Wu, H., Deng, M.: Robust adaptive control scheme for uncertain non-linear model reference adaptive control systems with time-varying delays. IET Control Theory Appl. **9**(8), 1181–1189 (2015)

4. Anfinsen, H., Aamo, O.M.: Model reference adaptive control of 2×2 coupled linear hyperbolic PDEs. IEEE Trans. Autom. Control **63**(8), 2405–2420 (2018)
5. Skorina, E.H., Luo, M., Tao, W., Chen, F., Fu, J., Onal, C.D.: Adapting to flexibility: model reference adaptive control of soft bending actuators. IEEE Robot. Autom. Lett. **2**(2), 964–970 (2017)
6. Cao, C., Hovakimyan, N.: Design and analysis of a novel L_1 adaptive controller, part I: control signal and asymptotic stability. In: Proceedings of American Control Conference, USA, pp. 3397–3402 (2006)
7. Cao, C., Hovakimyan, N.: Design and analysis of a novel L_1 adaptive controller, part II: guaranteed transient performance. In: Proceedings of American Control Conference, USA, pp. 3403–3408 (2006)
8. Cao, C., Hovakimyan, N.: Guaranteed transient performance with L_1 adaptive controller for systems with unknown time-varying parameters and bounded disturbances: part I. In: Proceedings of the 2007 American Control Conference, Marriott Marquis Hotel at Times Square, New York City, USA, pp. 3925–3930 (2007)
9. Cao, C., Hovakimyan, N.: Design and analysis of a novel L_1 adaptive control architecture with guaranteed transient performance. IEEE Trans. Autom. Control **53**(2), 586–591 (2008)
10. Hovakimyan, N., Cao, C.: L_1 Adaptive Control Theory Guaranteed Robustness with Fast Adaptation. SIAM, Philadelphia (2010)
11. Patel, V., Cao, C., Hovakimyan, N., Wise, A., Lavretsky, E.: L_1 adaptive controller for tailless unstable aircraft. In: Proceedings of the 2007 American Control Conference, USA, pp. 5272–5277 (2007)
12. Cao, C., Hovakimyan, N.: L_1 adaptive controller for systems with unknown time-varying parameters and disturbances in the presence of non-zero trajectory initialization error. Int. J. Control **81**(7), 1147–1161 (2008)
13. Cao, C., Hovakimyan, N.: L_1 adaptive controller for a class of systems with unknown nonlinearities: part I. In: Proceedings of American Control Conference, Seattle, Washington, pp. 4093–4098 (2008)
14. Cao, C., Hovakimyan, N.: L_1 adaptive controller for nonlinear systems in the presence of unmodelled dynamics: part II. In: Proceedings of American Control Conference, Seattle, Washingtion, pp. 4099–4104 (2008)
15. Maiti, R., Das Sharma, K., Sarkar, G.: PSO based parameter estimation and PID controller tuning for 2-DOF non-linear twin rotor MIMO system. Int. J. Autom. Control (Inderscience) **12**(4), 582–609 (2018)
16. Maiti, R., Das Sharma, K., Sarkar, G.: Optimal state feedback controller and observer design for twin rotor MIMO system. In: 2nd International Conference on Control, Instrumentation, Energy & Communication (CIEC16), Kolkata, India, pp. 30–34 (2016)
17. Astrom, K.J., Wittenmark, B.: Adaptive Control, 2nd edn. Addison-Wesley Longman Publishing Co., Boston, MA, USA (1994)
18. Landau, I.D., Lozano, R., Saad, M.M., Karimi, A.: Adaptive Control, 2nd edn. Springer, London, U.K. (2011)
19. Pomet, J., Praly, L.: Adaptive nonlinear regulation: estimation from the Lyapunov equation. IEEE Trans. Autom. Control **37**(6), 729–740 (1992)
20. Saccon, A., Hauser, J., Aguiar, A.P.: Optimal control on lie groups: the projection operator approach. IEEE Trans. Autom. Control **58**(9), 2230–2245 (2013)
21. Stefani, R.T., Shahian, B., Savant, C.J., Jr., Hostetter, G.H.: Design of Feedback Control Systems. Oxford University Press, New York (2002)
22. Yuan, Q., Yin, G.: Analyzing convergence and rates of convergence of particle swarm optimization algorithms using stochastic approximation methods. IEEE Trans. Autom. Control **60**(7), 1760–1773 (2015)
23. Maiti, R., Das Sharma, K., Sarkar, G.: lBest-HS algorithm based concurrent L_1 adaptive control for nonlinear systems. ISA Trans. **96**, 390–414 (2020)
24. Yoo, S.J., Hovakimyan, N., Cao, C.: Decentralized L_1 adaptive control for large-scale nonlinear systems with interconnected unmodelled dynamics. IET Control Theory Appl. **4**(10), 1972–1988 (2010)

25. Polekhin, I.: On topological obstructions to global stabilization of an inverted pendulum. Syst. Control Lett. **113**, 31–35 (2018)
26. Das, S.K., Paul, K.K.: Robust compensation of a Cart-Inverted Pendulum system using a periodic controller: experimental results. Automatica **47**, 2543–2547 (2011)
27. Ghommam, J., Mnif, F.: Predictor-based control for an inverted pendulum subject to networked time delay. ISA Trans. **67**, 306–316 (2017)
28. Ghosh, A., Krishnan, T.R., Subudhi, B.: Brief Paper—robust proportional-integral-derivative compensation of an inverted cart-pendulum system: an experimental study. IET Control Theory Appl. **6**(8), 1145–1152 (2012)
29. Ghosh, A., Krishnan, T.R., Subudhi, B.: Robust proportional-integral-derivative compensation of an inverted cart-pendulum system: an experimental study. IET Control Theory Appl. **6**(8), 1145–1152 (2012)
30. Bettayeb, M., Boussalem, C., Mansouri, R., Al-Saggaf, U.M.: Stabilization of an inverted pendulum-cart system by fractional PI-state feedback. ISA Trans. **53**, 508–516 (2014)
31. Lee, J., Mukherjee, R., Khalil, H.K.: Output feedback stabilization of inverted pendulum on a cart in the presence of uncertainties. Automatica **54**, 146–157 (2015)
32. Irfan, S., Mehmood, A., Razzaq, M.T., Iqbal, J.: Advanced sliding mode control techniques for inverted pendulum: modelling and simulation. Eng. Sci. Technol. Int. J. **21**(4), 753–759 (2018)
33. Shalaby, R., EL-Hossainy, M., ABO-Zalam, B.: Fractional order modeling and control for under-actuated inverted pendulum. Commun. Nonlinear Sci. Numer. Simul. **74**(30), 97–121 (2019)
34. Tian, X., Peng, H., Zhou, F., et al.: RBF-ARX model-based fast robust MPC approach to an inverted pendulum. ISA Trans. **93**, 255–267 (2019)
35. Mishra, S.-K., Chandra, D.: Stabilization and tracking control of inverted pendulum using fractional order PID controllers. J. Eng. **2014**, 1–9 (2014)
36. Nasir, N.K., Tokhi, M.O.: An improved spiral dynamic optimization algorithm with engineering application. IEEE Trans. Syst. Man Cybern. Syst. **45**(6), 943–954 (2015)
37. Mondal, S., Mahanta, C.: Adaptive second-order sliding mode controller for a twin rotor multi-input-multi-output system. IET Control Theory Appl. **6**(14), 2157–2167 (2012)
38. Pradhan, J.K., Ghosh, A.: Design and implementation of decoupled compensation for a twin rotor multiple-input and multiple-output system. IET Control Theory Appl. **7**(2), 282–289 (2013)
39. Dheeraj, K., Jacob, J., Nandakumar, M.P.: Direct adaptive neural control design for a class of nonlinear multi input multi output systems. IEEE Access **7**, 15424–15435 (2019)
40. Roman, R., Radac, M., Precup, R.: Multi-input–multi-output system experimental validation of model-free control and virtual reference feedback tuning techniques. IET Control Theory Appl. **10**(12), 1395–1403 (2016)
41. Tastemirov, A., Lecchini-Visintini, A., Morales-Viviescas, R.M.: Complete dynamic model of the Twin Rotor MIMO System (TRMS) with experimental validation. Control. Eng. Pract. **66**, 89–98 (2017)
42. Chalupa, P., Přikryl, J., Novák, J.: Modelling of twin rotor MIMO system. Procedia Eng. **100**, 249–258 (2015)
43. Rotondo, D., Nejjari, F., Puig, V.: Quasi-LPV modeling, identification and control of a twin rotor MIMO system. Control. Eng. Pract. **21**(6), 829–846 (2013)
44. Haruna, A., Mohamed, Z., Efe, M.Ö., Basri, M.A.M.: Dual boundary conditional integral backstepping control of a twin rotor MIMO system. J. Franklin Inst. **354**(15), 6831–6854 (2017)
45. Juang, J.-G., Liu, W.-K., Lin, R.-W.: A hybrid intelligent controller for a twin rotor MIMO system and its hardware implementation. ISA Trans. **50**(4), 609–619 (2011)
46. Juang, J.-G., Lin, R.-W., Liu, W.-K.: Comparison of classical control and intelligent control for a MIMO system. Appl. Math. Comput. **205**(2), 778–791 (2008)

47. Juang, J.-G., Lin, R.-W., Liu, W.-K.: Comparison of classical control and intelligent control for a MIMO system. Appl. Math. Comput. **205**, 778–791 (2008)
48. Juang, J.-G., Liu, W.-K., Lin, R.-W.: A hybrid intelligent controller for a twin rotor MIMO system and its hardware implementation. ISA Trans. **50**, 609–619 (2011)
49. Shaik, F.A., Purwar, S., Pratap, B.: Real-time implementation of Chebyshev neural network observer for twin rotor control system. Expert Syst. Appl. **38**, 13043–13049 (2011)

Part III
Design of Hybrid L_1 Adaptive Controller

Chapter 3
Hybrid L_1 Adaptive Controller-I: Stochastic Optimization and Metaheuristics Based Approach

Abstract This chapter emphasis on the designing of stochastic optimization and metaheuristics based hybrid L_1 adaptive controller. The parameter values of L_1 adaptive controller obtained from the mathematical calculation of L_1 norm condition do not provide optimal performance. Hence, the optimal parameter setting of L_1 adaptive controller is obtained by employing a variant of HS algorithm, viz., local best harmony search (lbest HS) algorithm. The lbest HS-L_1 adaptive controller is employed on four simulation case studies and compared with basic L_1 adaptive, PSO-L_1 adaptive and HS-L_1 adaptive controller. The results portray that, the designed method provides better transient as well as steady state tracking performance than all other control methodologies tested.

3.1 Introduction

L_1 adaptive controller comprises of predictor, adaptation laws, state feedback control signal, L_1 control law and low-pass filter. High adaptation gain is utilized to adapt the unknown constant, time varying uncertainties and time varying disturbances with the aim to achieve fast transient performance. This includes high frequency signal into the control channel which may cause instability of the system. Therefore, a low-pass filter is attached and the high frequency control signal is passed through it. To keep balance between the quick transient performance and high robustness, the parameters of L_1 adaptive controller have to be chosen sensibly. In basic L_1 adaptive controller, the parameter values are selected arbitrarily within the ranges obtained from the mathematical calculation of L_1 norm condition which does not provide optimal performance.

In this work, a variant of harmony search (HS) algorithm, viz., local best harmony search (lbest HS) algorithm is hybridized with L_1 adaptive controller to obtain optimal parameter setting of it. In this lbest HS based L_1 (lbest HS-L_1) adaptive controller, the parameter values of L_1 adaptive controller are selected optimally within the mathematical ranges obtained from the L_1 norm calculation. The values of adaptive parameters, viz., unknown constant, time varying uncertainties and time varying disturbances are obtained concurrently from the lbest HS algorithm and L_1 adaptation

R. Maiti et al., *Hybrid L_1 Adaptive Control*, Studies in Systems, Decision and Control 422, https://doi.org/10.1007/978-3-030-97102-1_3

laws. At first, these values are obtained from the lbest HS algorithm and then, they are fine-tuned utilizing the L_1 adaptation laws formulated from the Lyapunov stability criterion. The design procedure of the lbest HS-L_1 adaptive controller is described in this chapter. The exploration capability of HS depends on the harmony memory considering rate (HMCR), bandwidth (bw) and the exploitation phenomenon governs by the pitch adjustment rate (PAR). In this lbest HS algorithm, the HMCR and PAR values change dynamically as iteration increases to provide better performance. The total harmony memory is divided into finite number of groups. Optimal parameter setting is found out from this group and then, the global optima is obtained from these local optima. The bandwidth (bw) is proportional to the square root of expectation mean of the current population which improves the explorative power of the lbest HS algorithm. There is no such work to the best of author's knowledge where the stability analysis of the optimization technique along with controller is provided. In this chapter, the stability analysis of the concurrent lbest HS-L_1 adaptive controller is performed in terms of spectral radius convergence phenomenon. Mathematical analysis shows that, the designed method has good exploration and exploitation capabilities. The transient and steady state stable performance of the designed method are guaranteed analytically. The performance of the designed method is examined by employing it on four systems in simulation environment. The efficacy of the lbest HS-L_1 adaptive control methodology is compared with basic L_1 adaptive, particle swarm optimization based L_1 (PSO-L_1) adaptive and harmony search based L_1 (HS-L_1) adaptive controller.

3.2 Description of Optimization Techniques Used in This Book

In this book, particle swarm optimization (PSO), basic harmony search algorithm (HS) and a variant of HS, viz., local best harmony search algorithm (lbest HS) is utilized. There brief descriptions are provided in the next sections.

3.2.1 Particle Swarm Optimization: A Stochastic Optimization Approach

In the year of 1995, Kenedy and Eberhart developed a population based stochastic optimization technique, particle swarm optimization (PSO). This process is inspired from the food search process of bird flock. Individual bird is resemble with particle whereas the flock is resemble with swarm. Searching of food is similar with finding the best fitness value. Cognitive information sharing improves the knowledge of an individual from the past experience and new discovaries of all other companion during the search process. Each paricle of PSO is searching the best position with some randamized velocity. It keep tracks of its own previous positions to find the best position among all iterations which is generally known as personal best position (*pBest*). At the same time, it also have knowledge about the best *pBest* position of all the particles which provides global best (*gBest*) position. Both the *pBest* and *gBest* values are utilized to search better fitness value of each particle [1].

Let, the swarm size is given by S which consists of i ($1 \le i \le S$) number of particles, viz., trial solutions. Each particle consists of some parameters, viz., $j = 1, 2, \ldots, n$. At t^{th} time instant, in the search space, each particle has a current position, viz., $p_i(t) = [p_{i,1}(t), p_{i,2}(t), \ldots p_{i,j}(t), \ldots, p_{i,n}(t)]$, current velocity, viz., $v_i(t) = [v_{i,1}(t), v_{i,2}(t), \ldots v_{i,j}(t), \ldots, v_{i,n}(t)]$ and a personal best position, viz., $pBest_i(t) = [pBest_{i,1}(t), pBest_{i,2}(t), \ldots pBest_{i,j}(t), \ldots, pBest_{i,n}(t)]$. There is also a *gBest* position among all particles and all iterations. Now let the objective is to optimize (minimize/ maximize) the fitness function for i^{th} particle, viz., f_i. Then each particle of the swarm at each instant will move with the velocity as follows:

$$v_{i,j}(t+1) = w(t)v_{i,j}(t) + c_1 r_1 \big(pBest_{i,j}(t) - p_{i,j}(t)\big) + c_2 r_2 \big(gBest(t) - p_{i,j}(t)\big) \tag{3.1}$$

The acceleration coefficient constants c_1 and c_2, generally known as trust parametres (set as 2), along with the random values $r_1 \rightarrow \cup(0, 1)$ and $r_2 \rightarrow \cup(0, 1)$ produces stochastic nature respectively in the cognitive and social components of the velocity vector. The inertia weight of the particle $w(t)$ at t^{th} instant is given as:

$$w(t) = \frac{(iter_{\max} - t) * \left(w_{initial} - w_{final}\right)}{iter_{\max}} + w_{final} \tag{3.2}$$

where $iter_{\max}$ is the maximum number of iteration. $w(t)$ reduces linearly generally from $w_{initial} = 0.9$ to $w_{final} = 0.4$.

The position of each particle will be updated following the formula as:

$$p_{i,j}(t \mid 1) = p_{i,j}(t) + v_{i,j}(t+1) \tag{3.3}$$

Now the personal best position will be updated as:

$$pBest_i(t+1) = \begin{cases} p_i(t+1), & \text{if } fit(p_i(t+1)) \ge pBest_i(t) \\ pBest_i(t), & otherwise \end{cases} \tag{3.4}$$

The global best position will be updated following:

$$gBest(t+1) = \arg\ \min\left(fit(pBest_i(t+1))\right) \tag{3.5}$$

where $fit(\cdot)$ represents the fitness function.

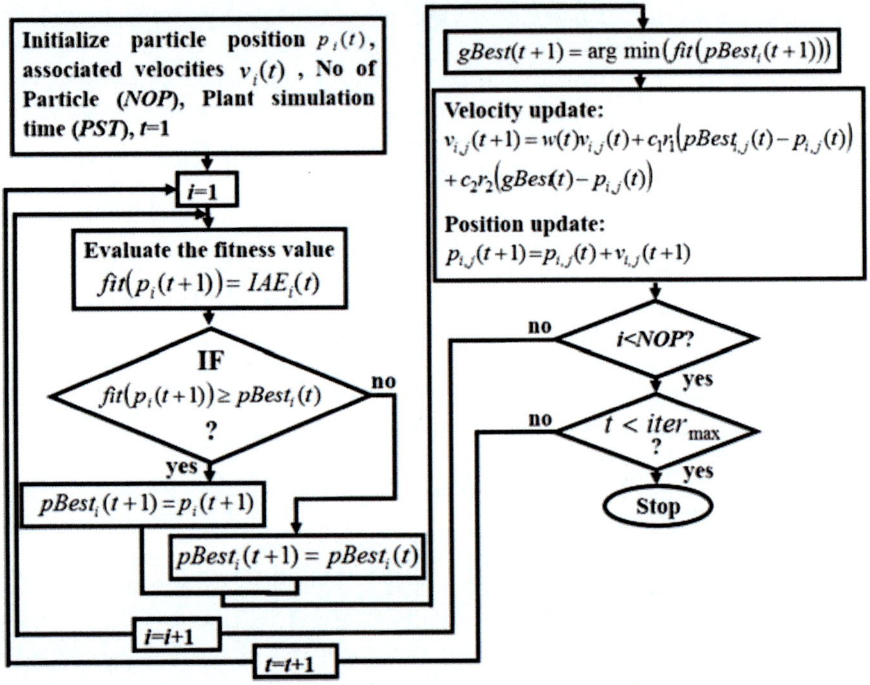

Fig. 3.1 Flowchart representation of PSO algorithm

At the end of the iteration, the *gBest* will be the particle with best fitness value. It's components will be optimized solution/ parameters of the optimization problem. The flowchart of the PSO algorithm is provided in Fig. 3.1.

3.2.2 *Harmony Search Algorithm: A Metaheuristics Approach*

Heuristic algorithims find the optimal solution based on the trial and error method. Thus it doen not guarantee the best or optimal solution. In contrary, meta-heuristic optimization is a higher level technique, consisting of lower level technique as well as exploration and exploitation of huge search spaces [2]. This process resembles with the production of aesthetic harmony. At the time of composition, musicians play different pitches randomly to produce a harmony. Then at the time of practice, they play some pitches from their memory or a completely new one. If they select a harmony from the memory, then they may fine tune it by adjusting it pitches slightly. The aim of this improvisation process is to generate a wounderful harmony altogether.

In this process, at first, a harmony memory (HM) matrix is to be considered. Each harmony consists of some pitches played by different musicians. The fitness value of each harmony is also kept at the HM matrix by adding one extra column. Then, to generate a new harmony, the pitches can be selected from the HM depending upon the harmony memory considering rate (HMCR) as given below:

$$\begin{array}{r} \text{Pitch will be selected from HM matrix,} \quad \text{if } HMCR \leq 0.5 \\ \text{A completely new Pitch will be selected,} \quad \text{if } HMCR > 0.5 \end{array} \tag{3.6}$$

If the pitch is selected from the harmony memory, then its value will be fine tuned by adding or substracting the bandwidth value depending upon the pitch adjustment rate (PAR) as follows:

$$\begin{array}{l} \text{Pitch } + \text{bandwidth, if } PAR \leq 0.5 \\ \text{Pitch } - \text{bandwidth, if } PAR > 0.5 \end{array} \tag{3.7}$$

Fitness value of the new harmony will be found out and it will be compared with the worst fitness value present in the memory. If the newly generated harmony has a better fitness value than that of the worst one, then the newly developed harmony will replace the worst harmony of the memory. Then, the whole harmony will be sorted in ascending/ descending order depending on the minimization/ maximization problem. At the end of the iteration of this improvisation process, the best/ optimal harmony will be obtained. The flowchart of HS algorithm is provided in Fig. 3.2.

3.2.3 *Local Best Harmony Search Algorithm: An Advanced Metaheuristics Approach*

Harmony search (HS) algorithm is a meta-heuristic search process or higher level technique which is the combination of lower level technique as well as exploration and exploitation of huge search spaces [1, 3]. The phenomenon of HS resembles with the composition of new harmony by combination of some pitches from memory and some new pitches randomly. At the time of practice, musicians play some pitches from memory with or without fine tuning of these. They may play a completely new harmony with the goal of obtaining better harmony. Motivated from this phenomenon, at first, a harmony memory (HM) matrix is considered whose rows are formed with some random harmonies. These harmonies, play by different musicians, resemble with candidate solution vectors of the optimization process. Each harmony comprises of some pitches which are similar as the candidates. Fitness values of the harmonies are included in the respective rows of those harmonies by adding an extra column in the HM matrix. Then, the harmonies are sorted in descending order of fitness values that the worst harmony places at the last row of HM matrix. To generate a new harmony, the musicians can play a pitch from the memory or randomly generate a new pitch depending upon the factor, viz., harmony memory considering rate (HMCR).

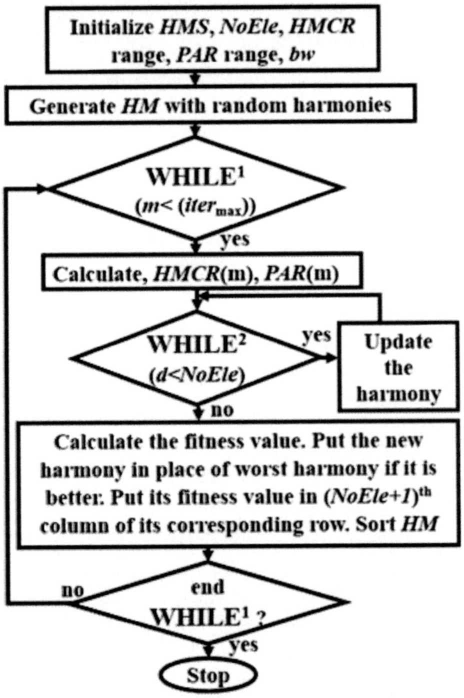

Fig. 3.2 Flowchart representation of HS algorithm

If the pitch is taken from the memory then, some fine tuning of that pitch may be performed depending upon pitch adjustment rate (PAR) where the amount of adjustment is dependent on the chosen bandwidth (bw). Fitness value of the newly generated harmony is executed and compared with the worst harmony placed at the last row. If the newly generated harmony has better fitness value then, it replaces the worst harmony otherwise the newly generated harmony is discarded. If the newly generated harmony is included at the HM matrix then, the whole HM matrix is sorted depending upon the fitness values. If the maximum iteration or specified error criterion is reached, the process stops and the best harmony placed at the first position of HM matrix is considered as the optimal parameter setting.

To improve the performance of basic HS algorithm, different modifications may be applied on it. HMCR, PAR and bw are very significant parameters in designing HS algorithm. With the increase of iteration, dependency upon memory increases as memory becomes rich in experience. At the same time, fine tuning of the harmony should also gradually increase as the pitches are chosen from the memory. Therefore, HMCR and PAR values increase linearly as iteration increases. In this work, the concept of local neighbourhood topology is employed to improvise the HM matrix with more newly generated harmonies in each iteration. In this concept, the HM matrix is divided into some groups and in each group harmony improvisation takes place in each iteration to increase exploitation capability. Each harmony of a particular group is improvised from its own group or others depending upon group memory

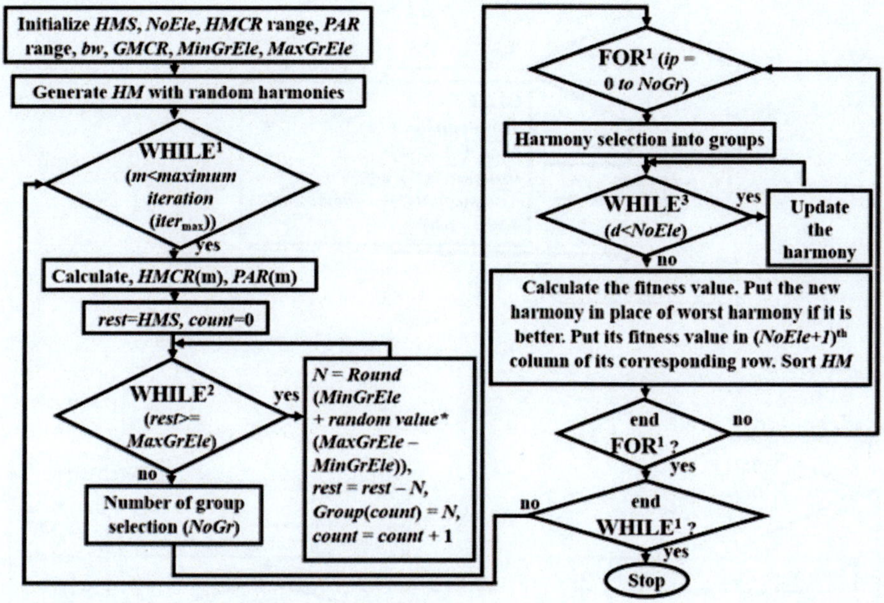

Fig. 3.3 Flowchart representation of lbest HS algorithm

considering rate (GMCR). This variant is termed as local best topology based HS or lbest HS in this book. This phenomenon resembles with the harmony improvisation of different groups by mutual cooperation and then combined performances in a philharmonic orchestra. Again, to improve exploration, the bandwidth (bw) is made linearly dependent upon the square root of the current population mean. Therefore, this search process possess exploration of huge search spaces with high exploitation, viz., convergence rate.

The flowchart of lbest HS algorithm is described in Figs. 3.3 and 3.4. At first, harmony memory size (HMS), number of elements (NoEle) in each harmony, bandwidth (bw), group memory considering rate (GMCR) are initialised. Ranges of harmony memory considering rate (HMCR), pitch adjustment rate (PAR), maximum and minimum number of elements may be considered in a group are set. Harmony memory is build up with random harmonies to initialize the process. With this randomly generated HM matrix, the optimization process enters into the iterative improvisation stages. HMCR and PAR are updated by following the formulas given as respectively:

$$HMCR(t) = HMCR_{\min} + \frac{(HMCR_{\max} - HMCR_{\min})}{iter_{\max}} \times t \quad (3.8)$$

$$PAR(t) = PAR_{\min} + \frac{(PAR_{\max} - PAR_{\min})}{iter_{\max}} \times t \quad (3.9)$$

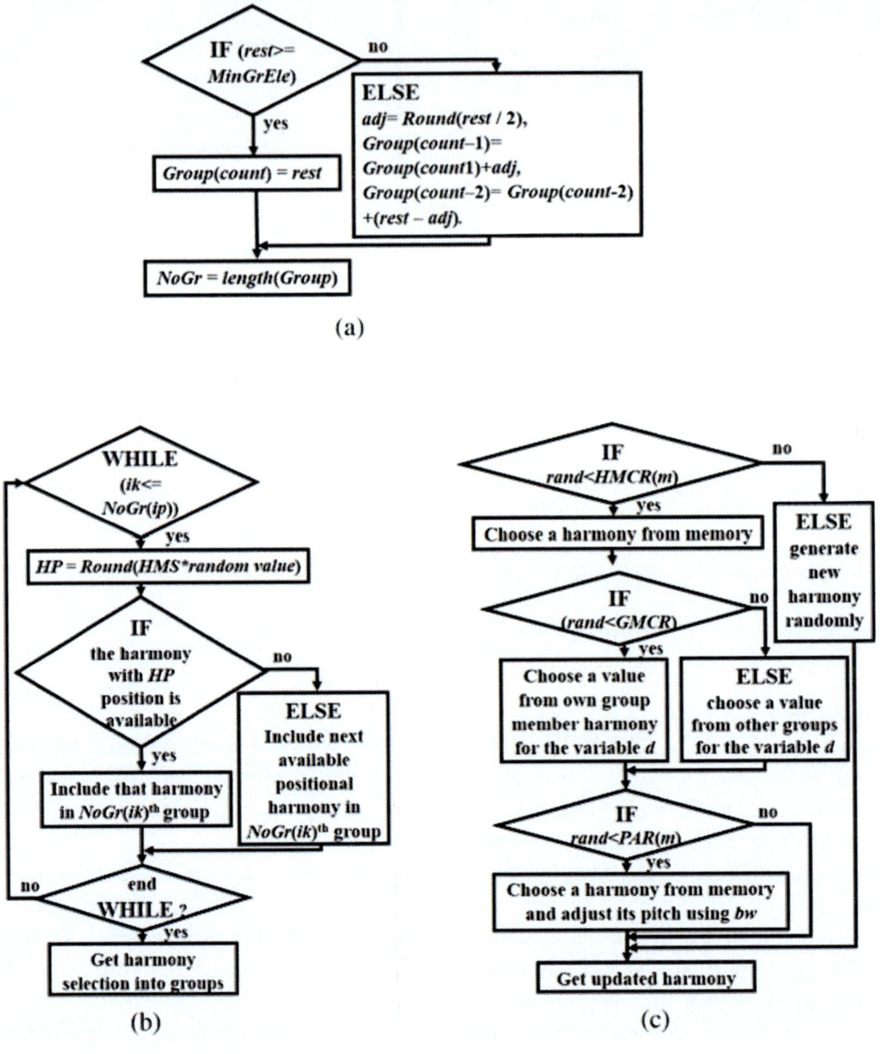

Fig. 3.4 Flowchart representation of **a** number of group selection, **b** harmony selection into groups, **c** update of harmony

where $HMCR_{\min}$ and $HMCR_{\max}$ are respectively the minimum and maximum value of $HMCR$. $PAR_{\min}$ and $PAR_{\max}$ are respectively the minimum and maximum value of PAR. $iter_{\max}$ is the maximum number of iterations.

In next step, number of groups (NoGr) are selected depending upon the ranges of maximum number of group elements (MaxGrEle) and minimum number of group elements (MinGrEle) that may present at one group. This procedure is provided at the **WHILE**[2] phase of Figs. 3.3 and 3.4a. Then, harmonies are selected in each group from the HM matrix following the method as shown in Fig. 3.4b. Next, harmony improvisation takes place in each group with the aim to obtain harmony with better fitness value as depicted in Fig. 3.4c.

In this harmony improvisation process, for each group, the new harmony may be chosen from the harmony memory matrix or random new harmony may be generated depending upon the harmony memory considering rate (HMCR). If the harmony is generated from the HM matrix then, the pitches of new harmony may be formulated from its own group or other groups depending upon group memory considering rate (GMCR). If the new harmony is generated from HM matrix then, fine tuning of its pitches may be done considering the pitch adjustment rate (PAR) by using the bandwidth (bw). After the end of group improvisation process, the fitness value of the newly generated harmony is calculated and compared with the worst harmony of that group. If the newly generated harmony is better than the worst harmony then, the new harmony replaces the worst harmony otherwise, the new harmony is discarded. After the end of improvisation of all the groups, the whole harmony memory is sorted. If the ending criterion reaches, then the harmony improvisation stops and the harmony with the best fitness value is considered as the optimal parameter setting.

lbest HS algorithm based hybrid L_1 adaptive controller design is elaborated in the next section.

3.3 Designing of Lbest HS-L_1 Adaptive Controller

In this section, a metaheuristic optimization based L_1 adaptive controller is designed to provide better control action to the system containing unknown constant, time varying uncertainties and time varying disturbances. The parameters of basic L_1 adaptive controller are obtained from the mathematical calculation of L_1 norm condition which does not guarantee optimal performance of L_1 adaptive controller. To select the optimal parameter setting of L_1 adaptive controller, metaheuristic optimization technique is employed. A variant of harmony search (HS) algorithm termed as local best harmony search (lbest HS) is utilized in this work to optimize the parameters of basic L_1 adaptive controller. This methodology is termed as lbest HS-L_1 adaptive controller in this book and shown in Fig. 3.5.

The parameters of L_1 adaptive controller are tuned with lbest HS algorithm [4] where the harmony vector or candidate solution vector (CSV) is formed as:

$$\underline{Z} = [k \ |k_g|K|\ \Gamma_C \ |\ \hat{\omega} \ |\ \underline{\hat{\theta}}\ |\ \underline{\hat{\sigma}}] \tag{3.10}$$

where k is the pre-filter gain, k_g is the feed-forward gain, K is the state feedback gain, Γ_C represents adaptation gain, $\hat{\omega}$, $\underline{\hat{\theta}}$ and $\underline{\hat{\sigma}}$ are respectively unknown constant related to system, time varying parameter uncertainties and time varying disturbances present in the system. The values of k, k_g, K and Γ_C are obtained from the lbest HS algorithm to design L_1 adaptive controller. These values are kept fixed when the system dynamics is simulated with Runge–Kutta (R-K) 4th order method. $\hat{\omega}$,

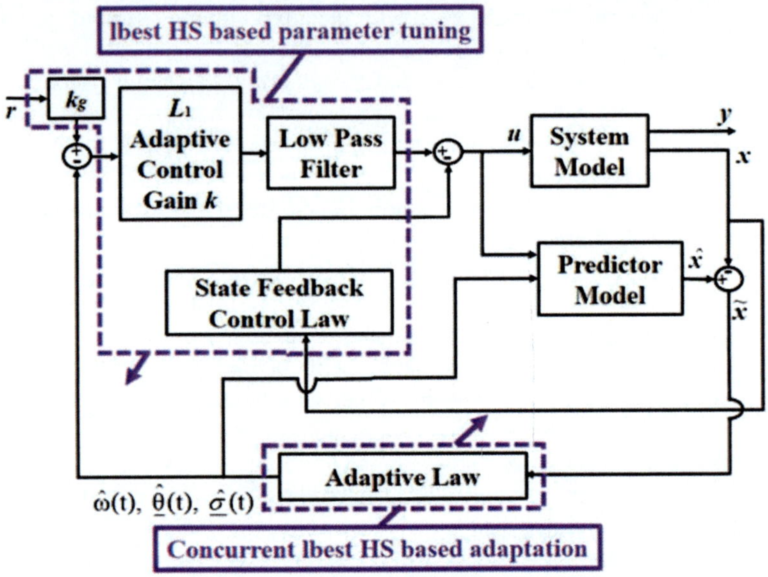

Fig. 3.5 Architecture of lbest HS-L_1 adaptive controller

$\underline{\hat{\theta}}$ and $\underline{\hat{\sigma}}$ values are also obtained from the lbest HS algorithm. Then, during the system dynamics simulation, these values are concurrently adapted following the L_1 adaptation laws. Thus, the harmony $\underline{Z}$ can be split into two parts, viz., static part ($\underline{\mu}$) and dynamic part ($\underline{\eta}$) as:

$$\underline{Z} = [\underline{\mu} \quad \Big| \quad \underline{\eta}] \tag{3.11}$$

where $\underline{\mu} = [k \ |k_g|K| \ \Gamma_C]$ and $\underline{\eta} = [\hat{\omega} \mid \underline{\hat{\theta}} \mid \underline{\hat{\sigma}}]$.

In basic L_1 adaptive controller as described in Chap. 2, the free parameters ($\underline{\mu}$) are selected within the ranges obtained from mathematical calculation of L_1 norm condition (2.56) and the adaptive parameters ($\underline{\eta}$) are adapted following the adaptation laws (2.42)–(2.44). In such case, the parameter values are selected arbitrarily which do not guarantee optimal controller. Therefore, in this chapter, $\underline{\mu}$ and $\underline{\eta}$ are obtained from the lbest HS algorithm. The $\underline{\eta}$, after obtained from the lbest HS, is adapted to reach the final fine-tuned values following L_1 adaptation laws. This concurrent adaptation is represented as:

$$\underline{Z}_C = \underline{Z} + \underline{\hat{Z}} \tag{3.12}$$

where adaptation part is $\underline{\hat{Z}} = [\underline{\hat{\mu}}| \ \underline{\hat{\eta}}] = [0| \ \underline{\hat{\eta}}]$. $\underline{\hat{\mu}}$ is the adaptation of $\underline{\mu}$ and $\underline{\hat{\eta}}$ is the adaptation of $\underline{\eta}$. As, there is no adaptation of $\underline{\mu}$, so, $\underline{\hat{\mu}} = 0$.

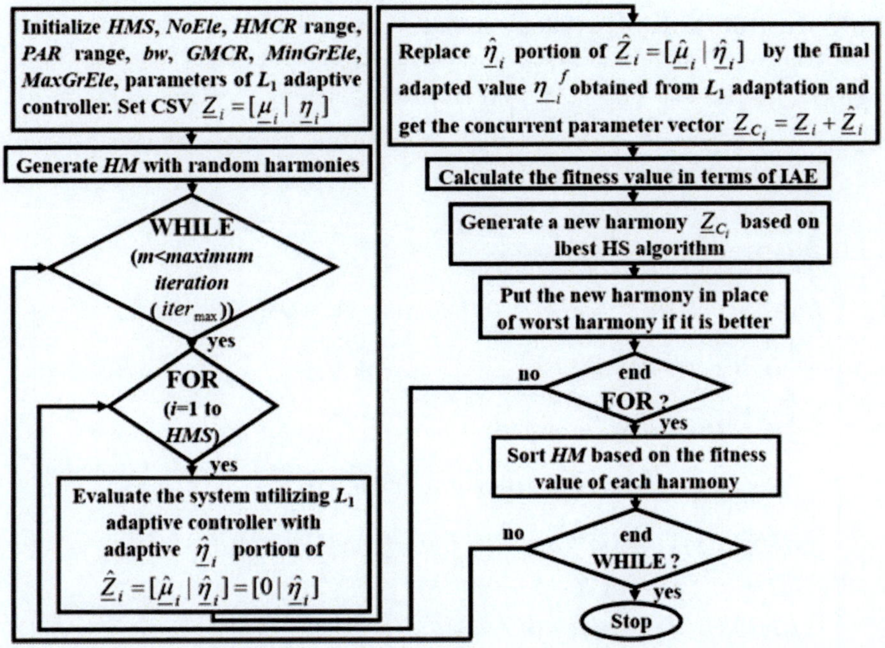

Fig. 3.6 Flowchart representation of lbest HS-L_1 adaptive controller

Therefore, the hybrid concurrent lbest HS-L_1 adaptive controller explores global search spaces and the adaptation part concentrates on local search phenomenon as described in Fig. 3.6.

The stability analysis of the lbest HS-L_1 adaptive method is given in the next section.

3.4 Stability Analysis of Lbest HS-L_1 Adaptive Controller

Multi agent search technique should poses proper balance between exploration and exploitation. Increasing population variance over the iteration provides good exploration capability of the search spaces where as convergence of the process within the spectral radius of iterative matrix conveys its explorative proficiency [5]. Exploration phenomenon of lbest HS-L_1 adaptive controller is analyzed in Theorem 3.1 and Lemma 3.1. The convergence phenomenon as exploitation is described in Theorem 3.2 and Lemma 3.2.

Theorem 3.1: Let, $\underline{Z}_C = \underline{Z} + \hat{\underline{Z}} = \left\{Z_{c_1}, Z_{c_2}, Z_{c_3}, \ldots, Z_{c_{HMS}}\right\}$ is the current population and $\underline{Y} = \{Y_1, Y_2, Y_3, \ldots, Y_{HMS}\}$ is some intermediate population obtained from new harmony improvisation step of lbest HS algorithm. Taking varying harmony memory considering rate (*HMCR*), viz., *HMCR*(t), varying pitch adjustment rate

(PAR), viz., $PAR(t)$, varying arbitrary distance bandwidth (bw), fixed group memory considering rate ($GMCR$), and the allowable range for the decision variables $(Z, \hat{Z})$ to be $\{Z_{\min}, Z_{\max}\}$ and $\left\{\hat{\underline{Z}}_{\min}, \hat{\underline{Z}}_{\max}\right\}$ respectively, where $\underline{Z}_{\min} = a_Z$, $\underline{Z}_{\max} = b_Z$, $\hat{\underline{Z}}_{\min} = c_Z$, $\hat{\underline{Z}}_{\max} = d_Z$ with $a_Z, b_Z, c_Z, d_Z \in \Re$ then, the expectation variance of current population can be given as:

$$E(Var(Y)) = \left(1 - \frac{1}{HMS}\right) \times \begin{bmatrix} HMCR(t) \cdot GMCR \cdot \mathrm{var}(Z_{C_g}) + HMCR(t) \cdot (1 - GMCR) \cdot \mathrm{var}(Z_{C_h}) \\ + \frac{1}{3} HMCR(t) \cdot PAR(t) \cdot bw^2 \\ + \frac{1}{3}(1 - HMCR(t)) \left(a_Z^2 + a_Z b_Z + b_Z^2 + c_Z^2 + c_Z d_Z + d_Z^2\right) \\ + HMCR(t) \cdot GMCR \cdot (1 - HMCR(t) \cdot GMCR) \cdot \overline{Z}_{C_g}^2 \\ + HMCR(t) \cdot (1 - GMCR) \cdot (1 - HMCR(t) \cdot (1 - GMCR)) \cdot \overline{Z}_{C_h}^2 \\ - 2HMCR^2(t) \cdot GMCR \cdot (1 - GMCR) \cdot \overline{Z}_{C_g} \cdot \overline{Z}_{C_h} \\ - HMCR(t) \cdot (1 - HMCR(t)) \cdot GMCR \cdot (a_Z + b_Z + c_Z + d_Z) \cdot \overline{Z}_{C_g} \\ - HMCR(t) \cdot (1 - GMCR) \cdot (1 - HMCR(t)) \cdot (a_Z + b_Z + c_Z + d_Z) \cdot \overline{Z}_{C_h} \\ - \frac{1}{4}(1 - HMCR(t))^2 \cdot (a_Z + b_Z + c_Z + d_Z)^2 \end{bmatrix} \tag{3.13}$$

Proof: Let, the current population of concurrent lbest HS-L_1 adaptive controller is given as $Z_C = \{Z_{c1}, Z_{c2}, Z_{c3}, \ldots, Z_{cHMS}\}$. Z_{C_g} and Z_{C_h} denotes current population generated from own group and other groups of harmony memory respectively.

Then, the population mean and quadratic population mean are given by $\overline{Z_C} = \left({}^1/_{HMS}\right) \sum_{i=1}^{HMS} Z_{ci}$ and $\overline{Z_C^2} = \left({}^1/_{HMS}\right) \sum_{i=1}^{HMS} Z_{ci}^2$ respectively, where HMS is the harmony memory size.

Population variance is $Var(Z_C) = \frac{1}{HMS} \sum_{i=1}^{HMS} \left(Z_{ci} - \overline{Z_C}\right)^2 = \overline{Z_C^2} - \overline{Z_C}^2$ and the expectation of the population variance is $E(Var(Z_C))$.

Again let, $\underline{Y} = \{Y_1, Y_2, Y_3, \ldots, Y_{HMS}\}$ is any intermediate population vector of harmony memory improvisation process. Then, each element of $\underline{Y}$ is formulated as [4]:

$$Y_i = \begin{cases} Z_{gr} + \hat{Z}_r\,, \text{ with probability } HMCR(t)\cdot GMCR\cdot(1-PAR(t)) \\ Z_{gr} + \hat{Z}_r + bw\cdot rand\,, \text{ with probability } 0.5HMCR(t)\cdot GMCR\cdot PAR(t) \\ Z_{gr} + \hat{Z}_r - bw\cdot rand\,, \text{ with probability } 0.5HMCR(t)\cdot GMCR\cdot PAR(t) \\ Z_{hr} + \hat{Z}_r\,, \text{ with probability } HMCR(t)\cdot(1-GMCR)\cdot(1-PAR(t)) \\ Z_{hr} + \hat{Z}_r + bw\cdot rand\,, \text{ with probability } 0.5HMCR(t)\cdot(1-GMCR)\cdot PAR(t) \\ Z_{hr} + \hat{Z}_r - bw\cdot rand\,, \text{ with probability } 0.5HMCR(t)\cdot(1-GMCR)\cdot PAR(t) \\ Z_{new} + \hat{Z}_r\,, \text{ with probability } (1-HMCR(t)) \end{cases} \tag{3.14}$$

Assumption 3.1: $Z_C = Z_r + \hat{Z}_r$, viz., $Z_{C_{gr}} = Z_{gr} + \hat{Z}_r$ and $Z_{C_{hr}} = Z_{hr} + \hat{Z}_r$ are random variables with uniformly distributed random index $r \in \{1, 2, 3, \ldots HMS\}$, with probability $p_k = P(r = k) = (1/HMS), k \in \{1, 2, 3, \ldots HMS\}$.

The expectation of Z_{C_r} and $Z_{C_r}^2$ are given as follows [4]:

$$E(Z_{C_r}) = \sum_{k=1}^{HMS} p_k \cdot Z_{C_k} = \frac{1}{HMS}\sum_{k=1}^{HMS} Z_{C_k} = \overline{Z_C}$$

and $E(Z_{C_r}^2) = \sum_{k=1}^{HMS} p_k \cdot Z_{C_k}^2 = \frac{1}{HMS}\sum_{k=1}^{HMS} Z_{C_k}^2 = \overline{Z_C^2}$.

Expectation of Y_i is provided as:

$$\begin{aligned} E(Y_i) &= HMCR(t)\cdot GMCR\cdot(1-PAR(t))\cdot E(Z_{C_{gr}}) \\ &+0.5\cdot HMCR(t)\cdot GMCR\cdot PAR(t)\cdot E(Z_{C_{gr}} + bw\cdot rand) \\ &+0.5\cdot HMCR(t)\cdot GMCR\cdot PAR(t)\cdot E(Z_{C_{gr}} - bw\cdot rand) \\ &+HMCR(t)\cdot(1-GMCR)\cdot(1-PAR(t))\cdot E(Z_{C_{hr}}) \\ &+0.5\cdot HMCR(t)\cdot(1-GMCR)\cdot PAR(t)\cdot E(Z_{C_{hr}} + bw\cdot rand) \\ &+0.5\cdot HMCR(t)\cdot(1-GMCR)\cdot PAR(t)\cdot E(Z_{C_{hr}} - bw\cdot rand) \\ &+(1-HMCR(t))\cdot E(Z_{new} + \hat{Z}_r) \end{aligned} \tag{3.15}$$

Then, expectation of Y_i^2 becomes:

$$\begin{aligned} E(Y_i^2) &= HMCR(t) \cdot GMCR \cdot (1 - PAR(t)) \cdot E(Z_{C_{gr}}^2) \\ &+0.5 \cdot HMCR(t) \cdot GMCR \cdot PAR(t) \cdot E(Z_{C_{gr}} + bw \cdot rand)^2 \\ &+0.5 \cdot HMCR(t) \cdot GMCR \cdot PAR(t) \cdot E(Z_{C_{gr}} - bw \cdot rand)^2 \\ &+HMCR(t) \cdot (1 - GMCR) \cdot (1 - PAR(t)) \cdot E(Z_{C_{hr}}^2) \\ &+0.5 \cdot HMCR(t) \cdot (1 - GMCR) \cdot PAR(t) \cdot E(Z_{C_{hr}} + bw \cdot rand)^2 \\ &+0.5 \cdot HMCR(t) \cdot (1 - GMCR) \cdot PAR(t) \cdot E(Z_{C_{hr}} - bw \cdot rand)^2 \\ &+(1 - HMCR(t)) \cdot E(Z_{new} + \hat{Z}_r)^2 \end{aligned} \tag{3.16}$$

where $Z_{new} = Z_{\min} + rand \cdot (Z_{\max} - Z_{\min})$ is the randomly selected variable between $Z_{\min} = a_Z,\ Z_{\max} = b_Z$. $\hat{Z}_r = \hat{Z}_{r_{\min}} + rand \cdot (\hat{Z}_{r_{\max}} - \hat{Z}_{r_{\min}})$ is the randomly selected variable between $\hat{Z}_{r_{\min}} = c_Z,\ \hat{Z}_{r_{\max}} = d_Z$, $rand$ is the random number between 0 and 1 satisfying a continuous uniform probability density function [4]:

$$\alpha(rand) = \begin{cases} 1, & rand \in [0,\ 1] \\ 0, & otherwise \end{cases}.$$

Now, it can be obtained that:

$$E(rand) = \int_0^1 rand \cdot \alpha(rand) \cdot drand = \left[\frac{rand^2}{2}\right]_0^1 = \frac{1}{2}$$

and $E(rand^2) = \int_0^1 rand^2 \cdot \alpha(rand) \cdot drand = \left[\frac{rand^3}{3}\right]_0^1 = \frac{1}{3}$.

These yield:

$$E(Z_{new}) = a_Z + E(rand) \cdot (b_Z - a_Z) = \frac{a_Z + b_Z}{2} \tag{3.17}$$

and

$$\begin{aligned} E\big((Z_{new})^2\big) &= a_Z^2 + 2 \cdot a_Z \cdot E(rand) \cdot (b_Z - a_Z) \\ &+E(rand^2) \cdot (b_Z - a_Z)^2 = \frac{a_Z^2 + a_Z b_Z + b_Z^2}{3} \end{aligned} \tag{3.18}$$

Similarly,

$$E(\hat{Z}_r) = \frac{c_Z + d_Z}{2} \tag{3.19}$$

and

$$E\left(\left(\hat{Z}_r\right)^2\right) = \frac{c_Z^2 + c_Z d_Z + d_Z^2}{3} \tag{3.20}$$

Putting the values of (3.17) and (3.18) in both (3.15) and (3.16) provide respectively:

$$\begin{aligned} &E(Y_i) = HMCR(t) \cdot GMCR \cdot E(Z_{C_g}) \\ &+HMCR(t) \cdot (1 - GMCR) \cdot E(Z_{C_h}) \\ &+\frac{1}{2}(1 - HMCR(t))(a_Z + b_Z + c_Z + d_Z) \end{aligned} \tag{3.21}$$

and

$$\begin{aligned} &E\left((Y_i)^2\right) = HMCR(t) \cdot GMCR \cdot \overline{Z_{C_g}^2} \\ &+\frac{bw^2}{3}HMCR(t) \cdot GMCR \cdot PAR(t) + HMCR(t) \cdot (1 - GMCR) \cdot \overline{Z_{C_h}^2} \\ &+\frac{bw^2}{3}HMCR(t) \cdot (1 - GMCR) \cdot PAR(t) \\ &+\frac{1}{3}(1 - HMCR(t))(a_Z^2 + a_Z b_Z + b_Z^2 + c_Z^2 + c_Z d_Z + d_Z^2) \end{aligned} \tag{3.22}$$

As, the mean of Y is $\overline{Y} = \frac{1}{HMS} \sum_{i=1}^{HMS} Y_i$.

Therefore, the expectation of $\overline{Y}$ is:

$$E\left(\overline{Y}\right) = \frac{1}{HMS} \sum_{i=1}^{HMS} E(Y_i)$$

or

$$E\left(\overline{Y}\right) = \frac{1}{HMS} \sum_{i=1}^{HMS} \begin{bmatrix} HMCR(t) \cdot GMCR \cdot E(Z_{C_g}) \\ +HMCR(t) \cdot (1 - GMCR) \cdot E(Z_{C_h}) \\ +\frac{1}{2}(1 - HMCR(t))(a_Z + b_Z + c_Z + d_Z) \end{bmatrix}$$

$$\begin{aligned} &E\left(\overline{Y}\right) == HMCR(t) \cdot GMCR \cdot E(\overline{Z_{C_g}}) \\ &+HMCR(t) \cdot (1 - GMCR) \cdot E(\overline{Z_{C_h}}) \\ &+\frac{1}{2}(1 - HMCR(t))(a_Z + b_Z + c_Z + d_Z) \end{aligned} \tag{3.23}$$

where $\frac{1}{HMS}\sum_{i=1}^{HMS} E(Z_{C_g}) = E(\overline{Z_{C_g}})$ and $\frac{1}{HMS}\sum_{i=1}^{HMS} E(Z_{C_h}) = E(\overline{Z_{C_h}})$.

In the same way $E(\overline{Y^2}) = \frac{1}{HMS}\sum_{i=1}^{HMS} (Y_i)^2$.

Therefore, the expectation of $\overline{Y^2}$ is:

$$E\left(\overline{Y^2}\right) = \frac{1}{HMS}\sum_{i=1}^{HMS} E\left((Y_i)^2\right) = \frac{1}{HMS} \times \sum_{i=1}^{HMS} \begin{bmatrix} HMCR(t) \cdot GMCR \cdot \overline{Z_{C_g}^2} \\ +\frac{bw^2}{3} HMCR(t) \cdot GMCR \cdot PAR(t) \\ +HMCR(t) \cdot (1 - GMCR) \cdot \overline{Z_{C_h}^2} \\ +\frac{bw^2}{3} HMCR(t) \cdot (1 - GMCR) \cdot PAR(t) \\ +\frac{1}{3}(1 - HMCR(t))(a_Z^2 + a_Z b_Z + b_Z^2 + c_Z^2 + c_Z d_Z + d_Z^2) \end{bmatrix}$$

or

$$\begin{aligned} E\left(\overline{Y^2}\right) &= HMCR(t) \cdot GMCR \cdot E(\overline{Z_{C_g}^2}) \\ &+\frac{bw^2}{3} HMCR(t) \cdot GMCR \cdot PAR(t) \\ &+HMCR(t) \cdot (1 - GMCR) \cdot E(\overline{Z_{C_h}^2}) \\ &+\frac{bw^2}{3} HMCR(t) \cdot (1 - GMCR) \cdot PAR(t) \\ &+\frac{1}{3}(1 - HMCR(t))(a_Z^2 + a_Z b_Z + b_Z^2 + c_Z^2 + c_Z d_Z + d_Z^2) \end{aligned} \tag{3.24}$$

$$\overline{Y}^2 = \left(\frac{1}{HMS}\right)^2 \left(\sum_{i=1}^{HMS} (Y_i)\right)^2.$$

Now,

$$E\left(\overline{Y}^2\right) = \left(\frac{1}{HMS}\right)^2 \left(\sum_{i=1}^{HMS} (Y_i)\right)^2$$

or

$$\begin{aligned} E\left(\overline{Y}^2\right) == \frac{1}{HMS}\cdot E\left[\frac{1}{HMS}\cdot\sum_{i=1}^{HMS}(Y_i)^2\right] \\ +\frac{(HMS-1)}{HMS}\cdot E\left[\frac{1}{HMS}\cdot\sum_{\substack{i,j=1\\ i\neq j}}^{HMS} Y_i\cdot Y_j\right] \end{aligned} \tag{3.25}$$

From the relation $E(Y_i\cdot Y_j)=E(Y_i)\cdot E(Y_j)$, it follows that:

$$E\left(\overline{Y}^2\right)=\frac{1}{HMS}\cdot E\left(\overline{Y^2}\right)+\frac{(HMS-1)}{HMS}\cdot\left[E\left(\overline{Y}\right)\right]^2 \tag{3.26}$$

Now,

$$\begin{aligned} E(Var(Y)) = E\left(\frac{1}{HMS}\sum_{i=1}^{HMS}(Y_i-\overline{Y})^2\right) = E\left(\overline{Y^2}\right)-E\left(\overline{Y}^2\right) \\ =\frac{HMS-1}{HMS}\cdot\left[E\left(\overline{Y^2}\right)-\left[E\left(\overline{Y}\right)\right]^2\right] \end{aligned} \tag{3.27}$$

Utilizing (3.23) and (3.24) in (3.27) yield:

$$E(Var(Y)) =$$

$$\left(1-\frac{1}{HMS}\right)\left[\begin{array}{l} HMCR(t)\cdot GMCR\cdot E(\overline{Z^2_{C_g}})+HMCR(t)\cdot(1-GMCR)\cdot E(\overline{Z^2_{C_h}}) \\ +\frac{1}{3}HMCR(t)\cdot PAR(t)\cdot bw^2 \\ +\frac{1}{3}(1-HMCR(t))\left(a_Z^2+a_Zb_Z+b_Z^2+c_Z^2+c_Zd_Z+d_Z^2\right) \\ -\left(\begin{array}{l} HMCR(t)\cdot GMCR\cdot E(\overline{Z_{C_g}})+HMCR(t)\cdot(1-GMCR)\cdot E(\overline{Z_{C_h}}) \\ +\frac{1}{2}(1-HMCR(t))(a_Z+b_Z+c_Z+d_Z) \end{array}\right)^2 \end{array}\right]$$

or

$$E(Var(Y)) = \left(1 - \frac{1}{HMS}\right) \times \left[\begin{array}{l} HMCR(t) \cdot GMCR \cdot Var(Z_{C_g}) + HMCR(t) \cdot (1 - GMCR) \cdot Var(Z_{C_h}) \\ + \frac{1}{3} HMCR(t) \cdot PAR(t) \cdot bw^2 + \frac{1}{3}(1 - HMCR(t)) \\ \times \left(a_Z^2 + a_Z b_Z + b_Z^2 + c_Z^2 + c_Z d_Z + d_Z^2\right) \\ + HMCR(t) \cdot GMCR \cdot E(\overline{Z}^2_{C_g}) + HMCR(t) \cdot (1 - GMCR) \cdot E(\overline{Z}^2_{C_h}) \\ - HMCR^2(t) \cdot GMCR^2 \cdot E(\overline{Z}^2_{C_g}) \\ - 2HMCR^2(t) \cdot GMCR \cdot (1 - GMCR) \cdot E(\overline{Z_{C_g}}) \cdot E(\overline{Z_{C_h}}) \\ - HMCR(t) \cdot (1 - HMCR(t)) \cdot GMCR \cdot E(\overline{Z_{C_g}})(a_Z + b_Z + c_Z + d_Z) \\ - HMCR^2(t) \cdot (1 - GMCR)^2 \cdot E(\overline{Z}^2_{C_h}) \\ - HMCR(t) \cdot (1 - HMCR(t)) \cdot (1 - GMCR) \cdot E(\overline{Z_{C_h}}) \\ \times (a_Z + b_Z + c_Z + d_Z) \\ - \frac{1}{4} \cdot (1 - HMCR(t))^2 \cdot (a_Z + b_Z + c_Z + d_Z)^2 \end{array}\right] \tag{3.28}$$

Equation (3.28) provides (3.13), which proofs the theorem. ■

Depending upon the variance, the exploration capability of the lbest HS-L_1 adaptive method is provided in Lemma 3.1.

Assumption 3.2: It is assumed that, the bandwidth, used to fine tune the newly generated harmony, is proportional to the square root of the expectation mean of population, viz., $bw \propto \sqrt{E(\overline{Z_{C_g}})} = \kappa\sqrt{E(\overline{Z_{C_g}})}$.

Lemma 3.1: If $HMCR(t)$ and $GMCR$ are chosen nearly equal to 1, and the bandwidth bw is chosen to vary depending upon $E(\overline{Z_{C_g}})$, viz., $bw \propto \sqrt{E(\overline{Z_{C_g}})} = \kappa\sqrt{E(\overline{Z_{C_g}})}$ then, the explorative power of the lbest HS-L_1 adaptive control algorithm increases with the increase of iteration.

Proof: If $HMCR(t) \approx 1$ and $GMCR \approx 1$, then $HMCR(t) \cdot GMCR \approx 1$, therefore, the terms related with $(1 - HMCR(t))$, $(1 - GMCR)$ and $(1 - HMCR(t) \cdot GMCR)$ are negligible. Then, (3.13) becomes:

$$E(Var(Y)) = \left(1 - \frac{1}{HMS}\right) \begin{pmatrix} HMCR(t) \cdot GMCR \cdot Var(Z_{C_g}) \\ + \frac{1}{3} HMCR(t) \cdot PAR(t) \cdot bw^2 \end{pmatrix}$$

or

$$E(Var(Y)) = \left(1 - \frac{1}{HMS}\right)\begin{pmatrix} HMCR(t) \cdot GMCR \cdot Var(Z_{C_g}) \\ +\frac{\kappa^2}{3} \cdot HMCR(t) \cdot PAR(t) \cdot E(\overline{Z_{C_g}}) \end{pmatrix} \tag{3.29}$$

At '*itr*' number of harmony improvisation, (3.29) takes a form:

$$\begin{aligned} E(Var(Y_{itr})) &= \left\{\left(1 - \frac{1}{HMS}\right) \cdot HMCR(t) \cdot GMCR\right\}^{itr} \cdot Var(Z_{C_g}) \\ &+ \left[\frac{1 - \left\{\left(1 - \frac{1}{HMS}\right) \cdot HMCR(t) \cdot GMCR\right\}^{itr}}{1 - \left\{\left(1 - \frac{1}{HMS}\right) \cdot HMCR(t) \cdot GMCR\right\}}\right] \\ &\times \left(1 - \frac{1}{HMS}\right)\left(\frac{\kappa^2}{3} \cdot HMCR(t) \cdot PAR(t)\right) \cdot E(\overline{Z_{C_g}}) \end{aligned} \tag{3.30}$$

Equation (3.30) shows that, as $\kappa > 1$, and $\left[1 - \left\{\left(1 - \frac{1}{HMS}\right) \cdot HMCR(t) \cdot GMCR\right\}^{itr}\right]$ increases with iteration therefore, the variance of the search process increases over the iteration of harmony improvisation process which implies that, the concurrent optimization based adaptive process has good exploration capability. (Proved) ■

Remarks 3.1: As the *bw* is increasing with the expectation mean of population and the expectation of the population variance is also increasing therefore, the lbest HS-L_1 adaptive controller has good exploration capability.

It is evident that, the lbest HS-L_1 adaptive controller is sufficiently explorative but it is also of great importance to have good exploitation phenomenon to converge within the spectral radius of the iterative matrix. The exploitation potentiality of this method is analyzed in Theorem 3.2 and Lemma 3.2.

Theorem 3.2: If the expectation of population variance is $E(Var(Z_{C_g}))$ and expectation of population mean is $E(\overline{Z_{C_g}})$ then, the iterative equation of lbest HS-L_1 adaptive controller is given as:

$$\begin{bmatrix} E(Var(Y)) \\ E(\overline{Y}) \end{bmatrix} = \begin{bmatrix} M_1 & M_2 \\ M_3 & M_4 \end{bmatrix} \cdot \begin{bmatrix} E(Var(Z_{C_g})) \\ E(\overline{Z_{C_g}}) \end{bmatrix} \tag{3.31}$$

where $M_1 = \left(1 - \frac{1}{HMS}\right) \cdot HMCR(t) \cdot GMCR$, $M_2 = \left(1 - \frac{1}{HMS}\right) \cdot \frac{\kappa^2}{3} \cdot HMCR(t) \cdot PAR(t)$, $M_3 = 0$ and $M_4 = HMCR(t) \cdot GMCR$.

Proof: As $HMCR(t) \approx 1$ and $GMCR \approx 1$, therefore $HMCR(t) \cdot GMCR \approx 1$ and the terms related to these quantities are negligible. Putting this condition and $bw \propto \sqrt{E(\overline{Z_C})} = \kappa\sqrt{E(\overline{Z_C})}$ in (3.13) and (3.23) yield respectively:

$$E(Var(Y)) = \left(1 - \frac{1}{HMS}\right)\begin{pmatrix} HMCR(t) \cdot GMCR \cdot Var(Z_{C_g}) \\ +\frac{1}{3} \cdot HMCR(t) \cdot PAR(t) \cdot bw^2 \end{pmatrix} \tag{3.32}$$

and

$$E(\overline{Y}) = HMCR(t) \cdot GMCR \cdot E(\overline{Z_{C_g}}) \tag{3.33}$$

Now, the iterative equation is formulated as:

$$\begin{bmatrix} E(Var(Y)) \\ E(\overline{Y}) \end{bmatrix} = \begin{bmatrix} M_1 & M_2 \\ M_3 & M_4 \end{bmatrix} \cdot \begin{bmatrix} E(Var(Z_C)) \\ E(\overline{Z_C}) \end{bmatrix} \tag{3.34}$$

Again, (3.34) can be written as, $Y = M \cdot Z_C'$,

where $Y = \begin{bmatrix} E(Var(Y)) \\ E(\overline{Y}) \end{bmatrix}$, $Z_C' = \begin{bmatrix} E(Var(Z_C)) \\ E(\overline{Z_C}) \end{bmatrix}$ and $M = \begin{bmatrix} M_1 & M_2 \\ M_3 & M_4 \end{bmatrix}$.

Therefore, the theorem is proved. ■

Lemma 3.2: Let, the spectral radius of the iterative matrix M of the lbest HS-L_1 adaptive method is $\rho(M)$ then, the iterative matrix will converge if $\rho(M) < 1$.

Proof: The characteristics roots of iterative matrix M are given as:

$\lambda_1 = \left(1 - \frac{1}{HMS}\right) \cdot HMCR(t) \cdot GMCR$ and $\lambda_2 = HMCR(t) \cdot GMCR$.

As $HMS > 1$, $\therefore \left(1 - \frac{1}{HMS}\right) < 1$, $HMCR(t) < 1$ and $GMCR < 1$, therefore $\lambda_1 < 1$.

Similarly, it can be stated that, the second root $\lambda_2 < 1$.

Again $\frac{\lambda_1}{\lambda_2} = \left(1 - \frac{1}{HMS}\right)$. As, $HMS > 1$, $\therefore \left(1 - \frac{1}{HMS}\right) < 1$, so, $\lambda_2 > \lambda_1$.

The spectral radius of M is $\rho(M) = MAX(\lambda_1, \lambda_2)$ which yields $\rho(M) = \lambda_2$.

As $\lambda_2 < 1$, $\therefore \rho(M) < 1$. Thus, the iterative matrix is convergent.

It demonstrates that the overall concurrent optimization based controller, viz., lbest HS-L_1 adaptive controller will converge. This proves the lemma. ■

Remarks 3.2: If the iterative matrix of the basic harmony search algorithm is M_{conv} then, the spectral radius of it is given as [6] $\rho(M_{conv}) = HMCR$. Now, in case of the lbest HS algorithm the spectral radius of iterative matrix is given as $\rho(M) = (HMCR(t) \cdot GMCR)$. Now as $HMCR(t) < 1$, $GMCR < 1$ then, $HMCR(t) \cdot GMCR < HMCR(t)$. Therefore, the convergence rate of lbest HS is better than the basic HS algorithm.

From these theorems and lemmas it is proved that, the lbest HS can find the parameter values of L_1 adaptive controller efficiently by exploring huge search spaces with good convergence phenomenon. Therefore, the concurrent hybrid lbest HS-L_1 adaptive controller provides fast transient performance with high robustness as well as satisfactory tracking performance.

3.5 Transient and Steady State Performance Analysis of Lbest HS-L_1 Adaptive Controller

The transient and steady state performance of the lbest HS-L_1 adaptive control method is similar as basic L_1 adaptive controller. Same system model (2.57) and (2.58) and reference system model (2.59) and (2.60) as considered in basic L_1 adaptive controller are also considered in this case. The only difference is that, in this method, the parameter values of L_1 adaptive controller are chosen optimally utilizing lbest HS algorithm.

The system model and control input in frequency domain are respectively:

$$\underline{x}(s) = H(s)(\,(1 - C(s))\underline{r}_2(s) + C(s)\,k_g\underline{r}(s) - C(s)\,\tilde{\underline{r}}_{ud}(s)\,) \tag{3.35}$$

$$\underline{u}(s) = \frac{C(s)}{\omega}\left[k_g\underline{r}(s) - \underline{r}_2(s) - \tilde{\underline{r}}_{ud}(s)\right] - K\underline{x}(s) \tag{3.36}$$

The closed loop LTI reference system model with control input are respectively:

$$\underline{x}_{ref}(s) = H(s)(\,(1 - C(s))\underline{r}_{2ref}(s) + C(s)k_g\underline{r}(s)\,) \tag{3.37}$$

$$\underline{u}_{ref}(s) = \frac{C(s)}{\omega}\left[k_g\underline{r}(s) - \underline{r}_{2ref}(s)\right] - K\underline{x}_{ref}(s) \tag{3.38}$$

The state error and control input error between the system and reference system model hold the following bounds respectively:

$$\left\|\underline{e}\right\|_{L_\infty} = \left\|\underline{x}_{ref} - \underline{x}\right\|_{L_\infty} \le val_1 \tag{3.39}$$

$$\left\|\underline{u} - \underline{u}_{ref}\right\|_{L_\infty} \le val_2 \tag{3.40}$$

where

$$val_1 = \frac{\|C(s)\|_{L_1}}{1 - \|H(s)(1 - C(s))\|_{L_1}\theta_{\max}}\sqrt{\frac{2\theta_m}{\lambda_{\min}(P)\Gamma_C}}$$

and

$$val_2 = \left(-\left\|\frac{C(s)\,\theta}{\omega} + K\right\|_{L_1} val_1 + \left\|\frac{C(s)}{\omega}\cdot\frac{1}{c_0^T H(s)}\cdot c_0^T\right\|_{L_1}\sqrt{\frac{2\theta_m}{\lambda_{\min}(P)\Gamma_C}}\right)$$

Remarks 3.3: These results (3.39) and (3.40) are similar as basic L_1 adaptive controller (2.61) and (2.62). Although the parameter values are optimal in this case (3.39) and (3.40). Hence, the lbest HS-L_1 adaptive controller provides better transient and steady state performance than basic L_1 adaptive controller.

3.6 Simulation Case Studies

In this chapter, lbest HS-L_1 adaptive controller is employed on four systems, viz., Duffing's oscillatory system, nonlinear spring mass damper system, inverted pendulum with cart and twin rotor MIMO system in simulation environment to validate its effectiveness. The efficiency of the lbest HS-L_1 adaptive control method is compared with other control strategies like basic L_1 adaptive, particle swarm optimization based L_1 (PSO-L_1) adaptive and basic harmony search based L_1 (HS-L_1) adaptive controller in terms of IAE. These methodologies are briefly described next.

Basic L_1 (Basic L_1) adaptive controller: In case of basic L_1 adaptive controller, the parameters, viz., $[k \; |k_g|K|\,\Gamma_C \mid \hat{\omega} \mid \hat{\underline{\theta}} \mid \hat{\underline{\sigma}}]$ are selected arbitrarily within the ranges getting from L_1 norm condition (2.56). The tracking performances are calculated in terms of integral absolute error (IAE).

Particle swarm optimization based L_1 (PSO-L_1) adaptive controller: The parameters of L_1 adaptive controller are tuned by utilizing particle swarm optimization (PSO). Candidate solution vector (CSV) is formed as $\underline{Z} = [k \; |k_g|K|\,\Gamma_C \mid \hat{\omega} \mid \hat{\underline{\theta}} \mid \hat{\underline{\sigma}}]$. $[\hat{\omega} \mid \hat{\underline{\theta}} \mid \hat{\underline{\sigma}}]$ values are adapted concurrently utilizing L_1 adaptation laws after obtained from the PSO. 30 particles of PSO are iterated for 200 times to obtain the optimal parameter setting of L_1 adaptive controller that, it can properly cancel out disturbances and provide good tracking performances. This whole process runs 10 times and from that minimum, viz., best, average and standard deviation of IAEs are tabulated for comparison.

Harmony search algorithm based L_1 (HS-L_1) adaptive controller: As a similar manner of the previous case, CSV $\underline{Z} = [k \; |k_g|K|\,\Gamma_C \mid \hat{\omega} \mid \hat{\underline{\theta}} \mid \hat{\underline{\sigma}}]$ is optimized utilizing HS algorithm to provide good tracking and transient performance by eliminating disturbances and uncertainties. To fine tune the values of $[\hat{\omega} \mid \hat{\underline{\theta}} \mid \hat{\underline{\sigma}}]$, they are concurrently adapted following L_1 adaptation rule after obtained from HS. 30 number of harmonies are improvised for 200 iterations with the objective to minimize the integral absolute error between reference and system output. The whole process runs for 10 times and minimum, viz., best, average and standard deviation of IAEs are tabulated for each system.

Local best harmony search algorithm based L_1 (lbest HS-L_1) adaptive controller: In this method, the CSV $\underline{Z} = [k \; |k_g|K|\,\Gamma_C \mid \hat{\omega} \mid \hat{\underline{\theta}} \mid \hat{\underline{\sigma}}]$ containing the parameter values of L_1 adaptive controller are optimized with the aim to provide better performance of L_1 adaptive controller. At first, the $[\hat{\omega} \mid \hat{\underline{\theta}} \mid \hat{\underline{\sigma}}]$ values are obtained from lbest HS and then, they are concurrently adapted following L_1 adaptation laws. In this case, *HMCR*, *PAR* values are varying from 0.65 to 0.95 and 0.25 to 0.65

respectively. *GMCR* is considered as 0.9 with group element varying within 10–15 number. *HMS* is considered as 30 and improvised for 200 times to incur optimal L_1 adaptive controller. This whole process run for 10 times to derive the minimum, viz., best, average and standard deviation of IAEs.

It is to be noted that, in a similar fashion of Chap. 2, in case of the controllers, designed utilizing any stochastic optimization and metaheuristics techniques, the statistical measures of performances, viz., best, average and standard deviation (std. dev.) of IAEs are taken. Whereas, for conventional adaptive controllers, these statistical measures are not required to be calculated. Thus, keeping that in mind, the table for comparative studies of different controllers in each case study is prepared.

These controllers are employed on the four systems described in Chap. 2. Tracking performances of the systems are evaluated for the references similar as given in Chap. 2.

3.6.1 Case Study-I: Duffing's Oscillatory System

In this case study, Duffing's oscillatory system with disturbance same as (2.63) is considered as [4]:

$$\begin{aligned} \dot{x}_1 &= x_2 \\ \dot{x}_2 &= -0.1x_2 - x_1^3 + 12\cos(t) + u(t) + d(t) \\ y &= x_1 \end{aligned} \tag{3.41}$$

With the same tracking objective as describes in Sect. 2.6.1, the Duffing's oscillatory system is controlled employing basic L_1 adaptive controller, PSO-L_1 adaptive, HS-L_1 adaptive and lbest HS-L_1 adaptive controller. Controllers are designed for the system with disturbance as shown in Fig. 2.2 and then tested for system with 10 dB white Gaussian noise as depicted in Fig. 2.3. The system runs for 10 s duration with time step 0.01 s and the results are tabulated in terms of integral absolute error in Table 3.1. System responses with disturbance for above mentioned four controllers

Table 3.1 Comparative study of different control methodologies for case study-I (Duffing's oscillatory system)

Control methodologies	IAE value			
	For disturbance			For white Gaussian noise
	Best	Average	Std. Dev	
Basic L_1 adaptive controller	0.2347	–	–	0.4559
PSO-L_1 adaptive controller	0.0129	0.0454	0.0208	0.0199
HS-L_1 adaptive controller	0.0121	0.0279	0.0128	0.0197
lbest HS-L_1 adaptive controller	0.0118	0.0119	0.0003	0.0186

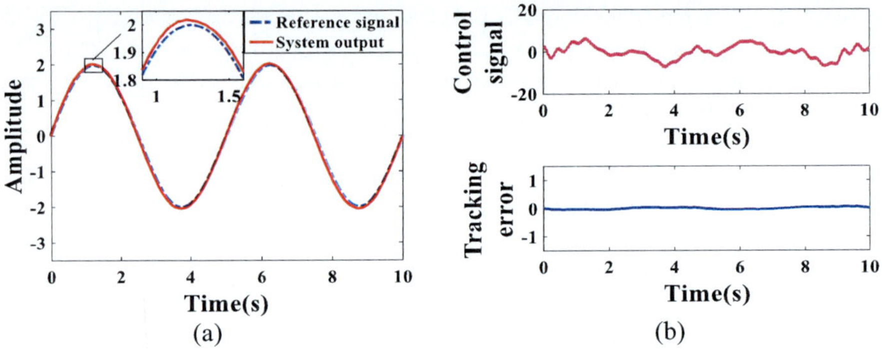

Fig. 3.7 Evaluation period **a** system response and corresponding **b** control signal with tracking error of case study-I (Duffing's oscillatory system) with disturbance for basic L_1 adaptive controller

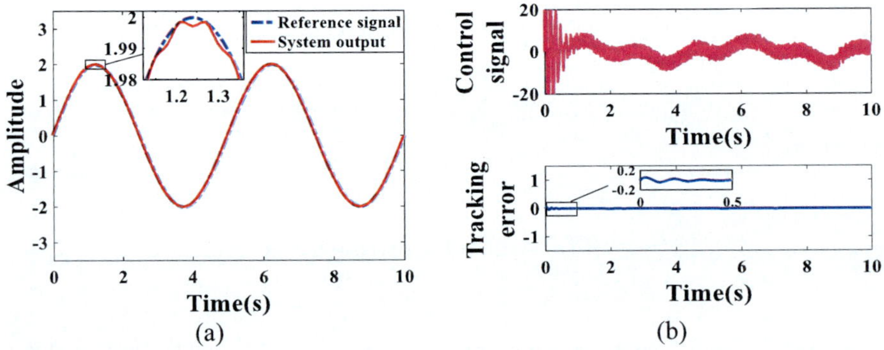

Fig. 3.8 Evaluation period **a** system response and corresponding **b** control signal with tracking error of case study-I (Duffing's oscillatory system) with disturbance for PSO-L_1 adaptive controller

are shown in Figs. 3.7a, 3.8a, 3.9a and 3.10a respectively. Respective control signal with error plots are provided in Figs. 3.7b, 3.8b, 3.9b and 3.10b. Performances of system with noise are depicted in Figs. 3.11, 3.12, 3.13 and 3.14. Results for system with disturbance show that, the lbest HS-L_1 adaptive controller possess fast transient response as well as better tracking response than the other methods tested in this case study. When tuned controllers are subjected to the unknown, time varying noise then the lbest HS-L_1 adaptive controller eliminates them efficiently and performs better than other controllers tested.

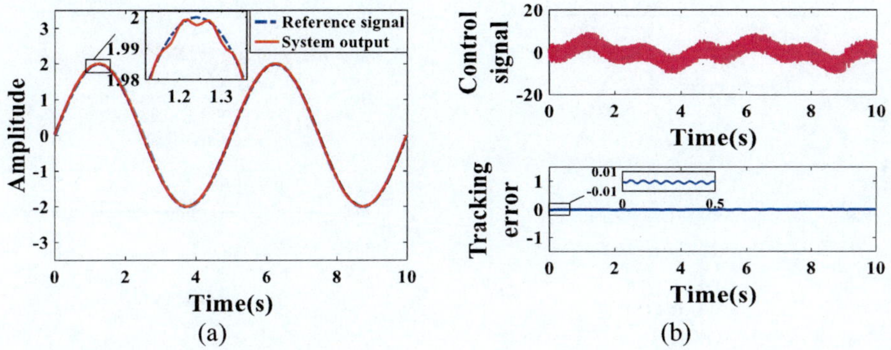

Fig. 3.9 Evaluation period **a** system response and corresponding **b** control signal with tracking error of case study-I (Duffing's oscillatory system) with disturbance for HS-L_1 adaptive controller

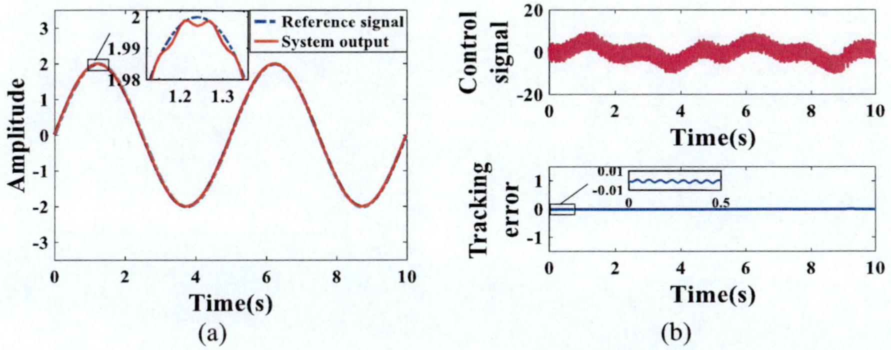

Fig. 3.10 Evaluation period **a** system response and corresponding **b** control signal with tracking error of case study-I (Duffing's oscillatory system) with disturbance for lbest HS-L_1 adaptive controller

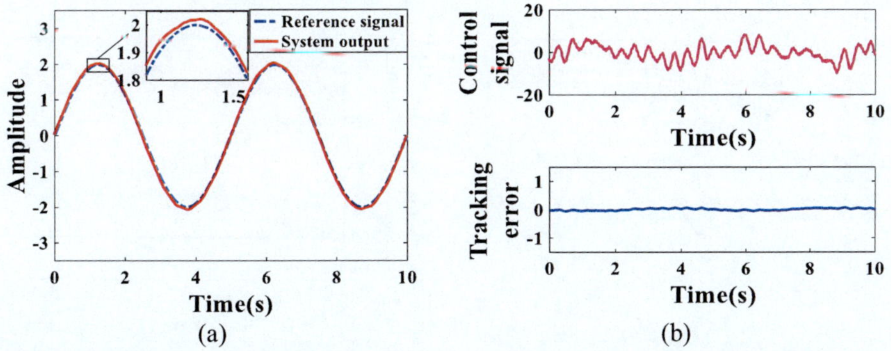

Fig. 3.11 Evaluation period **a** system response and corresponding **b** control signal with tracking error of case study-I (Duffing's oscillatory system) with white Gaussian noise for basic L_1 adaptive controller

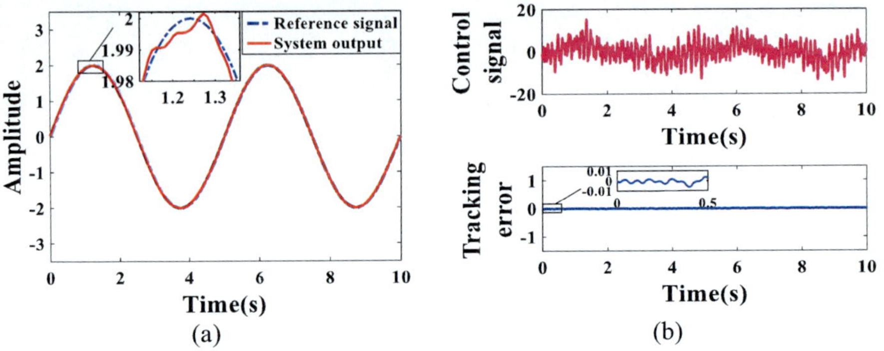

Fig. 3.12 Evaluation period **a** system response and corresponding **b** control signal with tracking error of case study-I (Duffing's oscillatory system) with white Gaussian noise for PSOL_1 adaptive controller

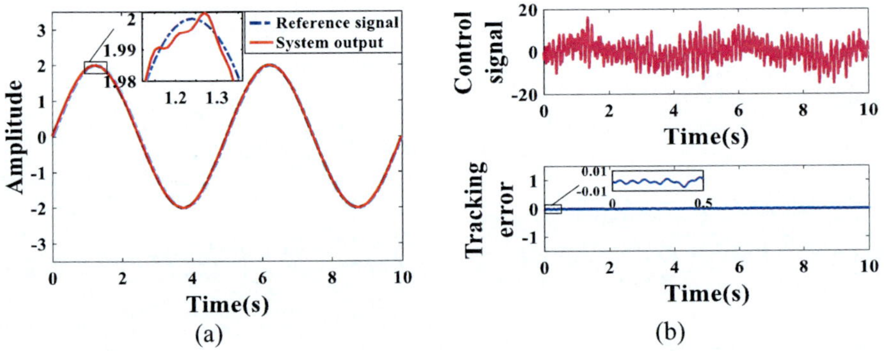

Fig. 3.13 Evaluation period **a** system response and corresponding **b** control signal with tracking error plot of case study-I (Duffing's oscillatory system) with white Gaussian noise for HS-L_1 adaptive controller

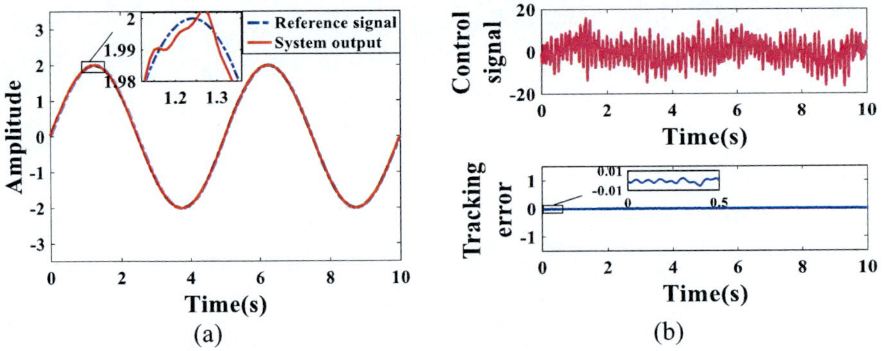

Fig. 3.14 Evaluation period **a** system response and corresponding **b** control signal with tracking error of case study-I (Duffing's oscillatory system) with white Gaussian noise for lbest HS-L_1 adaptive controller

3.6.2 Case Study-II: Nonlinear Spring Mass Damper System

The nonlinear spring mass damper system [4] with disturbance is considered similarly as (2.64):

$$\left.\begin{aligned} \dot{x}_1 &= x_2 \\ \dot{x}_2 &= -0.02x_1 - 0.67x_1^3 - 0.1x_2^3 + u(t) + d(t) \\ y &= x_1 \end{aligned}\right\} \tag{3.42}$$

Description of this system is provided in Sect. 2.6.2. Controllers are designed for system with disturbance as shown in Fig. 2.2 and then, the tuned controllers are examined for noise as given in Fig. 2.3. The performances of basic L_1 adaptive, PSO-L_1 adaptive, HS-L_1 adaptive and lbest HS-L_1 adaptive controller are investigated in terms of integral absolute error (IAE) as tabulated in Table 3.2. Responses of system with disturbance for above mentioned four controllers are provided in Figs. 3.15a, 3.16a, 3.17a and 3.18a respectively. The corresponding control signal with error

Table 3.2 Comparative study of different control methodologies for case study-II (nonlinear spring mass damper)

Control methodologies	IAE value			
	For disturbance			For white Gaussian noise
	Best	Average	Std. Dev	
Basic L_1 adaptive controller	0.0813	–	–	0.2694
PSO-L_1 adaptive controller	0.0311	0.4851	0.0175	0.1625
HS-L_1 adaptive controller	0.0305	0.0328	0.0016	0.0529
lbest HS-L_1 adaptive controller	0.0140	0.0207	0.0034	0.0359

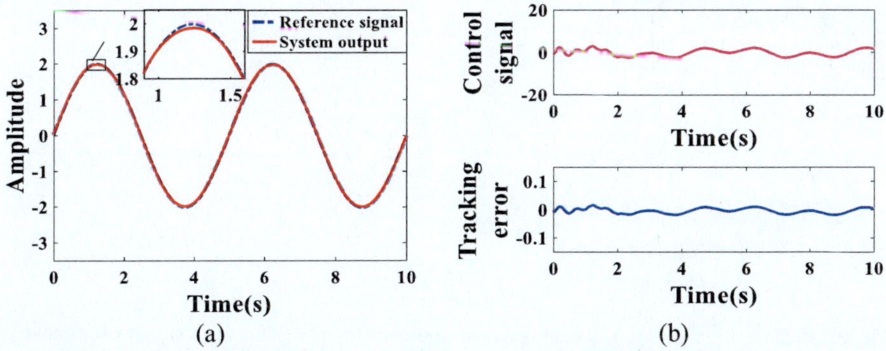

Fig. 3.15 Evaluation period **a** system response and corresponding **b** control signal with tracking error of case study-II (nonlinear spring mass damper) with disturbance for basic L_1 adaptive controller

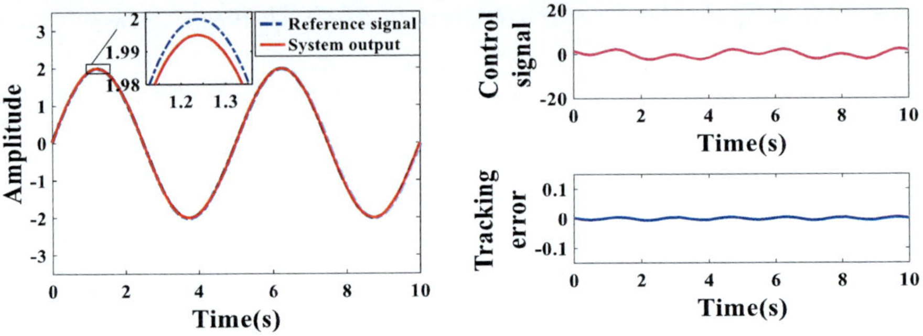

Fig. 3.16 Evaluation period **a** system response and corresponding **b** control signal with tracking error of case study-II (nonlinear spring mass damper) with disturbance for PSO-L_1 adaptive controller

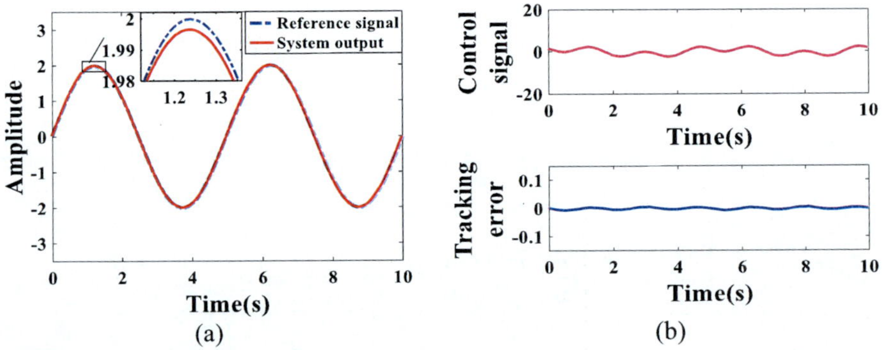

Fig. 3.17 Evaluation period **a** system response and corresponding **b** control signal with tracking error of case study-II (nonlinear spring mass damper) with disturbance for HS-L_1 adaptive controller

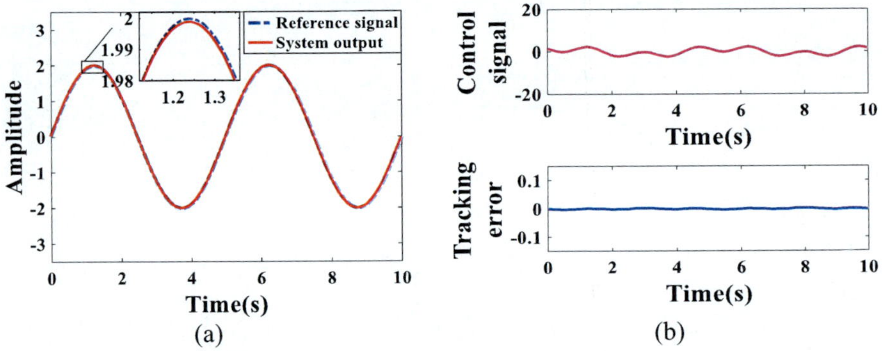

Fig. 3.18 Evaluation period **a** system response and corresponding **b** control signal with tracking error of case study-II (nonlinear spring mass damper) with disturbance for lbest HS-L_1 adaptive controller

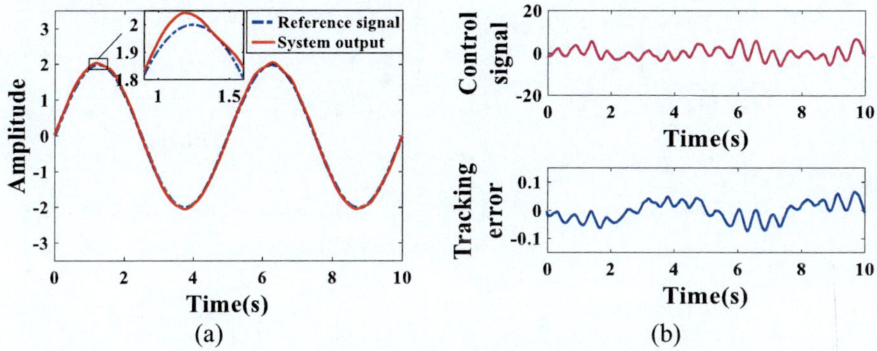

Fig. 3.19 Evaluation period **a** system response and corresponding **b** control signal with tracking error of case study-II (nonlinear spring mass damper) with white Gaussian noise for basic L_1 adaptive controller

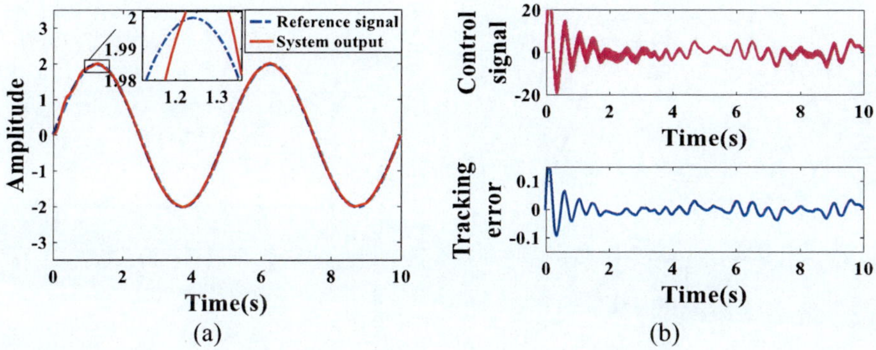

Fig. 3.20 Evaluation period **a** system response and corresponding **b** control signal with tracking error of case study-II (nonlinear spring mass damper) with white Gaussian noise for PSO-L_1 adaptive controller

signal plots are provided in Figs. 3.15b, 3.16b, 3.17b and 3.18b respectively. Whereas, the performances of system with 10 dB white Gaussian noise are shown in Figs. 3.19, 3.20, 3.21 and 3.22 for four controllers. From results it is evident that, the disturbance rejection phenomenon, transient and steady state performance of the lbest HS-L_1 adaptive controller is much better than the other control methods tested in this case study.

3.6.3 Case Study-III: Inverted Pendulum with Cart

The dynamics of the 4th order inverted pendulum with cart system (2.65) is given as follows [7]:

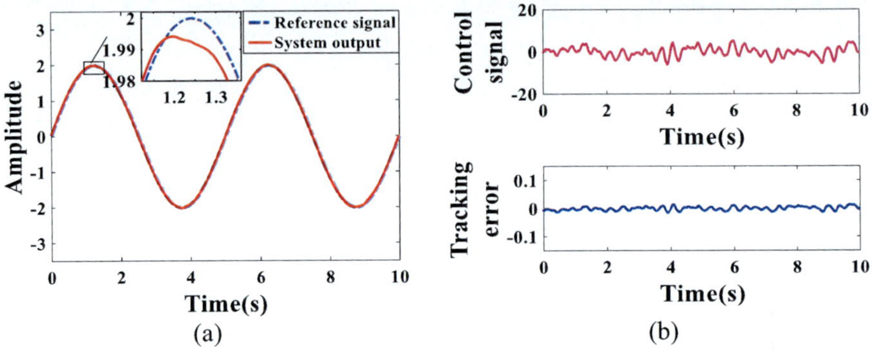

Fig. 3.21 Evaluation period **a** system response and corresponding **b** control signal with tracking error of case study-II (nonlinear spring mass damper) with white Gaussian noise for HS-L_1 adaptive controller

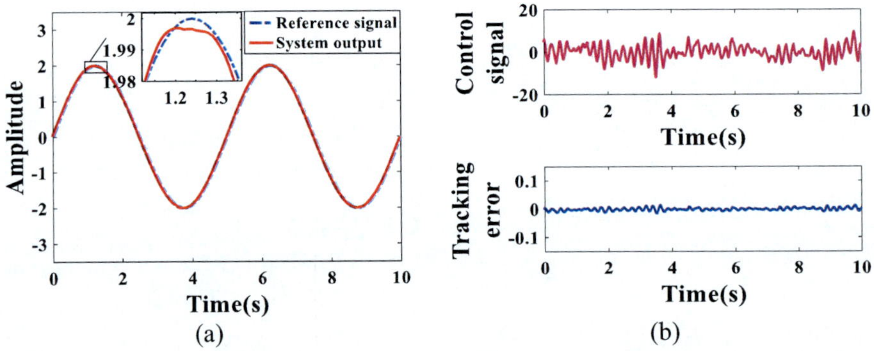

Fig. 3.22 Evaluation period **a** system response and corresponding **b** control signal with tracking error of case study-II (nonlinear spring mass damper) with white Gaussian noise for lbest HS-L_1 adaptive controller

$$\left.\begin{aligned}
\dot{x}_1 &= x_2 \\
\dot{x}_2 &= \frac{-mg\ \cos x_3\ \sin x_3 + ml \sin x_3\, x_4^2 + u}{M + m \sin^2 x_3} + d_1(t) \\
\dot{x}_3 &= x_4 \\
\dot{x}_4 &= \frac{-ml\ \cos x_3\ \sin x_3\, x_4^2 + (M+m) g \sin x_3\ - u \cos x_3}{Ml + ml \sin^2 x_3} + d_2(t)
\end{aligned}\right\} \tag{3.43}$$

Description of the system is provided in Sect. 2.6.3. Unstable, non-minimum phase inverted pendulum with cart is controlled employing basic L_1 adaptive, PSO-L_1 adaptive, HS-L_1 adaptive and lbest HS-L_1 adaptive controller with the aim to

Table 3.3 Comparative study of different control methodologies for case study-III (inverted pendulum with cart)

Control methodologies	IAE value			
	Best	Average	Std. Dev	With disturbance
Basic L_1 adaptive controller	0.0465	–	–	0.0502
PSO-L_1 adaptive controller	0.0019	0.0024	0.0005	0.0169
HS-L_1 adaptive controller	0.0018	0.0064	0.0058	0.0180
lbest HS-L_1 adaptive controller	0.0009	0.0033	0.0020	0.0078

track the reference trajectory $y = 0.1$ unit. System responses without disturbance for 10 s duration with 0.01 s time step are executed and the IAE values are tabulated in Table 3.3. The tuned controllers are then subjected to the disturbances as described in Sect. 2.6.3 and the performances are provided in Table 3.3. The evaluation period system responses without disturbance for above mentioned four controllers are shown in Figs. 3.23a, 3.24a, 3.25a and 3.26a respectively. The corresponding control signal with error plots are depicted in Figs. 3.23b, 3.24b, 3.25b and 3.26b respectively.

Results show that, the optimal parameter values of L_1 adaptive controller provide better transient as well as tracking performances. The responses of system with disturbances are depicted in Figs. 3.27, 3.28, 3.29 and 3.30 for four controllers. The performance of basic L_1 adaptive controller with disturbance degrades very much. On the other hand, the lbest HS-L_1 adaptive controller eliminates disturbances satisfactorily.

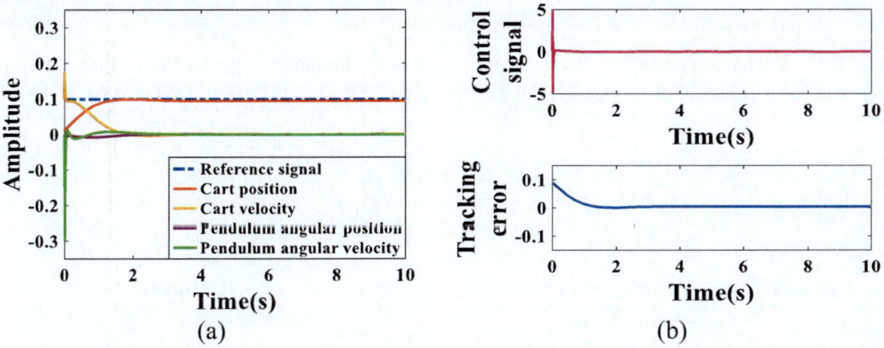

Fig. 3.23 Evaluation period **a** system response and corresponding **b** control signal with tracking error of case study-III (inverted pendulum with cart) without disturbance for basic L_1 adaptive controller

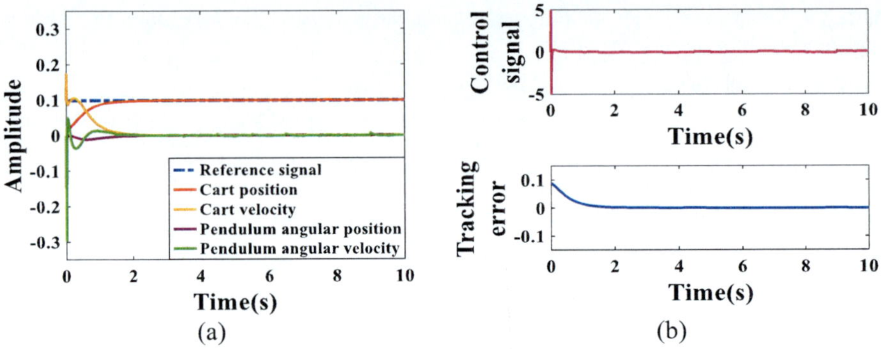

Fig. 3.24 Evaluation period **a** system response and corresponding **b** control signal with tracking error of case study-III (inverted pendulum with cart) without disturbance for PSO-L_1 adaptive controller

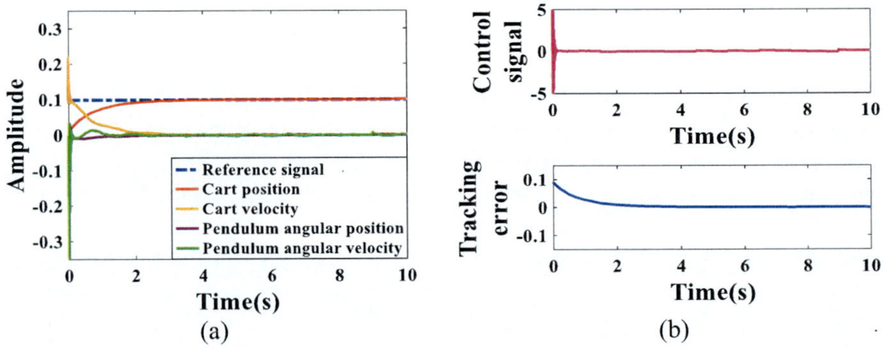

Fig. 3.25 Evaluation period **a** system response and corresponding **b** control signal with tracking error of case study-III (inverted pendulum with cart) without disturbance for HS-L_1 adaptive controller

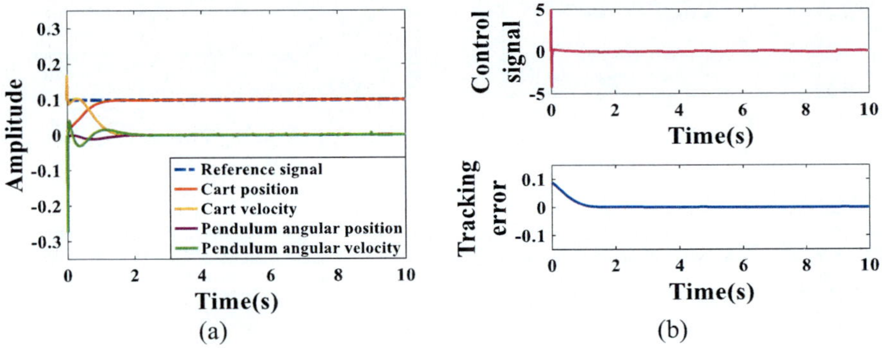

Fig. 3.26 Evaluation period **a** system response and corresponding **b** control signal with tracking error of case study-III (inverted pendulum with cart) without disturbance for lbest HS-L_1 adaptive controller

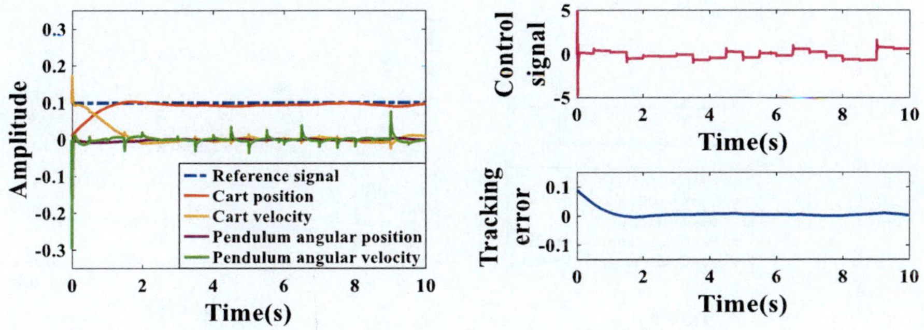

(a) (b)

Fig. 3.27 Evaluation period **a** system response and corresponding **b** control signal with tracking error of case study-III (inverted pendulum with cart) with disturbance for basic L_1 adaptive controller

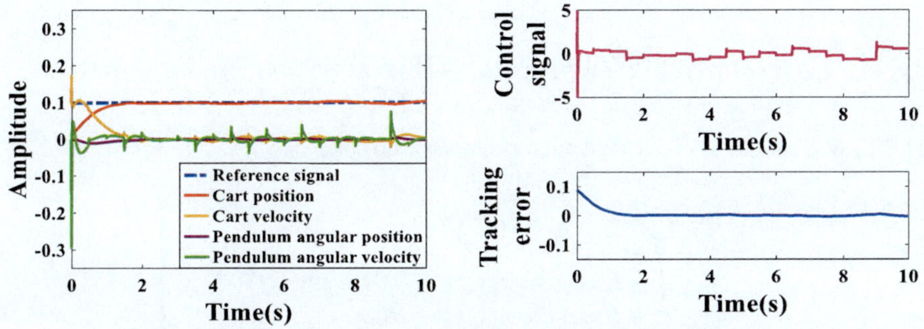

(a) (b)

Fig. 3.28 Evaluation period **a** system response and corresponding **b** control signal with tracking error of case study-III (inverted pendulum with cart) with disturbance for PSO-L_1 adaptive controller

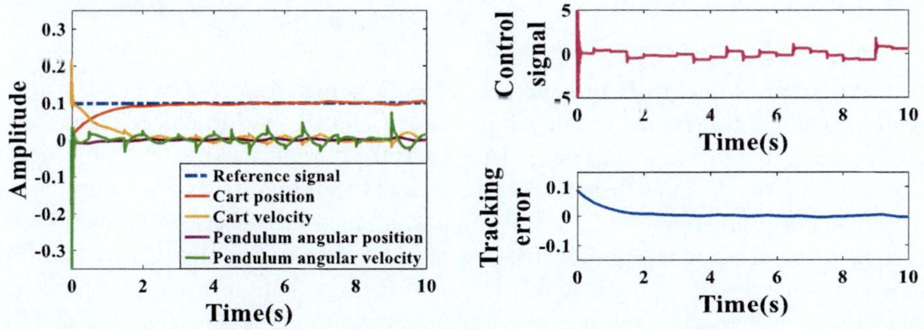

(a) (b)

Fig. 3.29 Evaluation period **a** system response and corresponding **b** control signal with tracking error of case study-III (inverted pendulum with cart) with disturbance for HS-L_1 adaptive controller

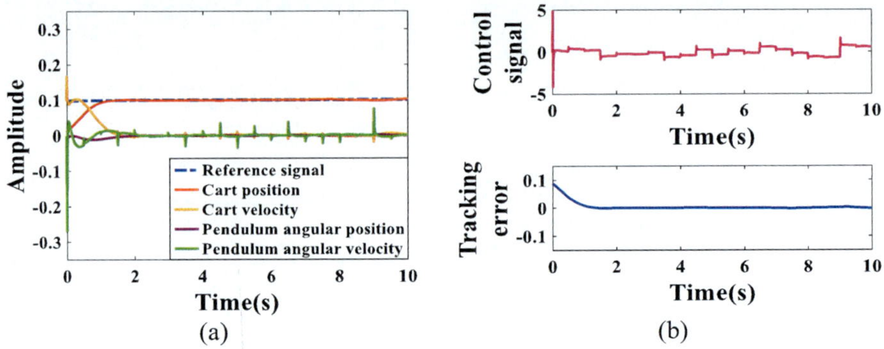

Fig. 3.30 Evaluation period **a** system response and corresponding **b** control signal with tracking error of case study-III (inverted pendulum with cart) with disturbance for lbest HS-L_1 adaptive controller

3.6.4 Case Study-IV: Twin Rotor MIMO System

The TRMS model [8] is described similar as (2.66):

$$\left.\begin{array}{c} \dot{x}_1 = x_2 \\ \dot{x}_2 = \frac{1}{I_1}\begin{pmatrix} a_1x_5^2 + b_1x_5 - M_g \sin x_1 \\ -K_{gy}b_1 \cos x_1x_4x_5 + 0.016 \ \sin 2x_1x_4^2 \\ -K_{gy}a_1 \cos x_1x_4x_5^2 - B_{1\phi}x_2 \end{pmatrix} \\ \dot{x}_3 = x_4 \\ \dot{x}_4 = \frac{1}{I_2}\begin{pmatrix} a_2x_6^2 + b_2x_6 - 1.75(K_ca_1x_5^2 + K_cb_1x_5) \\ +0.375(K_ca_1x_5^2 + K_cb_1x_5)\exp^{-0.5t} - B_{1\varphi}x_4 \end{pmatrix} \\ \dot{x}_5 = -\frac{T_{10}x_5}{T_{11}} + \frac{K_{11}u_1}{T_{11}} \\ \dot{x}_6 = -\frac{T_{20}x_6}{T_{21}} + \frac{K_{22}u_2}{T_{21}} \end{array}\right\} \tag{3.44}$$

Description of TRMS is provided in Sect. 2.6.4 and therefore avoided in this section. The controllers are employed on TRMS and the performance evaluations are provided in Table 3.4 in terms of IAE for 100 s duration with 0.01 s time step. The tabulated results show that, the performance of lbest HS-L_1 adaptive controller is better than the other methods tested. When Disturbance-I as described in Sect. 2.6.4 is subjected to the pitch rotor then, the yaw rotor performance is also affected due to cross-coupling as shown in Table 3.5. It is worth noting that, in case of lbest HS-L_1 adaptive controller, the yaw rotor performance without disturbance as provided in Table 3.4 does not vary much when disturbance is imposed on the pitch rotor as provided in Table 3.5. Therefore, it can be stated that, the disturbances and cross-coupling related uncertainties rejection phenomenon of the lbest HS-L_1 adaptive controller are very satisfactory. The performances of the system, viz., pitch angle

Table 3.4 Comparative study of different control methodologies for case study-IV (twin rotor MIMO system) without disturbance

Control methodologies	IAE value without disturbance					
	Pitch angle			Yaw angle		
	Best	Average	Std. Dev	Best	Average	Std. Dev
Basic L_1 adaptive controller	1.4914	–	–	4.0749	–	–
PSO-L_1 adaptive controller	0.8590	1.5621	0.4957	3.8614	4.1204	0.2648
HS-L_1 adaptive controller	0.8550	1.4235	0.5295	3.6819	3.9929	0.2868
lbest HS-L_1 adaptive controller	0.6037	1.3033	0.5577	2.6873	3.7344	0.5559

Table 3.5 Comparative study of different control methodologies for case study-IV (twin rotor MIMO system) with Disturbance-I

Control methodologies	IAE value with disturbance-I	
	Pitch angle	Yaw angle
Basic L_1 adaptive controller	1.5655	4.0783
PSO-L_1 adaptive controller	0.9401	3.8617
HS-L_1 adaptive controller	0.8691	3.6826
lbest HS-L_1 adaptive controller	0.7020	2.6873

and yaw angle without disturbance for four controllers are shown in Figs. 3.31, 3.32, 3.33 and 3.34. The system performances for four controllers with Disturbance-I are pictured in Figs. 3.35, 3.36, 3.37 and 3.38. Responses show that, the performance of lbest HS-L_1 adaptive controller is better than the other methods employed in this case study. The optimal parameter setting of the L_1 adaptive controller with the help of lbest HS algorithm provides satisfactory disturbance rejection as well as de-coupled performance.

3.7 Summary

In case of basic L_1 adaptive controller, the parameter values are chosen arbitrarily within the ranges obtained from L_1 norm condition which do not yield optimal performance. Hence, in this chapter, the parameter values of L_1 adaptive controller are tuned employing stochastic optimization and metaheuristics techniques to get optimized performance. In this work, a variant of HS algorithm, viz., lbest HS is utilized to tune the parameters of L_1 adaptive controller. It is proved analytically that, the exploration and exploitation capability of the lbest HS-L_1 adaptive controller is satisfactory. The lbest HS-L_1 adaptive controller is employed on four simulation case studies and the results are tabulated in terms of IAE. The results show that, optimal

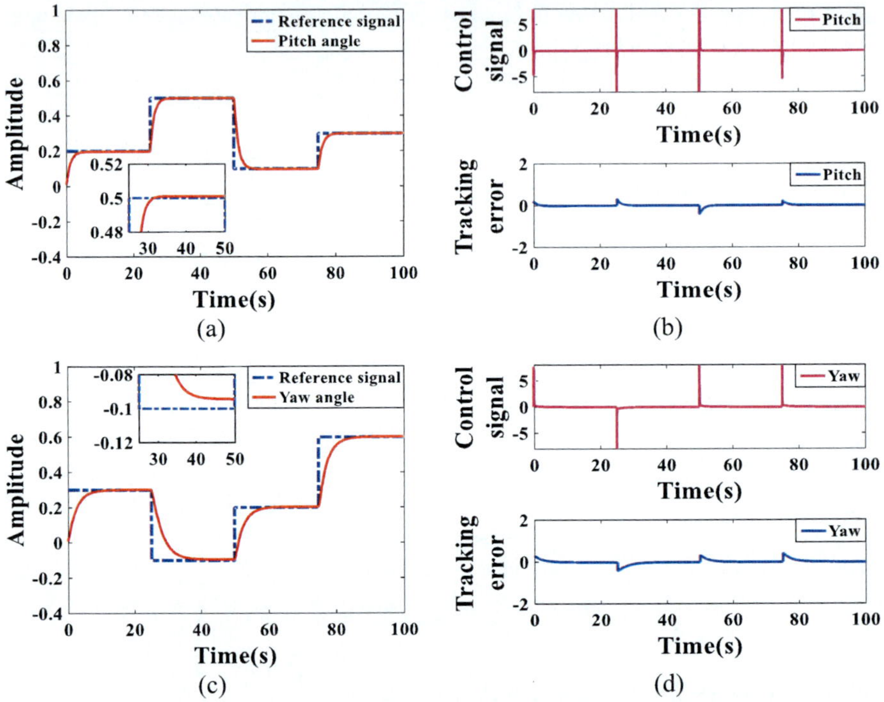

Fig. 3.31 Evaluation period **a** pitch angle and corresponding **b** control signal with tracking error, **c** yaw angle and corresponding **d** control signal with tracking error of case study-IV (twin rotor MIMO system) without disturbance for basic L_1 adaptive controller

parameter setting of L_1 adaptive controller provides better disturbance rejection, fast transient performance and satisfactory tracking phenomenon than the basic L_1 adaptive, PSO-L_1 adaptive and HS-L_1 adaptive controller.

The lbest HS-L_1 adaptive controller is efficient to tackle uncertainties and disturbances present in the system. However, separate consideration of nonlinearities is absent in this controller design. The performance of this controller may degrade when high nonlinearities present in the system. The components of the L_1 adaptive controller should be designed to handle uncertainties, disturbances as well as nonlinearities satisfactorily which is elaborated in the next chapter.

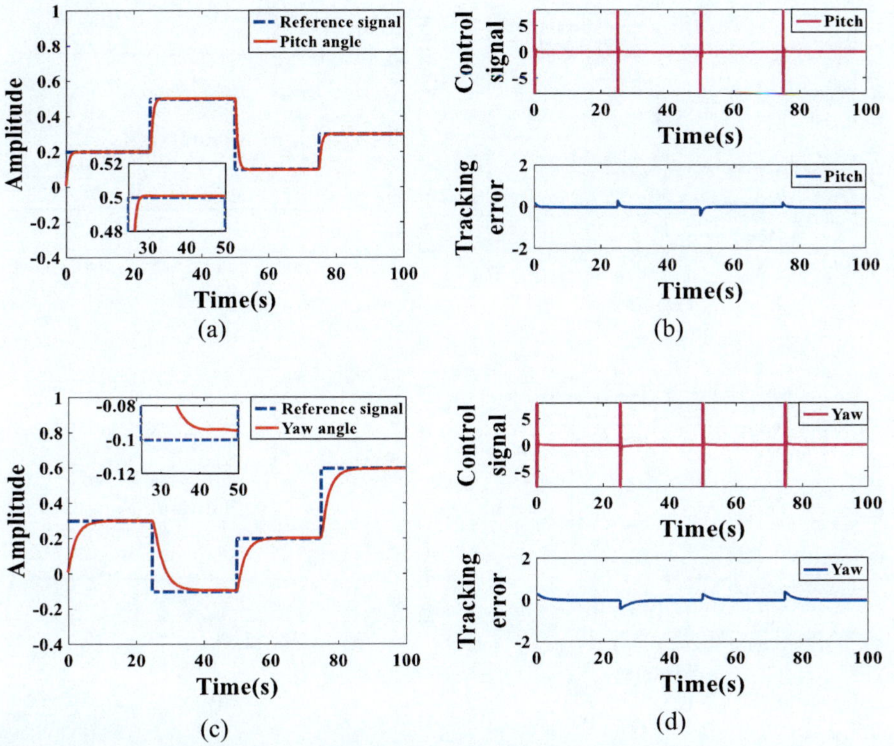

Fig. 3.32 Evaluation period **a** pitch angle and corresponding **b** control signal with tracking error, **c** yaw angle and corresponding **d** control signal with tracking error of case study-IV (twin rotor MIMO system) without disturbance for PSO-L_1 adaptive controller

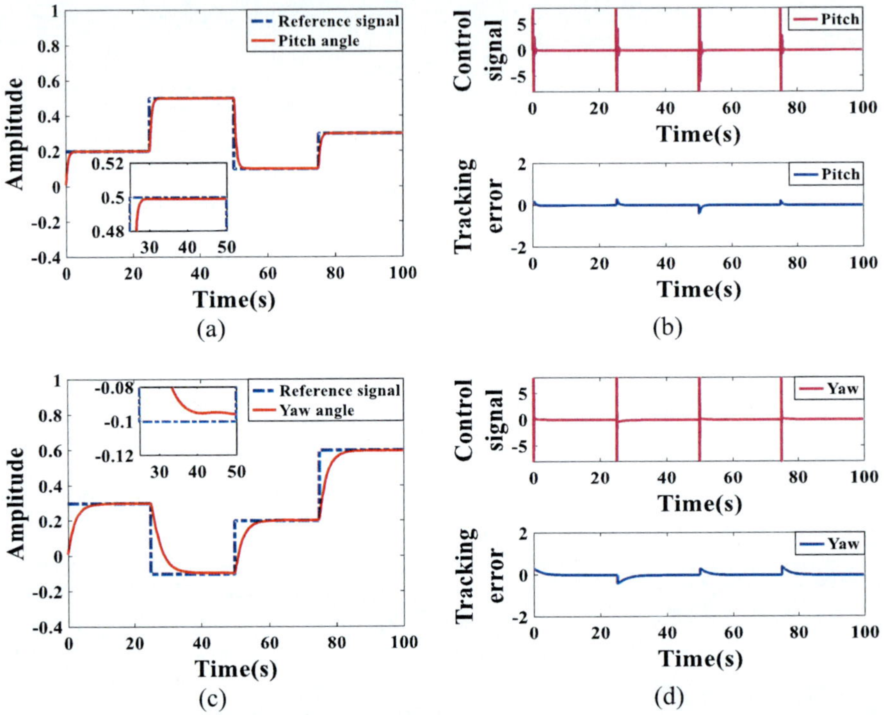

Fig. 3.33 Evaluation period **a** pitch angle and corresponding **b** control signal with tracking error, **c** yaw angle and corresponding **d** control signal with tracking error of case study-IV (twin rotor MIMO system) without disturbance for HS-L_1 adaptive controller

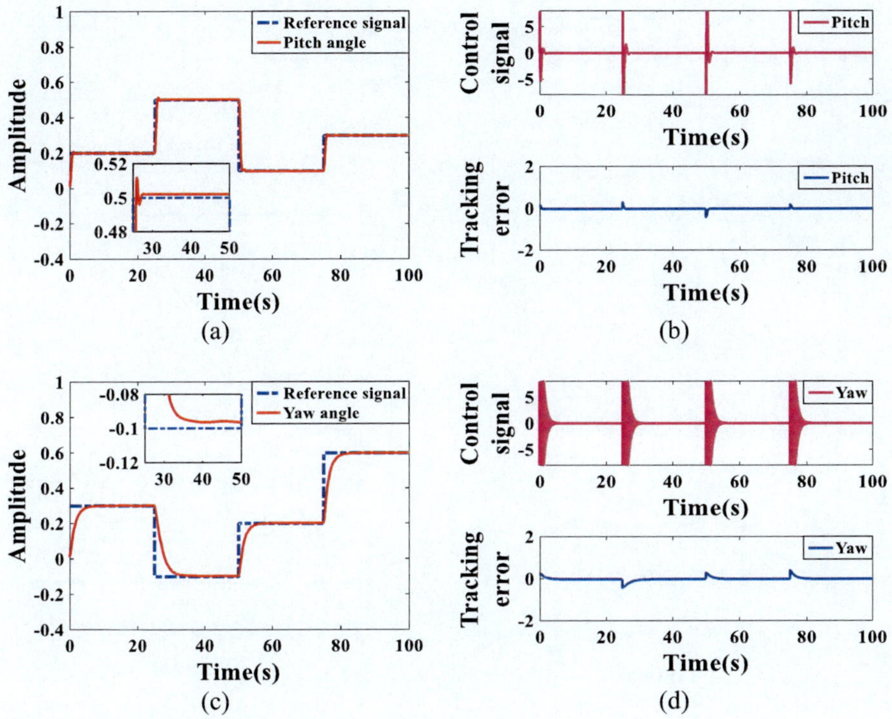

Fig. 3.34 Evaluation period **a** pitch angle and corresponding **b** control signal with tracking error, **c** yaw angle and corresponding **d** control signal with tracking error of case study-IV (twin rotor MIMO system) without disturbance for lbest HS-L_1 adaptive controller

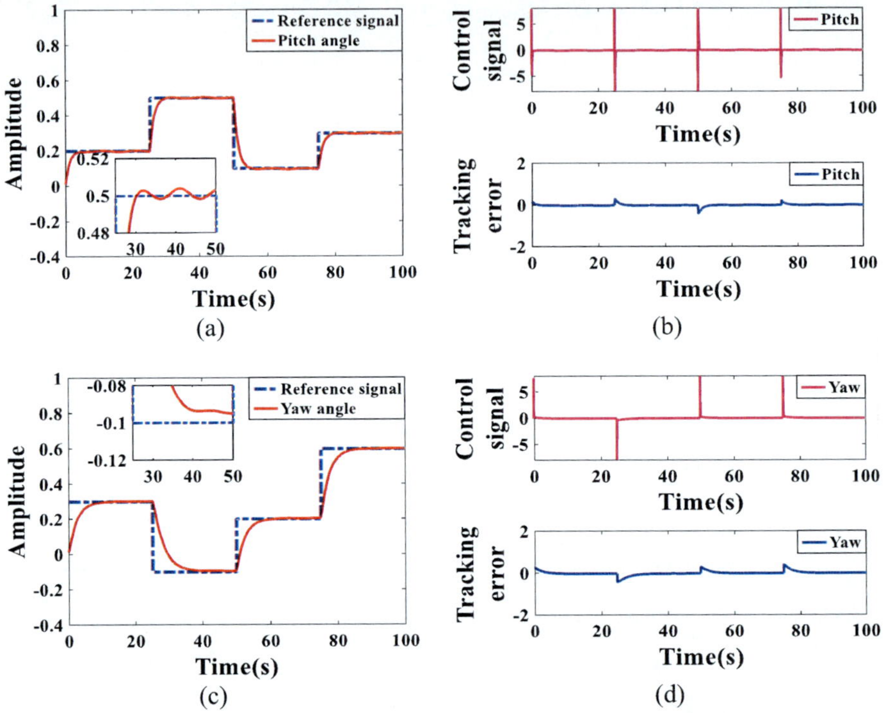

Fig. 3.35 Evaluation period **a** pitch angle and corresponding **b** control signal with tracking error, **c** yaw angle and corresponding **d** control signal with tracking error of case study-IV (twin rotor MIMO system) with Disturbance-I for basic L_1 adaptive controller

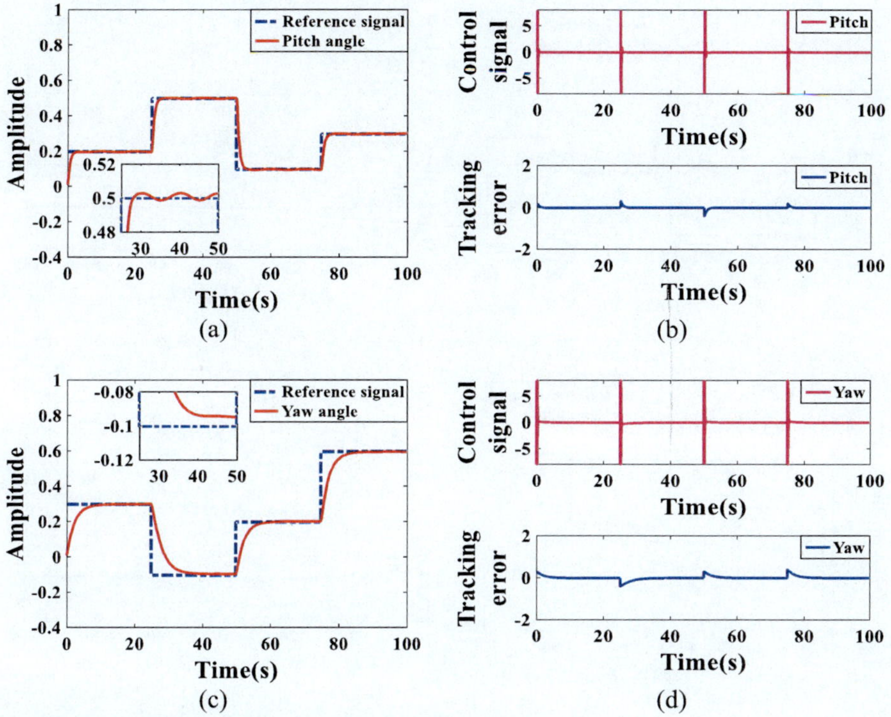

Fig. 3.36 Evaluation period **a** pitch angle and corresponding **b** control signal with tracking error, **c** yaw angle and corresponding **d** control signal with tracking error of case study-IV (twin rotor MIMO system) with Disturbance-I for PSO-L_1 adaptive controller

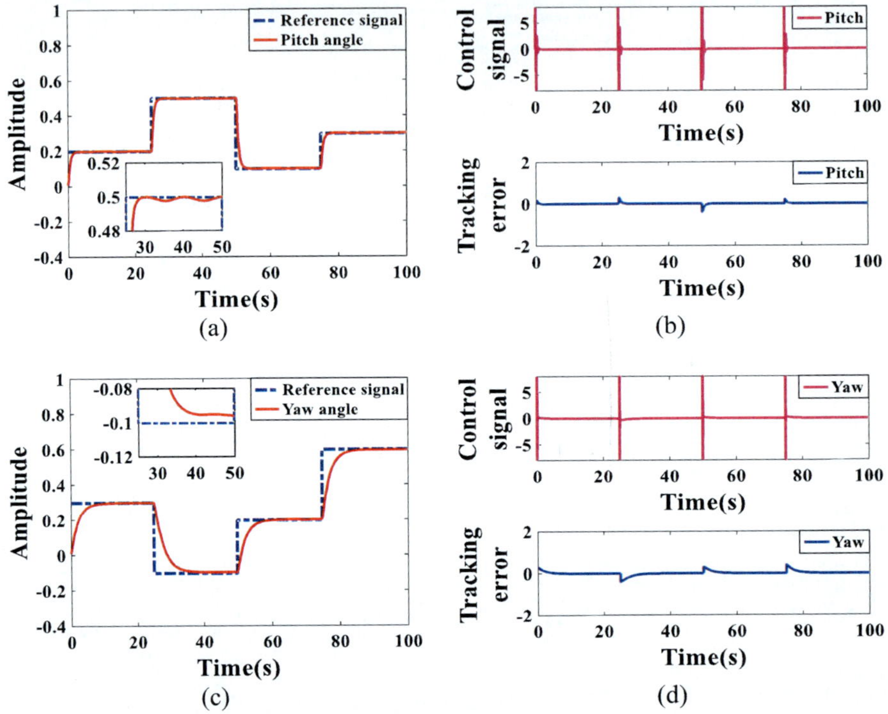

Fig. 3.37 Evaluation period **a** pitch angle and corresponding **b** control signal with tracking error, **c** yaw angle and corresponding **d** control signal with tracking error of case study-IV (twin rotor MIMO system) with Disturbance-I for HS-L_1 adaptive controller

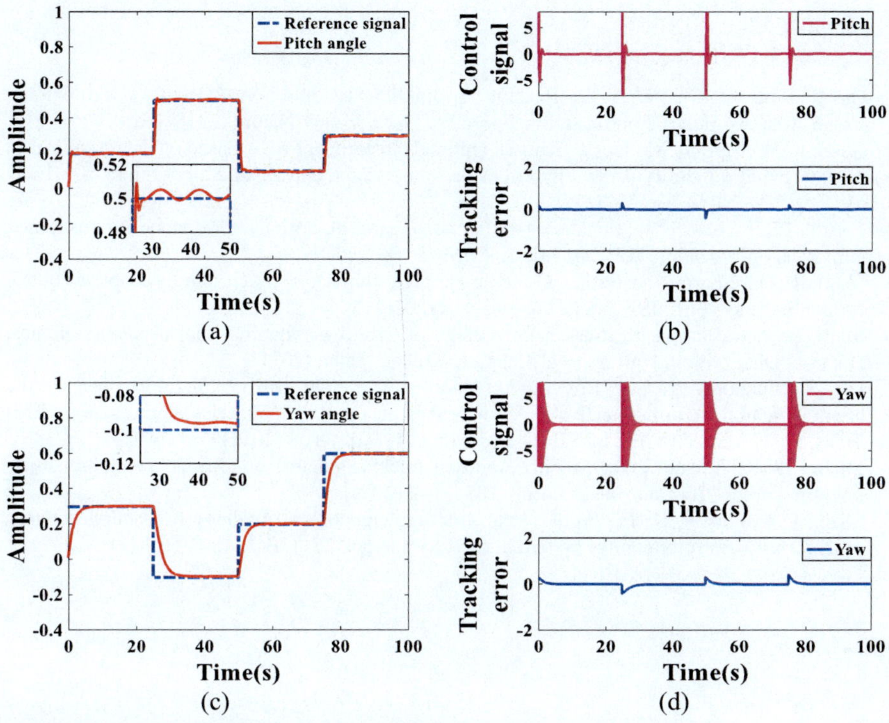

Fig. 3.38 Evaluation period **a** pitch angle and corresponding **b** control signal with tracking error, **c** yaw angle and corresponding **d** control signal with tracking error of case study-IV (twin rotor MIMO system) with Disturbance-I for lbest HS-L_1 adaptive controller

References

1. Das Sharma, K., Chatterjee, A., Rakshit, A.: Intelligent Control- A Stochastic Optimization Based Adaptive Fuzzy Approach. Singapore Pte Ltd: Springer Nature (2018)
2. Abedini, M., Moradi, M., Hosseinian, S.: Optimal clustering of MGs based on droop controller for improving reliability using a hybrid of harmony search and genetic algorithms. ISA Trans **61**, 119–128 (2016)
3. Portilla-Flores, E.A., et al.: Enhancing the harmony search algorithm performance on constrained numerical optimization. IEEE Access **5**, 25759–25780 (2017)
4. Maiti, R., Das Sharma, K., Sarkar, G.: lBest-HS algorithm based concurrent L_1 adaptive control for nonlinear systems. ISA Trans. **96**, 390–414 (2020)
5. Maiti, R., Das Sharma, K., Sarkar, G.: Design of hybrid L_1 adaptive control methodologies. Doctoral Dissertation, University of Calcutta, Kolkata, India (2021)
6. Das, S., Mukhopadhyay, A., Roy, A., Abraham, A., Panigrahi, B.K.: Exploratory power of the harmony search algorithm: analysis and improvements for global numerical optimization. IEEE Trans. Syst., Man, Cybern., Part B (Cybernetics) **41**(1), 89–106 (2011)
7. Mishra, S.-K., Chandra, D.: Stabilization and tracking control of inverted pendulum using fractional order PID controllers. J. Eng. **2014**, 1–9 (2014)
8. Shaik, F.A., Purwar, S., Pratap, B.: Real-time implementation of Chebyshev neural network observer for twin rotor control system. Expert Syst. Appl. **38**, 13043–13049 (2011)

Chapter 4
Hybrid L_1 Adaptive Controller-II: Fuzzy Parallel Distributed Compensation Based Approach

Abstract In this chapter, fuzzy PDC logic is augmented with fuzzy L_1 adaptive controller to tackle nonlinearities along with uncertainties and disturbances. The nonlinear system is considered as the combination of linear consequence rule based fuzzy systems. For each linearized zone, fuzzy L_1 adaptive controller with fuzzy PDC controller is designed. The fuzzy PDC-L_1 adaptive controller is employed on four simulation case studies. In this method, the fuzzy L_1 adaptive controller provides better rejection of uncertainties and disturbances as well as fuzzy PDC controller efficiently handles nonlinearities, delays, unmodelled dynamics to provide en bloc satisfactory transient and steady state tracking performance.

4.1 Introduction

L_1 adaptive controller can eliminate uncertainties and disturbances present in the system efficiently but handling large scale interconnected nonlinearities of complex systems are challenge to it [1]. On the other hand, fuzzy parallel distributed compensation (PDC) logic provides satisfactory controlling for nonlinear systems though time varying uncertainties and disturbances degrade its performance.

Hence, in this work, fuzzy PDC logic is hybridized with L_1 adaptive controller to cope up with nonlinearities as well as time varying uncertainties and disturbances. This methodology is termed as fuzzy PDC type L_1 (fuzzy PDC-L_1) adaptive controller. To handle nonlinearities, the nonlinear system is formulated by combining linear consequence rule based fuzzy systems following fuzzy PDC logic. Then, PDC augmented L_1 adaptive controller is formulated for each linearized zone and brought together to design the overall controller. The components of L_1 adaptive controller for each zone are developed following the adaptive fuzzy systems formulated from the fuzzy Lyapunov stability criteria. At first, zone-wise fuzzy predictors are designed and then combined to get the whole predictor model. Those zone-wise predictors estimate the uncertainties and disturbances present at that zone. Then, zonal adaptation laws are formulated and fuzzy blended to form total adaptation laws. After that, those adaptive estimates formulate the L_1 adaptive control law for each zone and combined to generate whole L_1 adaptive control action. High adaptation gain is used

R. Maiti et al., *Hybrid L_1 Adaptive Control*, Studies in Systems, Decision and Control 422, https://doi.org/10.1007/978-3-030-97102-1_4

to obtain fast transient performance which include high frequency components into the control signal. Thus, the control signal of each zone is filtered through respective zonal low-pass filter. Zone-wise developed low-pass filters are fuzzy blended to get the whole filter model. At the same time, fuzzy logic based state feedback controller is designed for each zone and then augmented together to get overall nonlinear state feedback control action. Such type of state feedback controller is known as fuzzy parallel distributed compensation (PDC) type controller. Filtered fuzzy L_1 adaptive control signal and fuzzy PDC control signal are hybridized to produce overall control action and given to the system under control. Zone-wise designing of this fuzzy PDC-L_1 adaptive controller helps to tackle nonlinearities, delays, unmodelled dynamics along with uncertainties and disturbances present in the system. The overall stability of this method is guaranteed as well as transient and steady state performances are analyzed. This method is employed on four simulation case studies, viz., Duffing's oscillatory system, nonlinear spring mass damper, 4th order inverted pendulum with cart and twin rotor MIMO system to show its effectiveness over basic L_1 adaptive controller and local best harmony search based L_1 (lbest HS-L_1) adaptive controller.

4.2 Linear Consequence Rule Based T-S Fuzzy System Design

4.2.1 Nonlinear System Approximation Utilizing T-S Fuzzy System with Linear Consequence

The nonlinear system is considered as the combination of linear Takagi–Sugeno (T-S) fuzzy systems. For each linear T-S fuzzy systems, *IF–THEN* rules are formulated with linear input–output relationships. Consider a system dynamics which is represented with W number of fuzzy *IF–THEN* rules. Mathematically, it can be expressed as:

For wth Rule (Rl^w):

Dynamic part:

$$\begin{aligned} &IF\ z_1(t) \text{ is } D_1^w, \ldots, z_q(t) \text{ is } D_q^w,\\ &THEN\ \dot{\underline{x}}(t) = A_w \underline{x}(t) + b_w \underline{u}(t), \end{aligned}$$

Output part:

$$\begin{aligned} &IF\ z_1(t) \text{ is } D_1^w, \ldots, z_q(t) \text{ is } D_q^w,\\ &THEN\ y(t) = c_w \underline{x}(t). \end{aligned}$$

where $w (w = 1, 2, \ldots, W)$ is the number of fuzzy *IF–THEN* rules, W represents the total number of fuzzy rules, $\underline{z}(t) = \left[z_1(t), z_2(t), \ldots, z_q(t)\right]^T \in \Re^q$ are the premise variables, q is the number of premise variables, D_i^w denotes the fuzzy set with $i = 1, 2, ...q$. $\dot{\underline{x}}(t) = A_w\underline{x}(t) + b_w\underline{u}(t)$ and $\underline{y}(t) = c_w\underline{x}(t)$ are the linear consequences of fuzzy systems. $\underline{x}(t)$ and $\underline{y}(t)$ are the system states and output respectively. $\underline{u}(t)$ represents system input. A_w, b_w, c_w are the parameter vectors for the wth rule. A_w is the system matrix, b_w and c_w are the input and output matrix respectively.

These rules are fuzzy blended to achieve the overall nonlinear system dynamics as:

$$\dot{\underline{x}}(t) = \sum_{w=1}^{W} \underline{\xi}_w(\underline{z})(A_w\underline{x}(t) + b_w\underline{u}(t)) \tag{4.1}$$

and the output of the system can be written as:

$$\underline{y}(t) = \sum_{w=1}^{W} \underline{\xi}_w(\underline{z})c_w\underline{x}(t) \tag{4.2}$$

Fuzzy basis function for *wth* zone is given as:

$$\underline{\xi}_w(\underline{z}) = \frac{\prod_{i=1}^{q} \mu_{D_i^w}(z_i(t))}{\sum_{w=1}^{W} \prod_{i=1}^{q} \mu_{D_i^w}(z_i(t))} = \frac{\mu_w(\underline{z})}{\sum_{w=1}^{W} \mu_w(\underline{z})} \tag{4.3}$$

where $\mu_{D_i^w}(z_i(t))$ is the grade of membership of $z_i(t)$ in $\mu_{D_i^w}$. $\mu(\underline{z})$ is the membership function satisfying the condition $\sum_{w=1}^{W} \mu_w(z) \neq 0$.

For each local zone, the degree of membership function follows the bound $\underline{\xi}_w \in [0, 1][0, 1]$ with following properties:

$$0 \leq \underline{\xi}_w \leq 1, \ \sum_{w=1}^{W} \underline{\xi}_w = 1 \text{ and } \sum_{w=1}^{W} \dot{\underline{\xi}}_w = 0 \tag{4.4}$$

4.2.2 *Fuzzy Parallel Distributed Compensation (PDC) Controller Design*

The premise part, utilized to design the system dynamics, is employed to formulate the state feedback controller in case of fuzzy PDC controller. The zone-wise designed state feedback controllers are combined to produce the overall state feedback controller which becomes nonlinear controller as a whole. The fuzzy PDC controller for W number of system dynamics are formulated as:

For *wth* Rule (Rl^w):

$$IF\ z_1(t)\ \text{is}\ D_1^w, \ldots, z_q(t)\ \text{is}\ D_q^w,$$

$$THEN u_{PDC}(t) = -\sum_{w=1}^{W} \underline{\xi}_w(x) K_w \underline{x}(t)$$

where K_w is the state feedback gain for *wth* zone.

In the next section, the components of the L_1 adaptive controller are designed by utilizing fuzzy logic.

4.3 Linear Consequence Rule Based Fuzzy PDC-L_1 Adaptive Controller Design

L_1 adaptive controller can eliminate uncertainties and disturbances from the system well enough though its performance degrades when nonlinearities, delays present in the system. On the other hand, fuzzy PDC controller is efficient for the nonlinear, delayed, unmodelled systems though it cannot properly tackle time varying uncertainties and disturbances. In this work, fuzzy PDC logic is conjugated with L_1 adaptive controller [2, 3] to provide better control action for the system with nonlinearities, delays, uncertainties, disturbances, etc., simultaneously. The state feedback controller of the L_1 adaptive controller is replaced with the fuzzy PDC controller. The architecture of the fuzzy PDC-L_1 adaptive control methodology is shown in Fig. 4.1.

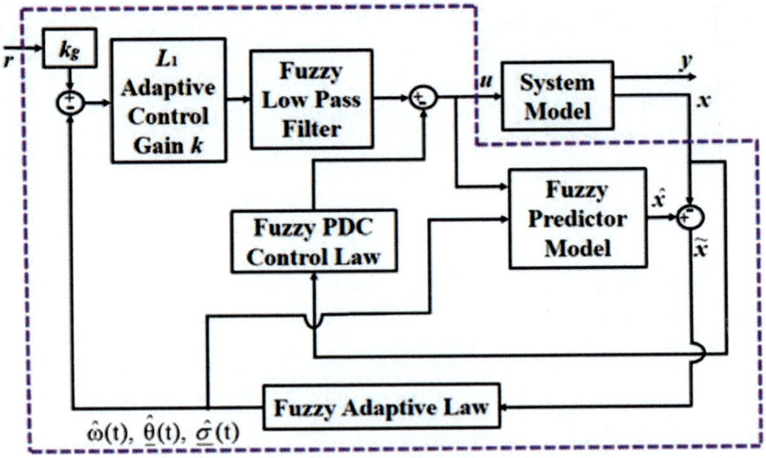

Fig. 4.1 Architecture of fuzzy PDC-L_1 adaptive controller

In this method, the nonlinear system is considered as the combination of linear consequence rule based fuzzy systems. For each linearized zone, predictor is formulated with the same premise part employed to design the nonlinear system. Linear consequence rule based zone-wise fuzzy predictors are combined to achieve the overall predictor model. Each linearized zone of the nonlinear system contains unknown constant, time varying uncertainties and time varying disturbances. These quantities are estimated from those zones and adapted following adaptation laws generated from the linear fuzzy systems with the same premise part utilized to develop the system. Zone-wise adaptation laws are combined to get the overall adaptive estimate. To eliminate unknown constant, time varying uncertainties and disturbances, their adaptive estimates are utilized to formulate zone-wise L_1 adaptive control signals. These control laws are accumulated to get the whole L_1 adaptive control law. The unknown constant, time varying uncertainties and disturbances are adapted utilizing high adaptation gain which includes high frequency components into the zone-wise developed L_1 adaptive control laws. Hence, the control signal of each zone is filtered through respective zone-wise designed low-pass filter. Zonal filters are formulated with the same premise part used to design the system and then they are fuzzy blended to generate the whole filter action. In this method, the conventional state feedback controller of the L_1 adaptive controller is designed utilizing fuzzy PDC logic to cope up with the nonlinearities present in the system. The same premise part employed to design the system is utilized to formulate the state feedback controller of each linearized zone, viz., fuzzy PDC controller. Then, zone-wise linear consequence rule based fuzzy PDC controllers are brought together to design the whole PDC controller which is nonlinear in nature. L_1 adaptive control law is augmented with fuzzy PDC controller to develop the overall control signal required to control the system.

The control action, formulated by appending fuzzy L_1 adaptive controller with fuzzy PDC controller, provides satisfactory disturbances and uncertainties rejection as well as efficient controlling for nonlinear system. The designing of the fuzzy linear consequence rule based system, predictor, adaptation laws, L_1 adaptive control law, PDC control law and low-pass filter are elaborated in the next sections.

4.3.1 *Fuzzy Logic Based Predictor Design*

In this work, predictor of L_1 adaptive controller is designed with the help of fuzzy logic system to obtain better estimation of unknown constant, time varying uncertainties and time varying disturbances [4]. Proper estimation and adequate cancellation of these quantities are essential to improve the performance of basic L_1 adaptive controller. To design the fuzzy predictor model at first, the system model is designed utilizing fuzzy PDC logic. The nonlinear system is represented by the combination of linear fuzzy systems. W number of linear fuzzy systems are formulated considering the error between reference and system output as the input of the fuzzy system. Combining those W number of fuzzy systems, the consequence part is formulated to obtain the overall system model as:

$$\underline{\dot{x}}(t) = \sum_{w=1}^{W} \underline{\xi}_w \left(A_{m_w}\underline{x}(t) + b_w(\omega_w \underline{u}_1(t) + \underline{\theta}_w(t) \left\|\underline{x}\right\|_{L_\infty} + \underline{\sigma}_w(t))\right) \tag{4.5}$$

and

$$\underline{y}(t) = \sum_{w=1}^{W} \underline{\xi}_w c_w \underline{x}(t) \tag{4.6}$$

where A_{m_w} is the closed loop system matrix of *wth* zone, b_w is the input matrix of that particular zone. $\underline{\xi}_w$ represents the premise part of that *wth* zone.

Again, (4.5) can be represented as:

$$\underline{\dot{x}}(t) = \sum_{w=1}^{W} (\alpha_{1_w} \underline{\xi}_w \underline{x}(t) + \alpha_{2_w} \underline{\xi}_w \underline{r}_{ud}(t)) \tag{4.7}$$

where $\alpha_{1_w} = A_{m_w}$, $\alpha_{2_w} = b_w$ and $\underline{r}_{ud}(t) = \omega_w \underline{u}_1(t) + \underline{\theta}_w(t) \left\|\underline{x}\right\|_{L_\infty} + \underline{\sigma}_w(t)$ for *wth* zone. In the above equation, α_{1_w} and α_{2_w} are unknown. Thus, the aim is to design (4.7) utilizing ideal fuzzy logic system. Let, the ideal fuzzy model of the system is considered as:

$$\underline{\dot{x}}_{id}^*(t) = \sum_{w=1}^{W} (\alpha_{1_w}^* \underline{\xi}_w \underline{x}_{id}^*(t) + \alpha_{2_w}^* \underline{\xi}_w \underline{r}_{ud}(t)) \tag{4.8}$$

where $\alpha_{1_w}^*$ is the system matrix and $\alpha_{2_w}^*$ is the input matrix of the ideal fuzzy system model for *wth* zone. $\underline{x}_{id}^*(t)$ is the state vector of the ideal fuzzy system model.

The values of α_{1_w} and α_{2_w} will be adapted to reach the values near about $\alpha_{1_w}^* \alpha_{1_w}^*$ and $\alpha_{2_w}^* \alpha_{2_w}^*$ respectively. To design the system model (4.7), the following theorem is provided.

Theorem 4.1 *The fuzzy system model (4.7) follows the ideal fuzzy system model (4.7), provided the system matrix and input matrix of the system are formulated following the adaptation laws given as respectively:*

$$\dot{\alpha}_{1_w} = -\Gamma_1 \underline{e}_{id}^T(t) P_w(\underline{\xi}) \underline{\xi}_w \underline{x}(t) \tag{4.9}$$

$$\dot{\alpha}_{2_w} = -\Gamma_2 \underline{e}_{id}^T(t) P_w(\underline{\xi}) \underline{r}_{ud}(t) \tag{4.10}$$

where Γ_1 and Γ_2 are the adaptation gains. $\underline{e}_{id}(t)$ is the error between system states and ideal system states. $P_w \underline{\xi}_w = P_w(\underline{\xi})$ is the solution of algebraic Lyapunov equation of the system, viz., $\frac{1}{2}\left(\alpha_{1_w}\underline{\xi}_w\right)^T \left(P_w\underline{\xi}_w\right) + \frac{1}{2}\left(P_w\underline{\xi}_w\right)^T \left(\alpha_{1_w}\underline{\xi}_w\right) = -Q_{m_w}\underline{\xi}_w$ for wth zone. It can be written that, $Q_{m_w}\underline{\xi}_w = Q_{m_w}(\underline{\xi})$. $Q_{m_w}(\underline{\xi}) = Q_{m_w}^T(\underline{\xi}) > 0$ be a positive definite arbitrary symmetric matrix.

Bounds on α_{1_w} *and* α_{2_w} *provides their stable adaptive estimates.*

Proof Let, for W number of zone, the fuzzy Lyapunov function is formulated as:

$$V_s(t) = \sum_{w=1}^{W} \left(\frac{1}{2} \underline{e}_{id}^T(t) P_w(\underline{\xi}) \underline{e}_{id}^T(t) + \frac{1}{2} \tilde{\alpha}_{1_w}^T \Gamma_1^{-1} \tilde{\alpha}_{1_w} + \frac{1}{2} \tilde{\alpha}_{2_w}^T \Gamma_2^{-1} \tilde{\alpha}_{2_w} \right) \tag{4.11}$$

where $\underline{e}_{id}(t) = \underline{x}(t) - \underline{x}_{id}^*(t)$ is the error between system states and ideal system states. $\tilde{\alpha}_{1_w} = \alpha_{1_w} - \alpha_{1_w}^*$ and $\tilde{\alpha}_{2_w} = \alpha_{2_w} - \alpha_{2_w}^*$.

Differentiation of (4.11) provides:

$$\dot{V}_s(t) = \sum_{w=1}^{W} \begin{pmatrix} \frac{1}{2} (\dot{\underline{e}}_{id}^T(t) P_w(\underline{\xi}) \underline{e}_{id}^T(t) + \underline{e}_{id}^T(t) P_w(\underline{\xi}) \dot{\underline{e}}_{id}^T(t)) \\ + \frac{1}{2} (\dot{\tilde{\alpha}}_{1_w}^T \Gamma_1^{-1} \tilde{\alpha}_{1_w} + \tilde{\alpha}_{1_w}^T \Gamma_1^{-1} \dot{\tilde{\alpha}}_{1_w}) \\ + \frac{1}{2} (\dot{\tilde{\alpha}}_{2_w}^T \Gamma_2^{-1} \tilde{\alpha}_{2_w} + \tilde{\alpha}_{2_w}^T \Gamma_2^{-1} \dot{\tilde{\alpha}}_{2_w}) \end{pmatrix} \tag{4.12}$$

Now, it can be written that, $\dot{\tilde{\alpha}}_{1_w} = \dot{\alpha}_{1_w} - \dot{\alpha}_{1_w}^*$ and $\dot{\tilde{\alpha}}_{2_w} = \dot{\alpha}_{2_w} - \dot{\alpha}_{2_w}^*$. $\alpha_{1_w}^*$ and $\alpha_{2_w}^*$ are constant, which provides respectively $\dot{\alpha}_{1_w}^* = 0$ and $\dot{\alpha}_{2_w}^* = 0$. Putting these on (4.12) yields:

$$\dot{V}_s(t) = \sum_{w=1}^{W} \left(\frac{1}{2} \dot{\underline{e}}_{id}^T(t) P_w(\underline{\xi}) \underline{e}_{id}^T(t) + \frac{1}{2} \underline{e}_{id}^T(t) P_w(\underline{\xi}) \dot{\underline{e}}_{id}^T(t) + \tilde{\alpha}_{1_w}^T \Gamma_1^{-1} \dot{\alpha}_{1_w} + \tilde{\alpha}_{2_w}^T \Gamma_2^{-1} \dot{\alpha}_{2_w} \right)$$

or

$$\dot{V}(t)_s = \sum_{w=1}^{W} \begin{pmatrix} -\underline{e}_{id}^T(t) Q_{m_w}(\underline{\xi}) \underline{e}_{id}(t) \\ + \begin{pmatrix} \underline{e}_{id}^T(t) P_w(\underline{\xi}) (\tilde{\alpha}_{1_w} \underline{\xi}_w \underline{x}(t) + \tilde{\alpha}_{2_w} \underline{\xi}_w \underline{r}_{ud}(t)) \\ + \tilde{\alpha}_{1_w}^T \Gamma_1^{-1} \dot{\alpha}_{1_w} + \tilde{\alpha}_{2_w}^T \Gamma_2^{-1} \dot{\alpha}_{2_w} \end{pmatrix} \end{pmatrix} \tag{4.13}$$

where $\dot{\underline{e}}_{id}(t) = \dot{\underline{x}}(t) - \dot{\underline{x}}_{id}^*(t) = \sum_{w=1}^{W} \left(\alpha_{1_w}^* \underline{\xi}_w \underline{e}_{id}(t) + \tilde{\alpha}_{1_w} \underline{\xi}_w \underline{x}(t) + \tilde{\alpha}_{2_w} \underline{\xi}_w \underline{r}_{ud}(t) \right)$.

Lyapunov stability criterion yields that, the system will be stable if $\dot{V}_s(t) = \sum_{w=1}^{W} -\underline{e}_{id}^T(t) Q_{m_w}(\underline{\xi}) \underline{e}_{id}(t) \leq 0$ condition holds.

This provides:

$$\underline{e}_{id}^T(t) P_w(\underline{\xi}) \tilde{\alpha}_{1_w} \underline{\xi}_w \underline{x}(t) + \tilde{\alpha}_{1_w}^T \Gamma_1^{-1} \dot{\alpha}_{1_w} = 0$$

$$\tilde{\alpha}_{1_w}^T \Gamma_1^{-1} (\Gamma_1 \underline{e}_{id}^T(t) P_w(\underline{\xi}) \underline{\xi}_w \underline{x}(t) + \dot{\alpha}_{1_w}) = 0 \tag{4.14}$$

and

$$\underline{e}_{id}^T(t)P_w(\underline{\xi})\tilde{\alpha}_{2_w}\underline{\xi}_w\underline{r}_{ud}(t) + \tilde{\alpha}_{2_w}^T\Gamma_2^{-1}\dot{\alpha}_{2_w} = 0$$

or

$$\tilde{\alpha}_{2_w}^T\Gamma_2^{-1}(\Gamma_2\underline{e}_{id}^T(t)P_w(\underline{\xi})\underline{\xi}_w\underline{r}_{ud}(t) + \dot{\alpha}_{2_w}) = 0 \tag{4.15}$$

Solution of (4.14) and (4.15) yield (4.9) and (4.10) respectively.

The bounds on α_{1_w} and α_{2_w} incur that, the adaptive estimates of these quantities are bounded. Hence, the designed linear consequence rule based fuzzy system follows the ideal system model with stable adaptive estimates. This proves the theorem. ■

The predictor model for this system is also formulated by combining zone-wise linear consequence rule based fuzzy systems. The same premise part, utilized to develop the system model, is used to design the predictor model. W number of linear consequences are fuzzy blended to design the predictor model as follows:

$$\dot{\hat{\underline{x}}}(t) = \sum_{w=1}^{W}\underline{\xi}_w\left(A_{m_w}\hat{\underline{x}}(t) + b_w(\hat{\omega}_w(t)\underline{u}_1(t) + \hat{\underline{\theta}}_w(t)\left\|\underline{x}\right\|_{L_\infty} + \hat{\underline{\sigma}}_w(t))\right) \tag{4.16}$$

$$\hat{\underline{y}}(t) = \sum_{w=1}^{W}\underline{\xi}_w c_w\hat{\underline{x}}(t) \tag{4.17}$$

where $\hat{\underline{x}}(t)$ is the state vector of the predictor, $\hat{\underline{y}}(t)$ is the output vector of the predictor. $[\hat{\omega}_w | \hat{\underline{\theta}}_w | \hat{\underline{\sigma}}_w]$ are the adaptive estimates of unknown constant, time varying uncertainties and time varying disturbances respectively for *wth* zone. These quantities are bounded to compact set given as $\hat{\omega}_w \in \Omega_w$, $\hat{\underline{\theta}}_w \in \underline{\Theta}_w$ and $\hat{\underline{\sigma}}_w \in \underline{\Sigma}_w$.

Again, (4.16) can be represented as:

$$\dot{\hat{\underline{x}}}(t) = \sum_{w=1}^{W}(\alpha_{1_w}\underline{\xi}_w\hat{\underline{x}}(t) + \alpha_{2_w}\underline{\xi}_w\hat{\underline{r}}_{ud}(t)) \tag{4.18}$$

where $\hat{\underline{r}}_{ud}(t) = \hat{\omega}_w(t)\underline{u}_1(t) + \hat{\underline{\theta}}_w(t)\left\|\underline{x}\right\|_{L_\infty} + \hat{\underline{\sigma}}_w(t)$.

4.3.2 *Fuzzy Logic Based Adaptive Laws Formulation for Unknown Constant, Time Varying Uncertainties and Time Varying Disturbances*

The nonlinear system is formulated by combining zone-wise linear consequence rule based fuzzy system models. Each zone of the system contains unknown constant, time varying uncertainties and time varying disturbances. Adaptation laws are formulated to adapt these quantities present in each linearized zone. The same premise part, used to design the nonlinear system, is applied to develop the adaptation laws. For *wth* zone, the estimated unknown constant, time varying uncertainties and time varying disturbances are represented respectively as $\hat{\omega}_w = \hat{\alpha}_{\omega_w} \underline{\xi}_w$, $\underline{\hat{\theta}}_w = \hat{\alpha}_{\theta_w} \underline{\xi}_w$ and $\underline{\hat{\sigma}}_w = \hat{\alpha}_{\sigma_w} \underline{\xi}_w$ with the bounds $\hat{\alpha}_{\omega_w} \in \alpha_{\Omega_w}, \hat{\alpha}_{\theta_w} \in \alpha_{\Theta_w}$ and $\hat{\alpha}_{\sigma_w} \in \alpha_{\Sigma_w}$. For W number of linear fuzzy zones, these adaptive estimates are formulated and then those consequence adaptive estimates are fuzzy blended to formulate the whole adaptation laws. The adaptation laws development for *wth* zone is described in the following theorem.

Theorem 4.2 *The unknown constant, time varying uncertainties and time varying disturbances, present in the wth zone of the system, are adapted following the fuzzy logic based bounded adaptation laws respectively as follows:*

$$\dot{\hat{\alpha}}_{\omega_w} = \Gamma_C \operatorname{Proj}\left(\hat{\alpha}_{\omega_w}, -\underline{\tilde{x}}^T(t) P_w(\underline{\xi}) b_w(\underline{\xi}) \underline{\xi}_w \underline{u}_1(t)\right) \tag{4.19}$$

$$\dot{\hat{\alpha}}_{\theta_w} = \Gamma_C \operatorname{Proj}\left(\hat{\alpha}_{\theta_w}, -\underline{\tilde{x}}^T(t) P_w(\underline{\xi}) b_w(\underline{\xi}) \underline{\xi}_w \left\|\underline{x}\right\|_{L_\infty}\right) \tag{4.20}$$

$$\dot{\hat{\alpha}}_{\sigma_w} = \Gamma_C \operatorname{Proj}\left(\hat{\alpha}_{\sigma_w}, -\underline{\tilde{x}}^T(t) P_w(\underline{\xi}) b_w(\underline{\xi}) \underline{\xi}_w\right) \tag{4.21}$$

where $\Gamma_C > 0$ is the adaptation gain and $\underline{\tilde{x}}(t) = \underline{\hat{x}}(t) - \underline{x}(t)$ is the error between predictor states and system states.

Proof Let, for W number of linear fuzzy zones, the fuzzy Lyapunov function is formulated as:

$$V(t) = \sum_{w=1}^{W} \frac{1}{2}\left(\underline{\tilde{x}}^T(t) P_w(\underline{\xi}) \underline{\tilde{x}}(t) + \tilde{\alpha}_{\omega_w}^T \Gamma_C^{-1} \tilde{\alpha}_{\omega_w} + \tilde{\alpha}_{\theta_w}^T \Gamma_C^{-1} \tilde{\alpha}_{\theta_w} + \tilde{\alpha}_{\sigma_w}^T \Gamma_C^{-1} \tilde{\alpha}_{\sigma_w}\right) \tag{4.22}$$

where $\tilde{\alpha}_{\omega_w} = \hat{\alpha}_{\omega_w} - \alpha_{\omega_w}, \tilde{\alpha}_{\theta_w} = \hat{\alpha}_{\theta_w} - \alpha_{\theta_w}, \tilde{\alpha}_{\sigma_w} = \hat{\alpha}_{\sigma_w} - \alpha_{\sigma_w}$.

Differentiation of (4.22) provides:

$$\dot{V}(t) = \sum_{w=1}^{W} \left(\begin{array}{l} \frac{1}{2}\left(\dot{\underline{\tilde{x}}}^T(t) P_w(\underline{\xi}) \underline{\tilde{x}}(t) + \underline{\tilde{x}}^T(t) P_w(\underline{\xi}) \dot{\underline{\tilde{x}}}(t)\right) \\ + \frac{1}{2}\left(\begin{array}{l} \dot{\tilde{\alpha}}_{\omega_w}^T \Gamma_C^{-1} \tilde{\alpha}_{\omega_w} + \tilde{\alpha}_{\omega_w}^T \Gamma_C^{-1} \dot{\tilde{\alpha}}_{\omega_w} + \dot{\tilde{\alpha}}_{\theta_w}^T \Gamma_C^{-1} \tilde{\alpha}_{\theta_w} + \tilde{\alpha}_{\theta_w}^T \Gamma_C^{-1} \dot{\tilde{\alpha}}_{\theta_w} \\ + \dot{\tilde{\alpha}}_{\sigma_w}^T \Gamma_C^{-1} \tilde{\alpha}_{\sigma_w} + \tilde{\alpha}_{\sigma_w}^T \Gamma_C^{-1} \dot{\tilde{\alpha}}_{\sigma_w} \end{array} \right) \end{array} \right) \tag{4.23}$$

Now, putting $\dot{\tilde{\alpha}}_{\omega_w} = \dot{\hat{\alpha}}_{\omega_w} - \dot{\alpha}_{\omega_w}, \dot{\tilde{\alpha}}_{\theta_w} = \dot{\hat{\alpha}}_{\theta_w} - \dot{\alpha}_{\theta_w}, \dot{\tilde{\alpha}}_{\sigma_w} = \dot{\hat{\alpha}}_{\sigma_w} - \dot{\alpha}_{\sigma_w}$ in (4.23) yield:

$$\dot{V}(t) = \sum_{w=1}^{W} \left(\begin{array}{l} \frac{1}{2}\left(\dot{\underline{\tilde{x}}}^T(t) P_w(\underline{\xi}) \underline{\tilde{x}}(t) + \underline{\tilde{x}}^T(t) P_w(\underline{\xi}) \dot{\underline{\tilde{x}}}(t)\right) \\ + \tilde{\alpha}_{\omega_w}^T \Gamma_C^{-1} \dot{\hat{\alpha}}_{\omega_w} - \tilde{\alpha}_{\omega_w}^T \Gamma_C^{-1} \dot{\alpha}_{\omega_w} \\ + \tilde{\alpha}_{\theta_w}^T \Gamma_C^{-1} \dot{\hat{\alpha}}_{\theta_w} - \tilde{\alpha}_{\theta_w}^T \Gamma_C^{-1} \dot{\alpha}_{\theta_w} + \tilde{\alpha}_{\sigma_w}^T \Gamma_C^{-1} \dot{\hat{\alpha}}_{\sigma_w} - \tilde{\alpha}_{\sigma_w}^T \Gamma_C^{-1} \dot{\alpha}_{\sigma_w} \end{array} \right) \tag{4.24}$$

where $\dot{\tilde{\underline{x}}}(t) = \dot{\hat{\underline{x}}}(t) - \dot{\underline{x}}(t) = \sum_{w=1}^{W} \left(\begin{array}{l} A_{m_w}(\underline{\xi})\,\tilde{\underline{x}}(t) \\ + b_w(\underline{\xi})\left(\tilde{\alpha}_{\omega_w}\underline{\xi}_w \underline{u}_1(t) + \tilde{\alpha}_{\theta_w}\underline{\xi}_w \|\underline{x}\|_{L_\infty} + \tilde{\alpha}_{\sigma_w}\underline{\xi}_w\right) \end{array} \right)$,

$A_{m_w}(\underline{\xi}) = A_{m_w}\underline{\xi}_w$, $b_w(\underline{\xi}) = b_w\underline{\xi}_w$. $\dot{\alpha}_{\omega_w} = 0$ as α_{ω_w} is constant, therefore, $\dot{\tilde{\alpha}}_{\omega_w} = \dot{\hat{\alpha}}_{\omega_w}$.

Putting the value of $\dot{\tilde{\underline{x}}}(t)$ and $\dot{\tilde{\alpha}}_{\omega_w} = \dot{\hat{\alpha}}_{\omega_w}$ in (4.24) yield:

$$\dot{V}(t) = \sum_{w=1}^{W} \left(\begin{array}{l} -\tilde{\underline{x}}^T(t) Q_{m_w}(\underline{\xi})\tilde{\underline{x}}(t) \\ +\tilde{\underline{x}}^T(t) P_w(\underline{\xi}) b_w(\underline{\xi}) \left(\tilde{\alpha}_{\omega_w}\underline{\xi}_w \underline{u}_1(t) + \tilde{\alpha}_{\theta_w}\underline{\xi}_w \|\underline{x}\|_{L_\infty} + \tilde{\alpha}_{\sigma_w}\underline{\xi}_w\right) \\ +\tilde{\alpha}_{\omega_w}^T \Gamma_C^{-1}\dot{\hat{\alpha}}_{\omega_w} + \tilde{\alpha}_{\theta_w}^T \Gamma_C^{-1}\dot{\hat{\alpha}}_{\theta_w} + \tilde{\alpha}_{\sigma_w}^T \Gamma_C^{-1}\dot{\hat{\alpha}}_{\sigma_w} \\ -\tilde{\alpha}_{\theta_w}^T \Gamma_C^{-1}\dot{\alpha}_{\theta_w} - \tilde{\alpha}_{\sigma_w}^T \Gamma_C^{-1}\dot{\alpha}_{\sigma_w} \end{array} \right) \tag{4.25}$$

where solution of algebraic Lyapunov equation of the system, viz., $\frac{1}{2}\left(A_{m_w}^T(\underline{\xi})\,P_w(\underline{\xi}) + P_w(\underline{\xi})\,A_{m_w}^T(\underline{\xi})\right) = -Q_{m_w}(\underline{\xi})$ is $P_w(\underline{\xi})$ for *wth* zone.

Applying Lyapunov stability condition on (4.25) incurs the system will be stable if following condition holds:

$$\dot{V}(t) = \sum_{w=1}^{W} \left(-\tilde{\underline{x}}^T(t) Q_{m_w}(\underline{\xi})\tilde{\underline{x}}(t) - \tilde{\alpha}_{\theta_w}^T \Gamma_C^{-1}\dot{\alpha}_{\theta_w} - \tilde{\alpha}_{\sigma_w}^T \Gamma_C^{-1}\dot{\alpha}_{\sigma_w} \right)$$

or

$$\dot{V}(t) = \sum_{w=1}^{W} \left(-\tilde{\underline{x}}^T(t) Q_{m_w}(\underline{\xi})\tilde{\underline{x}}(t) + \frac{1}{\Gamma_C}\left(\left|\tilde{\alpha}_{\theta_w}^T \dot{\alpha}_{\theta_w}\right| + \left|\tilde{\alpha}_{\sigma_w}^T \dot{\alpha}_{\sigma_w}\right|\right) \right) \tag{4.26}$$

From these Lyapunov stability criterion, the adaptive estimates for unknown constant, time varying uncertainties and time varying disturbances are formulated as follows.

The adaptive estimate of unknown constant is given as:

$$\tilde{\underline{x}}^T(t) P_w(\underline{\xi}) b_w(\underline{\xi}) \tilde{\alpha}_{\omega_w}\underline{\xi}_w \underline{u}_1(t) + \tilde{\alpha}_{\omega_w}^T \Gamma_C^{-1}\dot{\hat{\alpha}}_{\omega_w} = 0$$

or

$$\tilde{\alpha}_{\omega_w}^T \Gamma_C^{-1}\left(\Gamma_C \tilde{\underline{x}}^T(t) P_w(\underline{\xi}) b_w(\underline{\xi}) \underline{\xi}_w \underline{u}_1(t) + \dot{\hat{\alpha}}_{\omega_w}\right) = 0$$

or

$$\dot{\hat{\alpha}}_{\omega_w} = -\Gamma_C \tilde{\underline{x}}^T(t) P_w(\underline{\xi}) b_w(\underline{\xi}) \underline{\xi}_w \underline{u}_1(t) \tag{4.27}$$

Similarly, the adaptive estimate of time varying uncertainty is formulated as:

$$\tilde{\underline{x}}^T(t)P_w(\underline{\xi})b_w(\underline{\xi})\tilde{\alpha}_{\theta_w}\underline{\xi}_w\|\underline{x}\|_{L_\infty} + \tilde{\alpha}_{\theta_w}^T\Gamma_C^{-1}\dot{\hat{\alpha}}_{\theta_w} = 0$$

or

$$\tilde{\alpha}_{\theta_w}^T\Gamma_C^{-1}\left(\Gamma_C\tilde{\underline{x}}^T(t)P_w(\underline{\xi})b_w(\underline{\xi})\underline{\xi}_w\|\underline{x}\|_{L_\infty} + \dot{\hat{\alpha}}_{\theta_w}\right) = 0$$

or

$$\dot{\hat{\alpha}}_{\theta_w} = -\Gamma_C\tilde{\underline{x}}^T(t)P_w(\underline{\xi})b_w(\underline{\xi})\underline{\xi}_w\|\underline{x}\|_{L_\infty} \tag{4.28}$$

In a similar manner, the time varying disturbance is estimated following the adaptive rule as:

$$\tilde{\underline{x}}^T(t)P_w(\underline{\xi})b_w(\underline{\xi})\,\tilde{\alpha}_{\sigma_w}\underline{\xi}_w + \tilde{\alpha}_{\sigma_w}^T\Gamma_C^{-1}\dot{\hat{\alpha}}_{\sigma_w} = 0$$

or

$$\tilde{\alpha}_{\sigma_w}^T\Gamma_C^{-1}\left(\Gamma_C\tilde{\underline{x}}^T(t)P_w(\underline{\xi})b_w(\underline{\xi})\,\underline{\xi}_w + \dot{\hat{\alpha}}_{\sigma_w}\right) = 0$$

or

$$\dot{\hat{\alpha}}_{\sigma_w} = -\Gamma_C\tilde{\underline{x}}^T(t)P_w(\underline{\xi})b_w(\underline{\xi})\underline{\xi}_w \tag{4.29}$$

High adaptation gain Γ_C is applied to obtain fast adaptation of unknown constant, time varying uncertainties and time varying disturbances. To ensure the stable bounded adaptation of these parameters, projection operator [5] is used. With projection operator, the adaptive estimates of unknown constant (4.27), time varying uncertainties (4.28) and time varying disturbances (4.29) take the form of (4.19), (4.20) and (4.21) respectively. These prove the theorem. ■

Zone-wise developed respective adaptation laws of unknown constant, uncertainties and disturbances are combined correspondingly to obtain the overall adaptive estimates of those quantities. To eliminate these unknown constant, time varying uncertainties, disturbances and to produce efficient controlling for nonlinear system, fuzzy PDC control law augmented L_1 adaptive control signal is generated as elaborated next.

4.3.3 Fuzzy Logic Based Control Law Design

In this method, the essence of L_1 adaptive controller in eliminating uncertainties and disturbances is combined with the fuzzy PDC controller in dealing nonlinearities. The nonlinear system is represented as the combination of linear consequence rule based fuzzy systems. For each zone of the system, fuzzy rule based predictor is formulated. The unknown constant, uncertainties and disturbances present in each

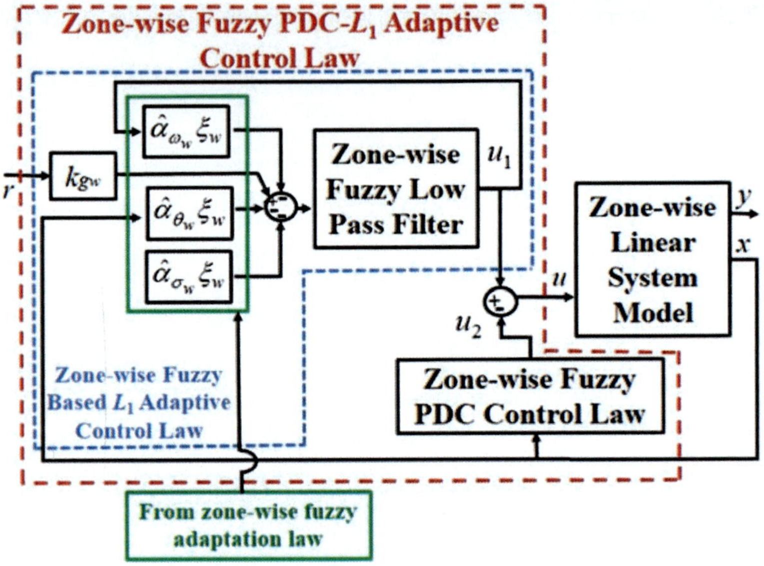

Fig. 4.2 Linear consequence rule based zone-wise fuzzy PDC-L_1 adaptive control law design

zone are estimated and adapted following adaptation laws. To eliminate these undesirable quantities, their zone-wise adaptive estimates are utilized to develop fuzzy L_1 adaptive control law for each zone. Due to the use of high adaptation gain, high frequency signal includes into the control channel which is nullified by means of low-pass filter dedicatedly designed for each zone. This filtered L_1 adaptive control signal is augmented with fuzzy PDC control law, developed for each linearized zone. The architecture of the zone-wise control law design is depicted in Fig. 4.2. The zone-wise developed fuzzy L_1 adaptive control laws with fuzzy PDC control signals are fuzzy blended to obtain the overall fuzzy PDC-L_1 adaptive control signal. The analytical formulations of the control laws are elaborated in the following sections. At first, the fuzzy logic based L_1 adaptive controller and fuzzy PDC controller design are described and then overall control action generation is discussed.

4.3.3.1 Fuzzy Logic Based L_1 Adaptive Control Law Formulation

Zone-wise corresponding adaptive estimates of unknown constant, time varying uncertainties and time varying disturbances are combined to get the respective entire adaptation laws. High frequency signal, introduced due to high adaptation gain, is eliminated by means of low-pass filter. The overall L_1 adaptive control law is formulated by combining W number of L_1 adaptive control laws of linear fuzzy zones as:

$$\underline{u}_1(s) = \sum_{w=1}^{W} \left(-k_w D(s)(\hat{\alpha}_{\omega_w \underline{\xi}_w} \underline{u}_1(s) + \hat{\alpha}_{\theta_w \underline{\xi}_w} \left\| \underline{x} \right\|_{L_\infty} + \hat{\alpha}_{\sigma_w \underline{\xi}_w} - k_{g_w} \underline{r}(s)) \right) \quad (4.30)$$

where for wth zone, k_w is the pre-filter gain and k_{g_w} represents feed-forward gain. $D(s)$ is the strictly proper stable transfer function related to the filter $C(s)$. This $D(s)$ should be designed in such a way that the filter $C(s)$ will become strictly proper stable transfer function as described in Sect. 4.3.4. The L_1 adaptive control signal, formulated in this Sect. 4.3.3.1, is consisted of stable adaptive estimates of unknown constant, time varying uncertainties and time varying disturbances, formulated in Sect. 4.3.2. Thus, the designed L_1 adaptive control signal is also stable.

The Stable fuzzy PDC controller design is described in Sect. 4.3.3.2. Then, stable fuzzy L_1 adaptive control law and stable fuzzy PDC control action are combined in Sect. 4.3.3.3 to develop the stable overall control action.

4.3.3.2 Fuzzy Logic Based PDC Control Law Formulation

In case of fuzzy PDC controller, the nonlinear system is divided into finite number of linear fuzzy systems and state feedback controller is developed for each of the linear system. Zone-wise state feedback controllers are brought together to formulate the overall control law which is nonlinear control action as a whole.

To formulate the premise part of PDC controller, the premise part designed for the system is utilized. The consequences, viz., the state feedback controllers designed for W number of linearized zones are fuzzy blended to produce whole fuzzy PDC control action as:

$$\underline{u}_2(t) = \sum_{w=1}^{W} -K_w^T \underline{\xi}_w \underline{x}(t) = \sum_{w=1}^{W} \left(-\alpha_{sf_w} \underline{\xi}_w \underline{x}(t) \right) \quad (4.31)$$

The system dynamics (4.5) in another form can be represented as:

$$\underline{\dot{x}}(t) = \sum_{w=1}^{W} \begin{pmatrix} A_{0_w}(\underline{\xi})\, \underline{x}(t) + b_w(\underline{\xi})(\alpha_{\omega_w} \underline{\xi}_w\, \underline{u}_1(t) \\ +\alpha_{\theta_w} \underline{\xi}_w \left\| \underline{x} \right\|_{L_\infty} + \alpha_{\sigma_w} \underline{\xi}_w) + b_w \underline{\xi}_w \underline{u}_2(t) \end{pmatrix} \quad (4.32)$$

where $A_{0_w}(\underline{\xi})$ is the open loop system matrix for wth zone.

The closed loop system matrix in (4.5) is formulated by putting the value of $\underline{u}_2(t)$ from (4.31) in (4.32) as $A_{m_w}(\underline{\xi}) = (A_{0_w}(\underline{\xi}) - b_w(\underline{\xi})\alpha_{sf_w}\underline{\xi}_w)$.

The system stability with fuzzy L_1 adaptive controller has already been examined in the previous section. In this section, the system stability considering only the fuzzy PDC control law is analyzed. Hence, the system (4.32) with only fuzzy PDC controller can be written as:

$$\underline{\dot{x}}(t) = \sum_{w=1}^{W}\left(A_{0_w}(\underline{\xi})\underline{x}(t) + b_w(\underline{\xi})\left(-\alpha_{sf_w}\underline{\xi}_w\underline{x}(t)\right)\right) \tag{4.33}$$

The value of α_{sf_w} has to be known to design the fuzzy PDC controller. Therefore an ideal fuzzy PDC controller model is considered and α_{sf_w} is adapted to follow that ideal model. The system model with ideal fuzzy PDC controller is considered as:

$$\underline{\dot{x}}^*_{PDC}(t) = \sum_{w=1}^{W}\left(A_{0_w}(\underline{\xi})\underline{x}^*_{PDC}(t) - b_w(\underline{\xi})\alpha^*_{sf_w}\underline{\xi}_w\underline{x}^*_{PDC}(t)\right) \tag{4.34}$$

where $\underline{x}^*_{PDC}$ is the state of the system with ideal fuzzy PDC controller. $\alpha^*_{sf_w}$ is the ideal singleton value which is to be adapted by α_{sf_w} following the theorem given next.

Theorem 4.3 *The system with fuzzy PDC controller (4.33) will be asymptotically stable provided the singleton value α_{sf_w} is designed following the stable bounded adaptive estimate as follows:*

$$\dot{\alpha}_{sf_w} = \Gamma_{sf}\underline{e}^T_{PDC}(t)P_w(\underline{\xi})b_w(\underline{\xi})\underline{\xi}_w\underline{x}(t) \tag{4.35}$$

where $\underline{e}_{PDC}(t) = x(t) - \underline{x}^*_{PDC}(t)$ is the state error between original system and the system with ideal fuzzy PDC controller. Γ_{sf} is the adaptation gain to adapt the value of singleton. Bounded adaptive estimates of α_{sf_w} yields stable PDC control law as well as stable system performance.

Proof For W number of zones, the fuzzy Lyapunov function is formulated as:

$$V_c(t) = \sum_{w=1}^{W}\left(\frac{1}{2}\underline{e}^T_{PDC}(t)P_w(\underline{\xi})\underline{e}_{PDC}(t) + \frac{1}{2}\tilde{\alpha}^T_{sf_w}\Gamma^{-1}_{sf}\tilde{\alpha}_{sf_w}\right) \tag{4.36}$$

where $\tilde{\alpha}_{sf_w} = \alpha_{sf_w} - \alpha^*_{sf_w}$ is the error between singleton value of fuzzy PDC controller and ideal fuzzy PDC controller.

Considering the derivative of (4.36) provides:

$$\dot{V}_c(t) = \sum_{w=1}^{W}\left(\frac{1}{2}\underline{\dot{e}}^T_{PDC}(t)P_w(\underline{\xi})\underline{e}_{PDC}(t) + \frac{1}{2}\underline{e}^T_{PDC}(t)P_w(\underline{\xi})\underline{\dot{e}}_{PDC}(t) + \tilde{\alpha}^T_{sf_w}\Gamma^{-1}_{sf}\dot{\tilde{\alpha}}_{sf_w}\right) \tag{4.37}$$

where $\dot{\tilde{\alpha}}_{sf_w} = \dot{\alpha}_{sf_w} - \dot{\alpha}^*_{sf_w}$. As, $\alpha^*_{sf_w}$ is a constant value therefore, $\dot{\alpha}^*_{sf_w} = 0$. This yields:

$$\dot{V}_c(t) = \sum_{w=1}^{W}\begin{pmatrix}\frac{1}{2}\underline{\dot{e}}^T_{PDC}(t)P_w(\underline{\xi})\underline{e}_{PDC}(t) + \frac{1}{2}\underline{e}^T_{PDC}(t)P_w(\underline{\xi})\underline{\dot{e}}_{PDC}(t) \\ +\tilde{\alpha}^T_{sf_w}\Gamma^{-1}_{sf}\dot{\alpha}_{sf_w}\end{pmatrix} \tag{4.38}$$

where $\dot{\underline{e}}_{PDC}(t) = \dot{\underline{x}}(t) - \dot{\underline{x}}^*_{PDC}(t) = \sum_{w=1}^{W} (A_{m_w}(\underline{\xi})\underline{e}_{PDC}(t) - b_w(\underline{\xi})\tilde{\alpha}_{sf_w}\underline{\xi}_w\underline{x}(t))$. Putting this value in (4.38) develops the following equation:

$$\dot{V}_c(t) = \sum_{w=1}^{W} \begin{pmatrix} \frac{1}{2}(A_{m_w}(\underline{\xi})\underline{e}_{PDC}(t) - b_w(\underline{\xi})\tilde{\alpha}_{sf_w}\underline{\xi}_w\underline{x}(t))^T P_w(\underline{\xi})\underline{e}_{PDC}(t) \\ +\frac{1}{2}\underline{e}^T_{PDC}(t)P_w(\underline{\xi})(A_{m_w}(\underline{\xi})\underline{e}_{PDC}(t) - b_w(\underline{\xi})\tilde{\alpha}_{sf_w}\underline{\xi}_w\underline{x}(t)) + \tilde{\alpha}^T_{sf_w}\Gamma^{-1}_{sf}\dot{\alpha}_{sf_w} \end{pmatrix}$$

or

$$\dot{V}_c(t) = \sum_{w=1}^{W} \begin{pmatrix} \frac{1}{2}\underline{e}^T_{PDC}(t)(A^T_{m_w}(\underline{\xi})P_w(\underline{\xi}) + P_w(\underline{\xi})A_{m_w}(\underline{\xi}))\underline{e}_{PDC}(t) \\ -\frac{1}{2}(b_w(\underline{\xi})\tilde{\alpha}_{sf_w}\underline{\xi}_w\underline{x}(t))^T P_w(\underline{\xi})\underline{e}_{PDC}(t) \\ -\frac{1}{2}\underline{e}^T_{PDC}(t)P_w(\underline{\xi})b_w(\underline{\xi})\tilde{\alpha}_{sf_w}\underline{\xi}_w\underline{x}(t) + \tilde{\alpha}^T_{sf_w}\Gamma^{-1}_{sf}\dot{\alpha}_{sf_w} \end{pmatrix}$$

or

$$\dot{V}_c(t) = \sum_{w=1}^{W} \begin{pmatrix} -\underline{e}^T_{PDC}(t)\,Q_{m_w}(\underline{\xi})\underline{e}_{PDC} - \underline{e}^T_{PDC}(t)P_w(\underline{\xi})b_w(\underline{\xi})\tilde{\alpha}_{sf_w}\underline{\xi}_w\underline{x}(t)) \\ +\tilde{\alpha}^T_{sf_w}\Gamma^{-1}_{sf}\dot{\alpha}_{sf_w} \end{pmatrix} \tag{4.39}$$

From Lyapunov stability criterion, it can be stated that, the system will be stable if following condition holds: $\dot{V}_c(t) = -\sum_{w=1}^{W} \underline{e}^T_{PDC}(t)\,Q_{m_w}(\underline{\xi})\underline{e}_{PDC}(t) \leq 0$. This develops the following equation as:

$$-\underline{e}^T_{PDC}(t)P_w(\underline{\xi})b_w(\underline{\xi})\tilde{\alpha}_{sf_w}\underline{\xi}_w\underline{x}(t) + \tilde{\alpha}^T_{sf_w}\Gamma^{-1}_{sf}\dot{\alpha}_{sf_w} = 0$$

or

$$\tilde{\alpha}^T_{sf_w}\Gamma^{-1}_{sf}(-\Gamma_{sf}\underline{e}^T_{PDC}(t)P_w(\underline{\xi})b_w(\underline{\xi})\underline{\xi}_w\underline{x}(t) + \dot{\alpha}_{sf_w}) = 0 \tag{4.40}$$

This provides (4.35).

If α_{sf_w} is bounded then, the adaptive estimates of the singleton is bounded and hence the system containing fuzzy PDC controller is stable.

This proves the theorem. ∎

After designing stable L_1 adaptive control law and stable fuzzy PDC control law, they are augmented to develop overall control action to control the system with uncertainties, disturbances and nonlinearities as described next.

4.3.3.3 Fuzzy PDC-L_1 Adaptive Control Law Formulation

The fuzzy PDC-L_1 adaptive control law is formulated by hybridized fuzzy L_1 adaptive control action and fuzzy PDC control law for W number of zones as:

$$\underline{u}(s) = \sum_{w=1}^{W}\left(-k_w D(s)\begin{pmatrix}\hat{\alpha}_{\omega_w}\underline{\xi}_w \underline{u}_1(t) + \hat{\alpha}_{\theta_w}\underline{\xi}_w \left\|\underline{x}\right\|_{L_\infty} \\ +\hat{\alpha}_{\sigma_w}\underline{\xi}_w - k_{g_w}\underline{r}(s)\end{pmatrix} - \alpha_{sf_w}^T \underline{\xi}_w \underline{x}(s)\right) \quad (4.41)$$

In this formulation, the nonlinear system is considered as the combination of linear fuzzy systems and for each linearized zone, fuzzy PDC-L_1 adaptive controller is formulated to cope up with uncertainties, disturbances, nonlinearities, etc., present at that zone. In this way, efficient uncertainties, disturbances rejection phenomenon is obtained with better tracking performances for nonlinear system.

High adaptation gain is utilized to facilitate fast transient performance by quickly adapting and rejecting the unknown constant, uncertainties and disturbances present in the system. The L_1 adaptive control law, formulated from the estimations of unknown constant, uncertainties, disturbances, contains high frequency component which is nullified employing low-pass filter. The fuzzy filter design is elaborated in the next section.

4.3.4 Fuzzy Logic Based Filter Design

High adaptation gain is utilized to obtain fast transient performance by quickly adapting unknown constant, time varying uncertainties and time varying disturbances. This introduces high frequencies into the control channel. The high frequency component from the control signal has to be eliminated efficiently to guarantee stable performance. Thus, the control signal is passed through a low-pass filter before applying to the system. During the estimation phase, the control signal should contain the adaptive estimates of time varying uncertainties and disturbances that in turn eliminate these undesired quantities from the actual system during the physical implementation phase. Furthermore, in basic L_1 adaptive controller, the filter structure is kept fixed which creates problems in two ways. Firstly, if the time varying uncertainties and disturbances have very high magnitude with lower frequency then, they will be passed through the low-pass filter making system response noisier. Secondly, if the time varying uncertainties and disturbances have high magnitude with high frequency then, their estimated values will be nullified from the control signal by the low pass filter. Hence, adequate control signal cannot be generated for the system during the physical implementation phase to eliminate time varying uncertainties and disturbances. To obtain proper estimate of uncertainties, disturbances and to nullify high frequency component at the same time, the low-pass filter is developed utilizing fuzzy logic. In this proposition, the filter structure varies depending upon the uncertainties and disturbances [6]. The zone-wise developed L_1 adaptive control law, consist of high frequency component, is filtered through dedicated low-pass filter of each zone.

The filter model $C(s)$ is dependent on a strictly proper stable transfer function $D(s)$. The aim is to design $D(s)$ in such a way that, the low-pass filter model $C(s)$ will become strictly proper stable transfer function. To design $D(s)$, the same premise part utilized to design the system is chosen. The output of $D(s)$ for total W number of zones are taken as the consequence parts. These consequences are fuzzy blended to develop the overall output of $D(s)$ as:

$$\underline{y}_f(t) = \sum_{w=1}^{W} \alpha_{f_w} \underline{\xi}_w \tag{4.42}$$

Here, the value of α_{f_w} is unknown. Therefore, an ideal $D(s)$ model is considered with ideal output as:

$$\underline{y}_f^*(t) = \sum_{w=1}^{W} \alpha_{f_w}^* \underline{\xi}_w \tag{4.43}$$

where $\alpha_{f_w}^*$ is the position of ideal fuzzy singleton value.

Putting the value of (4.42) in (4.30) incurs:

$$\underline{u}_1(s) = \sum_{w=1}^{W} \left(-k_w \alpha_{f_w} \underline{\xi}_w (\hat{\alpha}_{\omega_w} \underline{\xi}_w \underline{u}_1(s) + \hat{\alpha}_{\theta_w} \underline{\xi}_w \left\| \underline{x} \right\|_{L_\infty} + \hat{\alpha}_{\sigma_w} \underline{\xi}_w - k_{g_w} \underline{r}(s)) \right)$$

or

$$\underline{u}_1(s) = \sum_{w=1}^{W} -k_w \alpha_{f_w} \underline{\xi}_w (\hat{\underline{r}}_{ud}(s) - k_{g_w} \underline{r}(s)) \tag{4.44}$$

where $\hat{\underline{r}}_{ud}(s) = \hat{\alpha}_{\omega_w} \underline{\xi}_w \underline{u}_1(s) + \hat{\alpha}_{\theta_w} \underline{\xi}_w \left\| \underline{x} \right\|_{L_\infty} + \hat{\alpha}_{\sigma_w} \underline{\xi}_w$.

Putting the L_1 adaptive control law, viz., $\underline{u}_1(s)$ from (4.44) in (4.5) provides:

$$\dot{\underline{x}}(t) = \sum_{w=1}^{W} \begin{pmatrix} A_{m_w}(\underline{\xi})\, \underline{x}(t) + b_w(\underline{\xi})\, (\alpha_{\omega_w} \underline{\xi}_w k_w \alpha_{f_w} \underline{\xi}_w (-\hat{\underline{r}}_{ud}(t) + k_{g_w} \underline{r}(t)) \\ +\alpha_{\theta_w} \underline{\xi}_w \left\| \underline{x} \right\|_{L_\infty} + \alpha_{\sigma_w} \underline{\xi}_w) \end{pmatrix} \tag{4.45}$$

The system model with ideal output of $D(s)$, viz., (4.43) becomes:

$$\dot{\underline{x}}_f^*(t) = \sum_{w=1}^{W} \begin{pmatrix} A_{m_w}(\underline{\xi})\, \underline{x}_f^*(t) + b_w(\underline{\xi})(\alpha_{\omega_w} \underline{\xi}_w k_w \alpha_{f_w}^* \underline{\xi}_w (-\hat{\underline{r}}_{ud}(t) + k_{g_w} \underline{r}(t)) \\ +\alpha_{\theta_w} \underline{\xi}_w \left\| \underline{x} \right\|_{L_\infty} + \alpha_{\sigma_w} \underline{\xi}_w) \end{pmatrix} \tag{4.46}$$

The position of singleton α_{f_w} are designed following the adaptation rule to reach the value near about the ideal position of singleton $\alpha_{f_w}^*$ as analyzed in the next theorem.

Theorem 4.4 *The output of the strictly proper stable transfer function $D(s)$ will follow the output of the ideal $D(s)$ model when the position of the singleton α_{f_w} will be designed following the stable adaptive law given as:*

$$\dot{\alpha}_{f_w} = -\Gamma_f \underline{e}_f^T(t) P_w(\underline{\xi}) b_w(\underline{\xi}) \alpha_{\omega_w} \underline{\xi}_w k_w \underline{\xi}_w (-\underline{\hat{r}}_{ud}(t) + k_{g_w} \underline{r}(t)) \quad (4.47)$$

where Γ_f is the adaptation gain, $\underline{e}_f(t) = \underline{x}(t) - \underline{x}_f^*(t)$ is the error between system state and state of the system model with ideal $D(s)$. The value of α_{f_w} is bounded.

Proof Let, for W number of fuzzy *IF–THEN* rules, the fuzzy Lyapunov candidate vector is formed as:

$$V_f(t) = \sum_{w=1}^{W} \left(\frac{1}{2} \underline{e}_f^T(t) P_w(\underline{\xi}) \underline{e}_f(t) + \frac{1}{2} \tilde{\alpha}_{f_w}^T \Gamma_f^{-1} \tilde{\alpha}_{f_w} \right) \quad (4.48)$$

where $\tilde{\alpha}_{f_w} = \alpha_{f_w} - \alpha_{f_w}^*$ is the error between position of singleton to ideal position of singleton.

Derivative of (4.48) is given as:

$$\dot{V}_f(t) = \sum_{w=1}^{W} \begin{pmatrix} \frac{1}{2} \underline{\dot{e}}_f^T(t) P_w(\underline{\xi}) \underline{e}_f(t) + \frac{1}{2} \underline{e}_f^T(t) P_w(\underline{\xi}) \underline{\dot{e}}_f(t) \\ + \frac{1}{2} \dot{\tilde{\alpha}}_{f_w}^T \Gamma_f^{-1} \tilde{\alpha}_{f_w} + \frac{1}{2} \tilde{\alpha}_{f_w}^T \Gamma_f^{-1} \dot{\tilde{\alpha}}_{f_w} \end{pmatrix} \quad (4.49)$$

Replacing $\underline{\dot{e}}_f(t)$ in the above equation with its value.

$$\underline{\dot{e}}_f(t) = \underline{\dot{x}}(t) - \underline{\dot{x}}_f^*(t) = \sum_{w=1}^{W} \left(A_{m_w}(\underline{\xi}) \underline{e}_f(t) + b_w(\underline{\xi}) (\alpha_{\omega_w} \underline{\xi}_w k_w \tilde{\alpha}_{f_w} \underline{\xi}_w (-\underline{\hat{r}}_{ud}(t) + k_{g_w} \underline{r}(t))) \right)$$

yields:

$$\dot{V}_f(t) = \sum_{w=1}^{W} \frac{1}{2} \begin{pmatrix} \begin{pmatrix} A_{m_w}(\underline{\xi}) \underline{e}_f(t) \\ + b_w(\underline{\xi}) (\alpha_{\omega_w} \underline{\xi}_w k_w \tilde{\alpha}_{f_w} \underline{\xi}_w (-\underline{\hat{r}}_{ud}(t) + k_{g_w} \underline{r}(t))) \end{pmatrix}^T P_w(\underline{\xi}) \underline{e}_f(t) \\ + \underline{e}_f^T(t) P_w(\underline{\xi}) \begin{pmatrix} A_{m_w}(\underline{\xi}) \underline{e}_f(t) \\ + b_w(\underline{\xi}) (\alpha_{\omega_w} \underline{\xi}_w k_w \tilde{\alpha}_{f_w} \underline{\xi}_w (-\underline{\hat{r}}_{ud}(t) + k_{g_w} \underline{r}(t))) \end{pmatrix} \\ + \dot{\tilde{\alpha}}_{f_w}^T \Gamma_f^{-1} \tilde{\alpha}_{f_w} + \tilde{\alpha}_{f_w}^T \Gamma_f^{-1} \dot{\tilde{\alpha}}_{f_w} \end{pmatrix}$$

or

$$\dot{V}_f(t) = \sum_{w=1}^{W} \begin{pmatrix} \frac{1}{2} \underline{e}_f^T(t) \left(A_{m_w}^T(\underline{\xi}) P_w(\underline{\xi}) + P_w(\underline{\xi}) A_{m_w}(\underline{\xi}) \right) \underline{e}_f(t) \\ + \frac{1}{2} \left(b_w(\underline{\xi}) (\alpha_{\omega_w} \underline{\xi}_w k_w \tilde{\alpha}_{f_w} \underline{\xi}_w (-\underline{\hat{r}}_{ud}(t) + k_{g_w} \underline{r}(t))) \right)^T P_w(\underline{\xi}) \underline{e}_f(t) \\ + \frac{1}{2} \underline{e}_f^T(t) P_w(\underline{\xi}) \left(b_w(\underline{\xi}) (\alpha_{\omega_w} \underline{\xi}_w k_w \tilde{\alpha}_{f_w} \underline{\xi}_w (-\underline{\hat{r}}_{ud}(t) + k_{g_w} \underline{r}(t))) \right) \\ + \frac{1}{2} \dot{\tilde{\alpha}}_{f_w}^T \Gamma_f^{-1} \tilde{\alpha}_{f_w} + \frac{1}{2} \tilde{\alpha}_{f_w}^T \Gamma_f^{-1} \dot{\tilde{\alpha}}_{f_w} \end{pmatrix}$$

Now, $\dot{\tilde{\alpha}}_{f_w} = \dot{\alpha}_{f_w} - \dot{\alpha}^*_{f_w} = \dot{\alpha}_{f_w}$, as $\alpha^*_{f_w}$ is a constant value. Putting this provides:

$$\dot{V}_f(t) = \sum_{w=1}^{W} \begin{pmatrix} -\underline{e}_f^T(t)\, Q_{m_w}(\underline{\xi})\underline{e}_f(t) + \tilde{\alpha}_{f_w}^T \Gamma_f^{-1} \dot{\alpha}_{f_w} \\ +\left(\underline{e}_f^T(t) P_w(\underline{\xi}) b_w(\underline{\xi})\, \alpha_{\omega_w \underline{\xi}_w} k_w \tilde{\alpha}_{f_w \underline{\xi}_w} (-\hat{\underline{r}}_{ud}(t) + k_{g_w} \underline{r}(t))\right) \end{pmatrix} \tag{4.50}$$

The system will be stable with fuzzy logic based filter model if the following Lyapunov stability condition holds: $\dot{V}_f(t) = -\sum_{w=1}^{W} \underline{e}_f^T(t) Q_{m_w}(\underline{\xi})\underline{e}_f(t) \leq 0$. This leads to:

$$\underline{e}_f^T(t) P_w(\underline{\xi}) b_w(\underline{\xi})\, \alpha_{\omega_w \underline{\xi}_w} k_w \tilde{\alpha}_{f_w \underline{\xi}_w} (-\hat{\underline{r}}_{ud}(t) + k_{g_w} \underline{r}(t)) + \tilde{\alpha}_{f_w}^T \Gamma_f^{-1} \dot{\alpha}_{f_w} = 0$$

or

$$\tilde{\alpha}_{f_w}^T \Gamma_f^{-1} (\Gamma_f \underline{e}_f^T(t) P_w(\underline{\xi}) b_w(\underline{\xi})\, \alpha_{\omega_w \underline{\xi}_w} k_w \underline{\xi}_w \begin{pmatrix} -\hat{\underline{r}}_{ud}(t) \\ +k_{g_w} \underline{r}(t)) + \dot{\alpha}_{f_w} \end{pmatrix} = 0 \tag{4.51}$$

Solution of (4.51) for $\dot{\alpha}_{f_w}$ develops the adaptive estimate (4.47). Now, as α_{f_w} is bounded, therefore, its adaptive estimate is bounded and affords strictly proper stable transfer function $D(s)$. This proves the theorem. ■

The strictly proper stable transfer function $D(s)$ is utilized to design the filter model. Then, the filter model designed for W number of zones are combined to obtain the overall fuzzy filter model as:

$$C(s) = \sum_{w=1}^{W} \frac{\alpha_{\omega_w \underline{\xi}_w} k_w D(s)}{1 + \alpha_{\omega_w \underline{\xi}_w} k_w D(s)} \tag{4.52}$$

where k_w is the pre-filter gain for *wth* zone.

After developing all the components of the fuzzy PDC-L_1 adaptive controller, its stability analysis is provided in the subsequent section.

4.4 Stability Analysis of Fuzzy PDC-L_1 Adaptive Controller

In this method, the nonlinear system, predictor, adaptive estimates of unknown constant, time varying uncertainties, time varying disturbances, L_1 adaptive control law, low-pass filter, PDC control law are designed utilizing same premise part and corresponding linear consequence rule based fuzzy systems are combined to obtain the overall components. The bounded adaptive estimates of the position of singletons to design those fuzzy systems are derived from the stability condition of fuzzy Lyapunov function. Now, the stability analysis of the overall system with designed control methodology is investigated in the following theorem. The L_1 stability condition is also analyzed in the theorem given as follows.

Theorem 4.5 *The system with fuzzy PDC-L_1 adaptive controller will be stabilized if following conditions hold:*

(i) *The state error between predictor and system will gradually reaches zero as time tends to infinity, viz.,* $\lim_{t\to\infty}\left|\underline{\tilde{x}}(t)\right| \to 0$.

(ii) *The components of the fuzzy PDC-L_1 adaptive controller is formulated from the stable bounded adaptive estimates of the position of the singletons. Unknown constant, uncertainties, disturbances present in the system are also bounded.*

(iii) *L_1 gain requirement is satisfied, viz.,* $\|G(s)\|_{L_1}(\hat{\alpha}_{\theta_w}\underline{\xi}_w)_{\max} < 1$.

Proof Re-writing (4.26):

$$\dot{V}(t) = \sum_{w=1}^{W}\left(-\underline{\tilde{x}}^T(t)Q_{m_w}(\underline{\xi})\underline{\tilde{x}}(t) + \frac{1}{\Gamma_C}\left(\left|\tilde{\alpha}_{\theta_w}^T\dot{\alpha}_{\theta_w}\right| + \left|\tilde{\alpha}_{\sigma_w}^T\dot{\alpha}_{\sigma_w}\right|\right)\right)$$

Stability condition of the system incurs the upper bound as follows:

$$\sum_{w=1}^{W}\frac{1}{2}\lambda_{\min}(P_w(\underline{\xi}))\left\|\underline{\tilde{x}}(t)\right\|_{L_\infty}^2 \le \sum_{w=1}^{W}\frac{1}{2}\underline{\tilde{x}}^T(t)P_w(\underline{\xi})\underline{\tilde{x}}(t) \le V(t) \le V(0), \forall t \ge 0 \tag{4.53}$$

where $\lambda_{\min}(P_w(\underline{\xi}))$ denotes the minimum eigenvalue of $P_w(\underline{\xi})$.

Projection operator ensures that, the adaptive estimates of unknown constant, time varying uncertainties and time varying disturbances are bounded to compact set as $\left|\alpha_{\omega_w}\right| \le \alpha_\Omega$, $\left|\alpha_{\theta_w}\right| \le \alpha_\Theta$, $\left|\alpha_{\sigma_w}\right| \le \alpha_\Sigma$, $\left|\hat{\alpha}_{\omega_w}\right| \le \alpha_\Omega$, $\left|\hat{\alpha}_{\theta_w}\right| \le \alpha_\Theta$, $\left|\hat{\alpha}_{\sigma_w}\right| \le \alpha_\Sigma$ for all $t \ge 0$, which in turn develops the upper bound as:

$$\begin{aligned}
&\max_{t\ge0}\left(\frac{1}{\Gamma_C}(\tilde{\alpha}_{\omega_w}^T\tilde{\alpha}_{\omega_w} + \tilde{\alpha}_{\theta_w}^T\tilde{\alpha}_{\theta_w} + \tilde{\alpha}_{\sigma_w}^T\tilde{\alpha}_{\sigma_w})\right)\\
&= \frac{1}{\Gamma_C}\max_{t\ge0}\begin{pmatrix}(\hat{\alpha}_{\omega_w} - \alpha_{\omega_w})^T(\hat{\alpha}_{\omega_w} - \alpha_{\omega_w}) + (\hat{\alpha}_{\theta_w} - \alpha_{\theta_w})^T(\hat{\alpha}_{\theta_w} - \alpha_{\theta_w})\\ +(\hat{\alpha}_{\sigma_w} - \alpha_{\sigma_w})^T(\hat{\alpha}_{\sigma_w} - \alpha_{\sigma_w})\end{pmatrix}\\
&\le \frac{1}{\Gamma_C}\max_{t\ge0}\begin{pmatrix}(2\alpha_{\omega_w})^T(2\alpha_{\omega_w}) + (2\alpha_{\theta_w})^T(2\alpha_{\theta_w})\\ +(2\alpha_{\sigma_w})^T(2\alpha_{\sigma_w})\end{pmatrix}
\end{aligned}$$

or

$$\begin{aligned}
&\max_{t\ge0}\left(\frac{1}{\Gamma_C}(\tilde{\alpha}_{\omega_w}^T\alpha_{\omega_w} + \tilde{\alpha}_{\theta_w}^T\alpha_{\theta_w} + \tilde{\alpha}_{\sigma_w}^T\alpha_{\sigma_w})\right)\\
&\le \frac{4}{\Gamma_C}\begin{pmatrix}\max\limits_{\alpha_{\omega_w}\in\alpha_\Omega}\mathrm{tr}(\alpha_{\omega_w}^T\alpha_{\omega_w}) + \max\limits_{\alpha_{\theta_w}\in\alpha_\Theta}(\alpha_{\theta_w}^T\alpha_{\theta_w})\\ + \max\limits_{\alpha_{\sigma_w}\in\alpha_\Sigma}(\alpha_{\sigma_w}^T\alpha_{\sigma_w})\end{pmatrix}
\end{aligned} \tag{4.54}$$

and also:

$$(\tilde{\alpha}_{\theta_w}^T \dot{\alpha}_{\theta_w} + \tilde{\alpha}_{\sigma_w}^T \dot{\alpha}_{\sigma_w}) \le 2\left(\frac{\max}{\alpha_{\theta_w}} \|\alpha_{\theta_w}\| d_{\theta_w} + \frac{\max}{\sigma_w \in \sum_w} \|\alpha_{\sigma_w}\| d_{\sigma_w}\right) \tag{4.55}$$

Here, α_{θ_w} and α_{σ_w} are completely differentiable and the derivatives are bounded as $\dot{\alpha}_{\theta_w} \le d_{\theta_w} < \infty$ and $\dot{\alpha}_{\sigma_w} \le d_{\sigma_w} < \infty$ for all $t \ge 0$.

Therefore, it can be written that:

$$\begin{aligned}\underline{\tilde{x}}^T(t) Q_{m_w}(\underline{\xi}) \underline{\tilde{x}}(t) &\le \frac{1}{2} \frac{\lambda_{\min}(Q_{m_w}(\underline{\xi}))}{\lambda_{\max}(P_w(\underline{\xi}))} \underline{\tilde{x}}^T(t) P_w(\underline{\xi}) \underline{\tilde{x}}(t) \\ &\le 2\Gamma_C^{-1}\left(\max_{\alpha_{\theta_w} \in \alpha_\Theta} \|\alpha_{\theta_w}\| d_{\theta_w} + \max_{\sigma_w \in \Sigma_w} \|\alpha_{\sigma_w}\| d_{\sigma_w}\right)\end{aligned}$$

or

$$\underline{\tilde{x}}^T(t) P_w(\underline{\xi}) \underline{\tilde{x}}(t) \le \frac{4}{\Gamma_C} \cdot \frac{\lambda_{\max}(P_w(\underline{\xi}))}{\lambda_{\min}(Q_{m_w}(\underline{\xi}))} \begin{pmatrix} \max_{\alpha_{\theta_w} \in \alpha_\Theta} \|\alpha_{\theta_w}\| d_{\theta_w} \\ + \max_{\sigma_w \in \Sigma_w} \|\alpha_{\sigma_w}\| d_{\sigma_w} \end{pmatrix} \tag{4.56}$$

Putting the values of (4.54) and (4.56) in (4.22) incurs:

$$V(t) \le \sum_{w=1}^{W} \begin{pmatrix} \frac{4}{\Gamma_C} \cdot \frac{\lambda_{\max}(P_w(\underline{\xi}))}{\lambda_{\min}(Q_{m_w}(\underline{\xi}))} \left(\max_{\alpha_{\theta_w} \in \alpha_\Theta} \|\alpha_{\theta_w}\| d_{\theta_w} + \max_{\sigma_w \in \Sigma_w} \|\alpha_{\sigma_w}\| d_{\sigma_w}\right) \\ + \frac{2}{\Gamma_C} \begin{pmatrix} \max_{\alpha_{\omega_w} \in \alpha_\Omega} \operatorname{tr}(\alpha_{\omega_w}^T \alpha_{\omega_w}) + \max_{\alpha_{\theta_w} \in \alpha_\Theta} (\alpha_{\theta_w}^T \alpha_{\theta_w}) \\ + \max_{\alpha_{\sigma_w} \in \alpha_\Sigma} (\alpha_{\sigma_w}^T \alpha_{\sigma_w}) \end{pmatrix} \end{pmatrix}$$

or

$$V(t) \le \sum_{w=1}^{W} \frac{\theta_{m_w}}{\Gamma_C} \tag{4.57}$$

where

$$\begin{aligned}\theta_{m_w} = {} & 2\left(\max_{\alpha_{\omega_w} \in \alpha_\Omega} \operatorname{tr}(\alpha_{\omega_w}^T \alpha_{\omega_w}) + \max_{\alpha_{\theta_w} \in \alpha_\Theta} (\alpha_{\theta_w}^T \alpha_{\theta_w}) + \max_{\alpha_{\sigma_w} \in \alpha_\Sigma} (\alpha_{\sigma_w}^T \alpha_{\sigma_w})\right) \\ & + 4\frac{\lambda_{\max}(P_w(\xi))}{\lambda_{\min}(Q_w(\xi))}\left(\max_{\alpha_{\theta_w} \in \alpha_\Theta} \|\alpha_{\theta_w}\| d_{\theta_w} + \max_{\sigma_w \in \Sigma_w} \|\alpha_{\sigma_w}\| d_{\sigma_w}\right)\end{aligned}$$

Now, the system with fuzzy PDC-L_1 adaptive controller will be stable if the Lyapunov stability criterion holds, viz., $\dot{V}(t) \le 0$. Putting (4.57) in (4.53) provides:

$$\sum_{w=1}^{W} \frac{1}{2}\lambda_{\min}(P_w(\underline{\xi})) \left\| \underline{\tilde{x}}(t) \right\|_{L_\infty}^2 \le \sum_{w=1}^{W} \frac{\theta_{m_w}}{\Gamma_C}$$

or

$$\left\| \underline{\tilde{x}}(t) \right\|_{L_\infty} \le \sum_{w=1}^{W} \sqrt{\frac{2\theta_{m_w}}{\lambda_{\min}(P_w(\underline{\xi}))\Gamma_C}} \tag{4.58}$$

Therefore, $\left\| \underline{\tilde{x}}(t) \right\|_{L_\infty}$ is bounded. But it does not incur that, $\lim_{t\to\infty} \left| \underline{\tilde{x}}(t) \right| \to 0$.
Let,

$$\Psi = \dot{V}(t) \le \sum_{w=1}^{W} \left(-\underline{\tilde{x}}^T(t) Q_{m_w}(\underline{\xi}) \underline{\tilde{x}}(t) + \frac{1}{\Gamma_C} \left(\left| \tilde{\alpha}_{\theta_w}^T \dot{\alpha}_{\theta_w} \right| + \left| \tilde{\alpha}_{\sigma_w}^T \dot{\alpha}_{\sigma_w} \right| \right) \right) \le 0 \tag{4.59}$$

Integrating (4.59), it can be written that:

$$\int_0^t \Psi(\tau)\, d\tau \le (V(0) - V(t)) \tag{4.60}$$

Here, $V(0)$ and $V(t)$ are bounded. $V(t)$ is non-increasing yields $\lim_{t\to\infty} \int_0^t \Psi(\tau)\, d\tau < \infty$ and $\ddot{V}(t)$ is also bounded. Therefore, $\Psi(t) = \dot{V}(t)$ is uniformly continuous. Now, from Barbalat's lemma it can be stated that, as $V(t)$ is bounded and $\dot{V}(t)$ is bounded and uniformly continuous, therefore, $\lim_{t\to\infty} \Psi(t) \to 0$. This provides $\lim_{t\to\infty} \left| \underline{\tilde{x}}(t) \right| \to 0$.

Therefore, the state error between system and predictor tends to zero as time tends to infinity. Though, it does not incur the system state is bounded as system state and predictor state both may diverge at the same rate keeping error between them bounded. Therefore, the predictor state should be bounded to provide bounded system performance.

Now, putting value of $\underline{\tilde{x}}(t)$ on (4.58) incurs:

$$\left| \left\| \underline{\hat{x}}(t) \right\|_{L_\infty} - \left\| \underline{x}(t) \right\|_{L_\infty} \right| \le \sum_{w=1}^{W} \sqrt{\frac{2\theta_{m_w}}{\lambda_{\min}(P_w(\underline{\xi}))\Gamma_C}} \tag{4.61}$$

In frequency domain the predictor model (4.16) can be represented as [2]:

$$\underline{\hat{x}}(s) = \sum_{w=1}^{W} \left(H(s)C(s)k_{g_w}\underline{r}(s) + G(s)\underline{\hat{r}}_{ud}(s) \right) \tag{4.62}$$

where $\underline{r}(s)$ is the reference input,

$$H(s) = \sum_{w=1}^{W} \left((sI - A_{m_w \underline{\xi}_w})^{-1} b_{w \underline{\xi}_w} \right) = \sum_{w=1}^{W} \left((sI - A_{m_w}(\underline{\xi}))^{-1} b_w(\underline{\xi}) \right) \tag{4.63}$$

and

$$G(s) = H(s)(1 - C(s)) \tag{4.64}$$

Now, (4.62) will satisfy the following upper bound:

$$\left\|\hat{\underline{x}}\right\|_{L_\infty} \leq \sum_{w=1}^{W} \left(\left\|H(s)C(s)k_{g_w}\right\|_{L_1} \left\|\underline{r}\right\|_{L_\infty} + \|G(s)\|_{L_1} \left\|\hat{\underline{r}}_{ud}\right\|_{L_\infty} \right)$$

Putting, the values of $C(s)$ and $\hat{\underline{r}}_{ud}$ in the above equation yield:

$$\|\hat{\underline{x}}\|_{L_\infty} \leq \sum_{w=1}^{W} \left(\left\|H(s)\alpha_{f_w \underline{\xi}_w} k_{g_w}\right\|_{L_1} \|\underline{r}\|_{L_\infty} + \|G(s)\|_{L_1} \left\| \left(\hat{\alpha}_{\omega_w \underline{\xi}_w} \underline{u}_1 + \hat{\alpha}_{\theta_w \underline{\xi}_w} \underline{x} + \hat{\alpha}_{\sigma_w \underline{\xi}_w} \right) \right\|_{L_\infty} \right)$$

or

$$\|\hat{\underline{x}}\|_{L_\infty} \leq \sum_{w=1}^{W} \begin{pmatrix} \|H(s)\|_{L_1} \left\|\alpha_{f_w \underline{\xi}_w} k_{g_w}\right\|_{L_1} \left\|\underline{r}\right\|_{L_\infty} + \|G(s)\|_{L_1} \left\|\hat{\alpha}_{\omega_w \underline{\xi}_w}\right\|_{L_\infty} \left\|\underline{u}_1\right\|_{L_\infty} \\ + \|G(s)\|_{L_1} \left\|\hat{\alpha}_{\theta_w \underline{\xi}_w}\right\|_{L_\infty} \left\|\underline{x}\right\|_{L_\infty} + \|G(s)\|_{L_1} \left\|\hat{\alpha}_{\sigma_w \underline{\xi}_w}\right\|_{L_\infty} \end{pmatrix} \tag{4.65}$$

Now, putting the value of $\left\|\underline{x}\right\|_{L_\infty}$ from (4.61) in (4.65) provides:

$$\|\hat{\underline{x}}\|_{L_\infty} \leq \sum_{w=1}^{W} \begin{pmatrix} \|H(s)\|_{L_1} \left\|\alpha_{f_w \underline{\xi}_w} k_{g_w}\right\|_{L_1} \left\|\underline{r}\right\|_{L_\infty} + \|G(s)\|_{L_1} \left\|\hat{\alpha}_{\omega_w \underline{\xi}_w}\right\|_{L_\infty} \left\|\underline{u}_1\right\|_{L_\infty} \\ +\|G(s)\|_{L_1} \left\|\hat{\alpha}_{\theta_w \underline{\xi}_w}\right\|_{L_\infty} \left(\left\|\hat{\underline{x}}(t)\right\|_{L_\infty} + \sum_{w=1}^{W} \sqrt{\frac{2\theta_{m_w}}{\lambda_{\min}(P_w(\underline{\xi}))\Gamma_C}} \right) \\ +\|G(s)\|_{L_1} \left\|\hat{\alpha}_{\sigma_w \underline{\xi}_w}\right\|_{L_\infty} \end{pmatrix}$$

or

$$\|\hat{\underline{x}}\|_{L_\infty} \leq \sum_{w=1}^{W} \begin{pmatrix} \|H(s)\|_{L_1} \left\|\alpha_{f_w \underline{\xi}_w} k_{g_w}\right\|_{L_1} \left\|\underline{r}\right\|_{L_\infty} + \|G(s)\|_{L_1} \left\|\hat{\alpha}_{\omega_w \underline{\xi}_w}\right\|_{L_\infty} \left\|\underline{u}_1\right\|_{L_\infty} \\ +\|G(s)\|_{L_1} \left\|\hat{\alpha}_{\theta_w \underline{\xi}_w}\right\|_{L_\infty} \left\|\hat{\underline{x}}(t)\right\|_{L_\infty} + \|G(s)\|_{L_1} \left\|\hat{\alpha}_{\sigma_w \underline{\xi}_w}\right\|_{L_\infty} \\ +\|G(s)\|_{L_1} \left\|\hat{\alpha}_{\theta_w \underline{\xi}_w}\right\|_{L_\infty} \sum_{w=1}^{W} \sqrt{\frac{2\theta_{m_w}}{\lambda_{\min}(P_w(\underline{\xi}))\Gamma_C}} \end{pmatrix}$$

or

$$\left\|\underline{\hat{x}}\right\|_{L_\infty}\sum_{w=1}^{W}\left(1-\|G(s)\|_{L_1}\left\|\hat{\alpha}_{\theta_w}\underline{\xi}_w\right\|_{L_\infty}\right)$$

$$\leq\sum_{w=1}^{W}\left(\begin{array}{l}\|H(s)\|_{L_1}\left\|\alpha_{f_w}\underline{\xi}_w k_{g_w}\right\|_{L_1}\|\underline{r}\|_{L_\infty}+\|G(s)\|_{L_1}\left\|\hat{\alpha}_{\omega_w}\underline{\xi}_w\right\|_{L_\infty}\|\underline{u}_1\|_{L_\infty}\\+\|G(s)\|_{L_1}\left\|\hat{\alpha}_{\theta_w}\underline{\xi}_w\right\|_{L_\infty}\sum_{w=1}^{W}\sqrt{\frac{2\theta_{m_w}}{\lambda_{\min}(P_w(\underline{\xi}))\Gamma_C}}+\|G(s)\|_{L_1}\left\|\hat{\alpha}_{\sigma_w}\underline{\xi}_w\right\|_{L_\infty}\end{array}\right)$$

or

$$\left\|\underline{\hat{x}}\right\|_{L_\infty}\leq\frac{\sum_{w=1}^{W}\left(\begin{array}{l}\|H(s)\|_{L_1}\left\|\alpha_{f_w}\underline{\xi}_w k_{g_w}\right\|_{L_1}\|\underline{r}\|_{L_\infty}+\|G(s)\|_{L_1}\left\|\hat{\alpha}_{\omega_w}\underline{\xi}_w\right\|_{L_\infty}\|\underline{u}_1\|_{L_\infty}\\+\|G(s)\|_{L_1}\left\|\hat{\alpha}_{\theta_w}\underline{\xi}_w\right\|_{L_\infty}\sqrt{\frac{2\theta_{m_w}}{\lambda_{\min}(P_w(\underline{\xi}))\Gamma_C}}\\+\|G(s)\|_{L_1}\left\|\hat{\alpha}_{\sigma_w}\underline{\xi}_w\right\|_{L_\infty}\end{array}\right)}{\sum_{w=1}^{W}\left(1-\|G(s)\|_{L_1}\left\|\hat{\alpha}_{\theta_w}\underline{\xi}_w\right\|_{L_\infty}\right)}\tag{4.66}$$

Here, $\left\|\hat{\alpha}_{\omega_w}\underline{\xi}_w\right\|_{L_\infty}=\max_{i=1,2,\ldots,n,t\geq0}\left|\hat{\alpha}_{\omega_{w_i}}\underline{\xi}_{w_i}\right|$, $\left\|\hat{\alpha}_{\theta_w}\underline{\xi}_w\right\|_{L_\infty}=\max_{i=1,2,\ldots,n,t\geq0}\left|\hat{\alpha}_{\theta_{w_i}}\underline{\xi}_{w_i}\right|$ and $\left\|\hat{\alpha}_{\sigma_w}\underline{\xi}_w\right\|_{L_\infty}=\max_{i=1,2,\ldots,n,t\geq0}\left|\hat{\alpha}_{\sigma_{w_i}}\underline{\xi}_{w_i}\right|$. These values are bounded to the compact set as $\left\|\hat{\alpha}_{\omega_w}\underline{\xi}_w\right\|_{L_\infty}\in\Omega_w$, $\left\|\hat{\alpha}_{\theta_w}\underline{\xi}_w\right\|_{L_\infty}\in\underline{\Theta}_w$ and $\left\|\hat{\alpha}_{\sigma_w}\underline{\xi}_w\right\|_{L_\infty}\in\underline{\Sigma}_w$. Again, it can be stated that, $\|H(s)\|_{L_1}$, $\|G(s)\|_{L_1}$, $\left\|\alpha_{f_w}\underline{\xi}_w\right\|_{L_1}$, $\|k_{g_w}\|_{L_1}$, $\|\underline{u}_1\|_{L_\infty}$ are designed as a way that they are finite, input $\|\underline{r}\|_{L_\infty}$ is finite. The filter $C(s)$ is designed as a strictly proper stable transfer function. $\left\|\underline{\hat{x}}\right\|_{L_\infty}<\infty$ follows if the following L_1 norm condition holds:

$$\|G(s)\|_{L_1}(\alpha_{\theta_w}\underline{\xi}_w)_{\max}<1\tag{4.67}$$

These bounded quantities and L_1 norm condition incur (4.66) is stable. Therefore, $\left\|\underline{\hat{x}}\right\|_{L_\infty}$ is finite and the predictor state, viz., $\underline{\hat{x}}(t)$ is bounded. Again, bounds on $\underline{\tilde{x}}(t)$ as stated in (4.61) provides $\|\underline{x}\|_{L_\infty}$ is finite and the system state, viz., $\underline{x}(t)$ is also bounded. Hence, the system with the fuzzy PDC-L_1 adaptive controller is stable. ■

After the stability proof, the transient and steady state performances are analyzed in the next section.

4.5 Transient and Steady State Performance Analysis of Fuzzy PDC-L_1 Adaptive Controller

To analyze the transient and steady state performance of the system, a reference system model is considered. The closed loop system has to follow the responses of closed loop reference system at transient and steady state as elaborated in this section.

The system model (4.5) in frequency domain can be expressed as:

$$\underline{x}(s) = \sum_{w=1}^{W} H(s)\left(\alpha_{\omega_w}\underline{\xi}_w \underline{u}_1(s) + \alpha_{\theta_w}\underline{\xi}_w \left\|\underline{x}\right\|_{L_\infty} + \alpha_{\sigma_w}\underline{\xi}_w\right) \tag{4.68}$$

Fuzzy L_1 adaptive control law (4.30) can be written as:

$$\underline{u}_1(s)\sum_{w=1}^{W}\left(1 + k_w D(s)\alpha_{\omega_w}\underline{\xi}_w\right) = \sum_{w=1}^{W}\begin{pmatrix} -k_w D(s)(\hat{\alpha}_{\omega_w}\underline{\xi}_w\underline{u}_1(s) - \alpha_{\omega_w}\underline{\xi}_w\underline{u}_1(s) \\ +\hat{\alpha}_{\theta_w}\underline{\xi}_w\left\|\underline{x}\right\|_{L_\infty} - \alpha_{\theta_w}\underline{\xi}_w\left\|\underline{x}\right\|_{L_\infty} \\ +\alpha_{\theta_w}\underline{\xi}_w\left\|\underline{x}\right\|_{L_\infty} \\ +\hat{\alpha}_{\sigma_w}\underline{\xi}_w - \alpha_{\sigma_w}\underline{\xi}_w + \alpha_{\sigma_w}\underline{\xi}_w - k_{g_w}\underline{r}(s)) \end{pmatrix}$$

or

$$\underline{u}_1(s) = \sum_{w=1}^{W}\frac{\begin{pmatrix} -k_w D(s)(\tilde{\alpha}_{\omega_w}\underline{\xi}_w\underline{u}_1(s) + \tilde{\alpha}_{\theta_w}\underline{\xi}_w\left\|\underline{x}\right\|_{L_\infty} + \tilde{\alpha}_{\sigma_w}\underline{\xi}_w \\ +\alpha_{\theta_w}\underline{\xi}_w\left\|\underline{x}\right\|_{L_\infty} + \alpha_{\sigma_w}\underline{\xi}_w - k_{g_w}\underline{r}(s)) \end{pmatrix}}{(1 + k_w D(s)\alpha_{\omega_w}\underline{\xi}_w)}$$

or

$$\underline{u}_1(s) = \sum_{w=1}^{W}\left(-\frac{k_w D(s)}{(1 + \alpha_{\omega_w}\underline{\xi}_w k_w D(s))}(\underline{r}_2(s) - k_{g_w}\underline{r}(s) + \underline{\tilde{r}}_{ud}(s))\right) \tag{4.69}$$

where $\underline{r}_2(s)$ is the Laplace transform of $\underline{r}_2(t) = \sum_{w=1}^{W}\left(\alpha_{\theta_w}\underline{\xi}_w\left\|\underline{x}\right\|_{L_\infty} + \alpha_{\sigma_w}\underline{\xi}_w\right)$ and $\underline{\tilde{r}}_{ud}(s)$ is the Laplace transform of $\underline{\tilde{r}}_{ud}(t) = \sum_{w=1}^{W}\left(\tilde{\alpha}_{\omega_w}\underline{\xi}_w\underline{u}_1(t) + \tilde{\alpha}_{\theta_w}\underline{\xi}_w\left\|\underline{x}\right\|_{L\infty} + \tilde{\alpha}_{\sigma_w}\underline{\xi}_w\right)$.

Utilizing the fuzzy filter model (4.52) in (4.69) yields:

$$\underline{u}_1(s) = \sum_{w=1}^{W}\left(-\frac{C(s)}{\alpha_{\omega_w}\underline{\xi}_w}(\alpha_{\theta_w}\underline{\xi}_w\left\|\underline{x}\right\|_{L_\infty} + \alpha_{\sigma_w}\underline{\xi}_w - k_{g_w}\underline{r}(s) + \underline{\tilde{r}}_{ud}(s))\right)$$

or

$$\underline{u}_1(s) = \sum_{w=1}^{W} \left(-\frac{C(s)}{\alpha_{\omega_w}\underline{\xi}_w} (\underline{r}_2(s) - k_{g_w}\underline{r}(s) + \tilde{\underline{r}}_{ud}(s)) \right) \tag{4.70}$$

Overall control law formulated by conjugating fuzzy L_1 adaptive control law and fuzzy PDC control law is given by:

$$\underline{u}(s) = \sum_{w=1}^{W} \left(\frac{C(s)}{\alpha_{\omega_w}\underline{\xi}_w} (k_{g_w}\underline{r}(s) - \underline{r}_2(s) - \tilde{\underline{r}}_{ud}(s)) - \alpha_{sf_w}\underline{\xi}_w \underline{x}(s) \right) \tag{4.71}$$

The system model (4.68) with the control input (4.71) can be expressed as:

$$\underline{x}(s) = \sum_{w=1}^{W} \left(H(s)((1 - C(s))\underline{r}_2(s) + C(s)\, k_{g_w}\underline{r}(s) - C(s)\, \tilde{\underline{r}}_{ud}(s)) \right) \tag{4.72}$$

Let, the closed loop LTI reference system model is considered as:

$$\dot{\underline{x}}_{ref}(t) = \sum_{w=1}^{W} \begin{pmatrix} A_{m_w}(\underline{\xi})\underline{x}_{ref}(t) + b_w(\underline{\xi})(\alpha_{\omega_w}\underline{\xi}_w \underline{u}_{ref_1}(t) \\ +\alpha_{\theta_w}\underline{\xi}_w \left\| \underline{x}_{ref} \right\|_{L_\infty} + \alpha_{\sigma_w}\underline{\xi}_w) \end{pmatrix} \tag{4.73}$$

The control input to the reference system is:

$$\underline{u}_{ref}(s) = \sum_{w=1}^{W} \left(\frac{C(s)}{\alpha_{\omega_w}\underline{\xi}_w} \left[k_{g_w}\underline{r}(s) - \underline{r}_{2ref}(s) \right] - \alpha_{sf_w}\underline{\xi}_w \underline{x}_{ref}(s) \right) \tag{4.74}$$

where $\underline{r}_{2ref}(s)$ is the Laplace transform of $\underline{r}_{2ref}(t) = \sum_{w=1}^{W} \left(\alpha_{\theta_w}\underline{\xi}_w \left\| \underline{x}_{ref} \right\|_{L_\infty} + \alpha_{\sigma_w}\underline{\xi}_w \right)$.

$\underline{r}_{2ref}(t)$ satisfies the following boundary condition.

$\left\| \underline{r}_{2ref} \right\|_{L_\infty} \leq \sum_{w=1}^{W} \left(\left\| \alpha_{\theta_w}\underline{\xi}_w \right\|_{L_1} \left\| \underline{x}_{ref} \right\|_{L_\infty} + \left\| \alpha_{\sigma_w}\underline{\xi}_w \right\|_{L_\infty} \right)$.

The closed loop LTI reference system model (4.73) in frequency domain has a form:

$$\underline{x}_{ref}(s) = \sum_{w=1}^{W} H(s) \left((1 - C(s))\underline{r}_{2ref}(s) + C(s)k_{g_w}\underline{r}(s) \right) \tag{4.75}$$

The state and control input of the system model have to follow the state and control input of the reference system model respectively at transient and steady state condition. This phenomenon is evaluated in the theorem described as follows.

Theorem 4.6 *The state error and control input error between the reference system (4.75) and system (4.72) hold the following bounds respectively:*

$$\left\|\underline{e}\right\|_{L_\infty} = \left\|\underline{x}_{ref} - \underline{x}\right\|_{L_\infty} \le val_1 \tag{4.76}$$

$$\left\|\underline{u}_{ref} - \underline{u}\right\|_{L_\infty} \le val_2 \tag{4.77}$$

where

$$val_1 = \sum_{w=1}^{W}\left(\frac{\left\|\frac{\alpha_{\omega w}\underline{\xi}_w k_w D(s)}{1+\alpha_{\omega w}\underline{\xi}_w k_w D(s)}\right\|_{L_1}\sqrt{\frac{2\theta_{mw}}{\lambda_{\min}(P_w(\underline{\xi}))\Gamma_C}}}{1-\left\|H(s)\left(1-\left(\frac{\alpha_{\omega w}\underline{\xi}_w k_w D(s)}{1+\alpha_{\omega w}\underline{\xi}_w k_w D(s)}\right)\right)\right\|_{L_1}(\alpha_{\theta_w}\underline{\xi}_w)_{\max}}\right)$$

and

$$val_2 = \sum_{w=1}^{W}\left(\begin{array}{l} -\left\|\dfrac{C(s)\,\alpha_{\theta_w}\underline{\xi}_w}{\alpha_{\omega_w}\underline{\xi}_w} + \alpha_{sf_w}\underline{\xi}_w\right\|_{L_1} val_1 \\ +\left\|\dfrac{C(s)}{\alpha_{\omega_w}\underline{\xi}_w}\cdot\dfrac{1}{c_0^T H(s)}\cdot c_0^T\right\|_{L_1}\sqrt{\dfrac{2\theta_{m_w}}{\lambda_{\min}(P_w(\underline{\xi}))\Gamma_C}}\end{array}\right)$$

Proof The state error between reference system (4.75) and system (4.72) is denoted as:

$$\underline{e}(t) = \underline{x}_{ref}(t) - \underline{x}(t)$$

In frequency domain it can be written as:

$$\underline{e}(s) = \sum_{w=1}^{W}\left(H(s)\left((1 - C(s))(\underline{r}_{2ref}(s) - \underline{r}_2(s)) + C(s)\,\underline{\tilde{r}}_{ud}(s)\right)\right) \tag{4.78}$$

Putting the values of $\underline{r}_2(s)$ and $\underline{r}_{2ref}(s)$ in (4.78) yield:

$$\underline{e}(s) = \sum_{w=1}^{W}\left(H(s)(\,(1 - C(s))\left(\alpha_{\theta_w}\underline{\xi}_w\left(\left\|\underline{x}_{ref}\right\|_{L_\infty} - \left\|\underline{x}\right\|_{L_\infty}\right)\right) + C(s)\,\underline{\tilde{r}}_{ud}(s)\,)\right)$$

or

$$\underline{e}(s) = \sum_{w=1}^{W}\left(H(s)(\,(1 - C(s))(\alpha_{\theta_w}\underline{\xi}_w\left\|\underline{e}\right\|_{L_\infty}) + C(s)\,\underline{\tilde{r}}_{ud}(s)\,)\right) \tag{4.79}$$

Equation (4.79) holds the following bound:

$$\left\|\underline{e}\right\|_{L_\infty} \le \sum_{w=1}^{W}\left(\|H(s)(1-C(s))\|_{L_1}\left\|\alpha_{\theta_w}\underline{\xi}_w\underline{e}\right\|_{L_\infty} + \|H(s)C(s)\|_{L_1}\left\|\underline{\tilde{r}}_{ud}\right\|_{L_\infty}\right) \tag{4.80}$$

The dynamics of $\underline{\tilde{x}}(t)$, viz., error between predictor state and system state $\underline{\tilde{x}}(t) = \underline{\hat{x}}(t) - \underline{x}(t)$, in frequency domain can be written as $\underline{\tilde{x}}(s) = H(s)\underline{\tilde{r}}_{ud}(s)$. Replacing this value in (4.80) provides:

$$\left\|\underline{e}\right\|_{L_\infty} \le \sum_{w=1}^{W}\left(\|H(s)\,(1-C(s))\|_{L_1}(\alpha_{\theta_w}\underline{\xi}_w)_{\max}\left\|\underline{e}\right\|_{L_\infty} + \|C(s)\|_{L_1}\left\|\underline{\tilde{x}}\right\|_{L_\infty}\right) \tag{4.81}$$

Putting value of $\left\|\underline{\tilde{x}}\right\|_{L_\infty}$ from (4.58) in the above equation yields:

$$\left\|\underline{e}\right\|_{L_\infty} \le \sum_{w=1}^{W}\left(\frac{\left\|\frac{\alpha_{\omega_w}\underline{\xi}_w k_w D(s)}{1+\alpha_{\omega_w}\underline{\xi}_w k_w D(s)}\right\|_{L_1}\sqrt{\frac{2\theta_{m_w}}{\lambda_{\min}(P_w(\underline{\xi}))\Gamma_C}}}{1-\left\|H(s)\left(1-\left(\frac{\alpha_{\omega_w}\underline{\xi}_w k_w D(s)}{1+\alpha_{\omega_w}\underline{\xi}_w k_w D(s)}\right)\right)\right\|_{L_1}(\alpha_{\theta_w}\underline{\xi}_w)_{\max}}\right) \tag{4.82}$$

This provides (4.76).

Difference between the control input to the reference system (4.74) and system (4.71) is given as:

$$\left\|\underline{u}_{ref}-\underline{u}\right\|_{L_\infty} \le \sum_{w=1}^{W}\left(\begin{array}{l} -\left\|\dfrac{C(s)\,\alpha_{\theta_w}\underline{\xi}_w}{\alpha_{\omega_w}\underline{\xi}_w}+\alpha_{sf_w}\underline{\xi}_w\right\|_{L_1}\left\|\underline{x}_{ref}-\underline{x}\right\|_{L_\infty} \\ +\left\|\dfrac{C(s)}{\alpha_{\omega_w}\underline{\xi}_w}\right\|_{L_1}\left\|\underline{\tilde{r}}_{ud}\right\|_{L_\infty} \end{array}\right)$$

or

$$\left\|\underline{u}_{ref}-\underline{u}\right\|_{L_\infty} \le \sum_{w=1}^{W}\left(\begin{array}{l} -\left\|\dfrac{C(s)\,\alpha_{\theta_w}\underline{\xi}_w}{\alpha_{\omega_w}\underline{\xi}_w}+\alpha_{sf_w}\underline{\xi}_w\right\|_{L_1}\left\|\underline{e}\right\|_{L_\infty} \\ +\left\|\dfrac{C(s)}{\alpha_{\omega_w}\underline{\xi}_w}\right\|_{L_1}\left\|\underline{\tilde{r}}_{ud}\right\|_{L_\infty} \end{array}\right) \tag{4.83}$$

Again, it can be stated that:

$$\left\|\frac{C(s)}{\alpha_{\omega_w}\underline{\xi}_w}\right\|_{L_1}\left\|\underline{\tilde{r}}_{ud}\right\|_{L_\infty} = \frac{C(s)}{\alpha_{\omega_w}\underline{\xi}_w}\cdot\frac{1}{c_0^T H(s)}\cdot c_0^T H(s)\cdot\underline{\tilde{r}}_{ud}(s) = \frac{C(s)}{\alpha_{\omega_w}\underline{\xi}_w}\cdot\frac{1}{c_0^T H(s)}\cdot c_0^T\cdot\underline{\tilde{x}}(s)$$

Putting this on (4.83) yields:

$$\left\|\underline{u}_{ref}-\underline{u}\right\|_{L_\infty}\leq\sum_{w=1}^{W}\left(\begin{array}{l}-\left\|\dfrac{C(s)\,\alpha_{\theta_w}\underline{\xi}_w}{\alpha_{\omega_w}\underline{\xi}_w}+\alpha_{sf_w}\underline{\xi}_w\right\|_{L_1}\left\|\underline{e}\right\|_{L_\infty}\\+\left\|\dfrac{C(s)}{\alpha_{\omega_w}\underline{\xi}_w}\cdot\dfrac{1}{c_0^TH(s)}\cdot c_0^T\right\|_{L_1}\left\|\underline{\tilde{x}}\right\|_{L_\infty}\end{array}\right)\tag{4.84}$$

Putting value of $\left\|\underline{\tilde{x}}\right\|_{L_\infty}$ from (4.58) in (4.84) provides:

$$\left\|\underline{u}_{ref}-\underline{u}\right\|_{L_\infty}\leq\sum_{w=1}^{W}\left(\begin{array}{l}-\left\|\dfrac{C(s)\,\alpha_{\theta_w}\underline{\xi}_w}{\alpha_{\omega_w}\underline{\xi}_w}+\alpha_{sf_w}\underline{\xi}_w\right\|_{L_1}\left\|\underline{e}\right\|_{L_\infty}\\+\left\|\dfrac{C(s)}{\alpha_{\omega_w}\underline{\xi}_w}\cdot\dfrac{1}{c_0^TH(s)}\cdot c_0^T\right\|_{L_1}\sqrt{\dfrac{2\theta_{m_w}}{\lambda_{\min}(P_w(\underline{\xi}))\Gamma_C}}\end{array}\right)\tag{4.85}$$

This proves (4.77). Therefore, the theorem is proved. ■

Theorem 4.6 shows that, the system states and control signal follow respectively the states and control signal of the reference system model. The state error and control signal error between them is bounded and tends to zero with increase of the adaptation gain Γ_C.

To design reference system model, the unknown terms α_{ω_w},α_{θ_w} and α_{σ_w} have to be designed properly. Hence, an ideal system model has to be considered as:

$$\dot{\underline{x}}_{ideal}=\sum_{w=1}^{W}\left(A_{m_w}(\underline{\xi})\underline{x}_{ideal}(t)+b_w(\underline{\xi})k_{g_w}\underline{r}(t)\right)\tag{4.86}$$

$$y_{ideal}(t)=\sum_{w=1}^{W}c_w(\underline{\xi})x_{ideal}(t)\tag{4.87}$$

with ideal control input given as:

$$\underline{u}_{ideal}(t)=\sum_{w=1}^{W}\left(\frac{k_{g_w}\underline{r}(t)-\theta_w(t)\left\|\underline{x}_{ideal}(t)\right\|_{L_\infty}-\sigma_w(t)}{\omega_w}-K_w^T\underline{x}_{ideal}(t)\right)\tag{4.88}$$

Lemma 4.1 *The ideal system model (4.86) will be stable if the feed-forward gain k_{g_w} for all W zones and input to the system $r(t)$ are finite and bounded.*

Proof In frequency domain, the ideal system model (4.86) takes the form:

$$x_{ideal}(s)=\sum_{w=1}^{W}H_{ideal}(s)k_{g_w}r(s)\tag{4.89}$$

where $H_{ideal}(s) = \sum_{w=1}^{W} \left(sI - A_{m_w}\underline{\xi}_w\right)^{-1} b_w\underline{\xi}_w$ is a strictly proper closed loop stable transfer function. $r(s)$ is the Laplace transform of reference input $r(t)$.

Equation (4.89) holds the following bound:

$$\|x_{ideal}\|_{L_\infty} \le \sum_{w=1}^{W} \|H_{ideal}(s)\|_{L_1} \left\|k_{g_w}\right\|_{L_1} \|r\|_{L_\infty} \tag{4.90}$$

Here $\|r\|_{L_\infty} = \max_{i=1,2,\ldots,n,t\ge0} |r_i|$. Therefore, if k_{g_w} and reference input to the system $r(t)$ are finite for each and every zone, then for whole region $\|x_{ideal}\|_{L_\infty}$ is also finite. The stability of the ideal model is evaluated in terms of fuzzy Lyapunov stability. Let, the fuzzy Lyapunov candidate is considered as:

$$V_{ideal}(t) = \sum_{w=1}^{W} \frac{1}{2} x_{ideal}^T(t) P_w(\underline{\xi}) x_{ideal}(t) \tag{4.91}$$

The time derivative of (4.91) provides:

$$\dot{V}_{ideal}(t) = \sum_{w=1}^{W} \frac{1}{2}\left(\dot{x}_{ideal}^T(t) P_w(\underline{\xi}) x_{ideal}(t) + x_{ideal}^T(t) P_w(\underline{\xi}) \dot{x}_{ideal}(t)\right) \tag{4.92}$$

The ideal system will be stable if the following condition holds.

$\dot{V}_{ideal}(t) = \sum_{w=1}^{W} -x_{ideal}^T(t) Q_{m_w}(\underline{\xi}) x_{ideal}(t) \le 0$, where $Q_{m_w}(\underline{\xi})$ is positive definite matrix.

It in turn provides $\dot{V}_{ideal}(t)$ is negative semi definite, viz., $V_{ideal}(t) \le V_{ideal}(0)$. Now, as k_{g_w} and $r(t)$ are finite. So, $V_{ideal}(t)$ is bounded and non-increasing and $V_{ideal}(0)$, $\dot{V}_{ideal}(t)$ and $\ddot{V}_{ideal}(t)$ are also bounded. Therefore, from Lyapunov stability criterion and Barbalat's lemma it is proved that the ideal system (4.86) is stable. ■

Remarks 4.1 The input (4.88), given to the ideal system model (4.86), is provided to the reference system model (4.73) after passing through low-pass filter (4.74). Hence, the input to the reference system model and ideal system model are different. However, the reference system model is designed as a way that it follows the ideal system model. Now, as the ideal system model is stable as proved in Lemma 4.1 therefore, the reference system model provides stable performance. Equations (4.76) and (4.77) show that, the actual system model follows the stable reference system model. Hence, it can be stated that, the system performance is also stable.

The designed fuzzy PDC-L_1 adaptive control methodology is employed on four systems to show its effectiveness as elaborated in the next section.

4.6 Simulation Case Studies

In this chapter, the fuzzy PDC-L_1 adaptive controller is employed on four systems as described in the Chap. 2 in simulation environment. The effectiveness of this method is compared with basic L_1 adaptive controller and lbest HS-L_1 adaptive controller in the basis of integral absolute error (IAE). The detailed architecture as well as implementation of basic L_1 adaptive controller and lbest HS-L_1 adaptive controller are provided in Chaps. 2 and 3 respectively. Therefore, the implementation of these two methods are briefly described in this section. Whereas, the implementation procedure of fuzzy PDC-L_1 adaptive controller is described elaborately in this section.

***Basic L_1 (Basic L_1) adaptive controller*:** In this case, the parameters of L_1 adaptive controller $\underline{Z} = [k\ |k_g|\ K\ |\ \Gamma_C\ |\ \hat{\omega}\ |\ \underline{\hat{\theta}}\ |\ \underline{\hat{\sigma}}]$ are chosen arbitrarily within the ranges obtained from the mathematical calculation of L_1 norm condition.

***lbest HS based L_1 (lbest HS-L_1) adaptive controller*:** In Chap. 3, it has been shown that, the lbest HS-L_1 adaptive controller performs superior than the PSO-L_1 adaptive and basic HS-L_1 adaptive controller. Therefore, in this chapter, only the performance of lbest HS-L_1 adaptive controller is taken for the comparison purpose. The parameters of L_1 adaptive controller $\underline{Z} = [k\ |k_g|\ K\ |\ \Gamma_C\ |\ \hat{\omega}\ |\ \underline{\hat{\theta}}\ |\ \underline{\hat{\sigma}}]$ are chosen as the CSV of the lbest HS-L_1 adaptive controller. The ranges of the parameter values are selected from the mathematical calculation of the L_1 norm condition. The optimized parameter setting of L_1 adaptive controller are obtained from the lbest HS algorithm by using 30 particles and 200 iteration of it. This whole process runs for 10 times each and minimum, viz., best, average, standard deviation of IAEs are tabulated.

***Fuzzy PDC-L_1 (Fuzzy PDC-L_1) adaptive controller*:** In this method, the nonlinear system with unknown constant, time varying uncertainties and time varying disturbances is designed by combining linear fuzzy systems. The premise part $\underline{\xi}_w$ is designed for each *wth* zone of total W number of rules. W number of linear consequence rule based systems are fuzzy blended to develop the overall system model (4.7). With the same premise part $\underline{\xi}_w$ and augmenting W number of linear consequence rules, the predictor model is designed (4.18). Difference between the predictor state and system state of each zone is utilized to estimate the unknown constant, time varying uncertainties and time varying disturbances present at that zone. Then, these quantities are adapted following the adaptation laws (4.19)–(4.21) formulated with the same premise variable and the adaptive estimates of W number of zones are combined to obtain the overall estimation. L_1 adaptive control law is formulated for each zone and the consequences are accumulated to form the whole L_1 adaptive control action (4.30). Use of the high adaptation gain to adapt the unknown constant, uncertainties, disturbances includes high frequency into the control channel which is adequately nullified with the help of low-pass filter. The premise part used to design the system is utilized to develop the zone-wise filter models for W number of zones. These zone-wise filter models are fuzzy augmented to produce the overall low-pass filter model (4.52). For W number of zones, fuzzy PDC controllers are designed with the premise part $\underline{\xi}_w$ and then augmented to generate the whole fuzzy PDC control action (4.31). Filtered fuzzy L_1 adaptive control law is combined with fuzzy PDC control law to develop the overall fuzzy PDC-L_1 adaptive controller (4.41). To design different components of the L_1 adaptive controller, the fuzzy systems are formulated with five number of triangular membership functions for each input. The positions of the output singletons of all the fuzzy systems are obtained from the designed adaptation laws. Then, linear consequence fuzzy systems are accumulated to formulate the components of the L_1 adaptive controller.

It is to be noted that, in a similar fashion of Chap. 2, in case of the controllers, designed utilizing any stochastic optimization and metaheuristics techniques, the statistical measures of performances, viz., best, average and standard deviation (std. dev.) of IAEs are taken. Whereas, for conventional adaptive controllers as well as fuzzy adaptive systems, these statistical measures are not required to be calculated. Thus, keeping that in mind, the table for comparative studies of different controllers in each case study is prepared.

These methodologies are employed on four systems, viz., Duffing's oscillatory system, nonlinear spring mass damper system, 4th order inverted pendulum with cart system and twin rotor MIMO system. The descriptions of these systems are elaborated in the Chap. 2, therefore omitted in this chapter.

4.6.1 Case Study-I: Duffing's Oscillatory System

The dynamics of Duffing's oscillatory system with disturbance same as (2.63) is given as [7]:

$$\left.\begin{aligned} \dot{x}_1 &= x_2 \\ \dot{x}_2 &= -0.1x_2 - x_1^3 + 12\cos(t) + u(t) + d(t) \\ y &= x_1 \end{aligned}\right\} \tag{4.93}$$

The disturbance, shown in Fig. 2.2, is subjected to the system at the time of designing the controllers. Tuned controllers are then tested with 10 dB white Gaussian noise as depicted in Fig. 2.3 without further tuning. The control objective is to track the desired trajectory, described in Sect. 2.6.1, quickly and properly. The performance evaluation is tabulated in Table 4.1 in terms of integral absolute error (IAE). The responses of the system with disturbance and corresponding control signal with error signal plots for basic L_1 adaptive, lbest HS-L_1 adaptive and fuzzy PDC-L_1 adaptive controller are depicted in Figs. 4.3, 4.4, 4.5 respectively. The results show that, for fuzzy PDC-L_1 adaptive controller the system performance is better than the other controllers tested. The transient performance of this controller is fast due to proper designing of fuzzy L_1 adaptive controller and satisfactory tracking performance is also obtained by augmenting fuzzy PDC controller. The tuned controllers are then subjected to the 10 dB white Gaussian noise of higher frequency and the performances of the system are shown in Figs. 4.6, 4.7, 4.8. It can be stated from

Table 4.1 Comparative study of different control methodologies for case study-I (Duffing's oscillatory system)

Control methodologies	IAE value			
	For disturbance			For white Gaussian noise
	Best	Average	Std. Dev.	
Basic L_1 adaptive controller	0.2347	–	–	0.4559
lbest HS-L_1 adaptive controller	0.0118	0.0119	0.0003	0.0186
Fuzzy PDC-L_1 adaptive controller	0.0115	–	–	0.0157

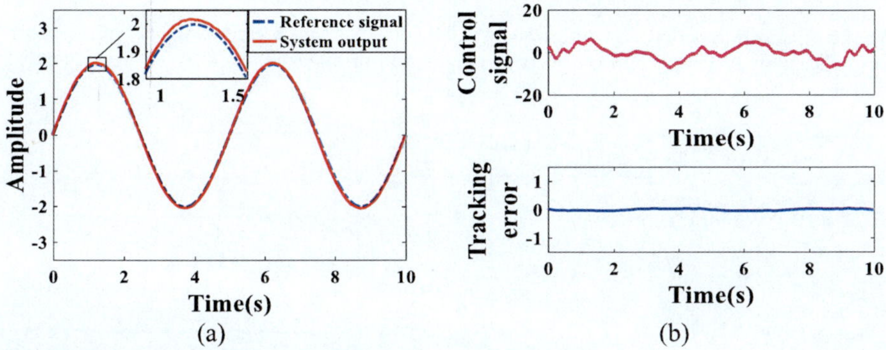

Fig. 4.3 Evaluation period **a** system response and corresponding **b** control signal with tracking error of case study-I (Duffing's oscillatory system) with disturbance for basic L_1 adaptive controller

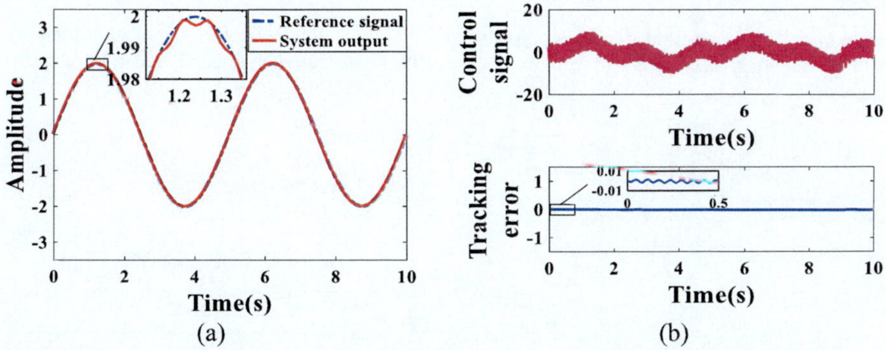

Fig. 4.4 Evaluation period **a** system response and corresponding **b** control signal with tracking error of case study-I (Duffing's oscillatory system) with disturbance for lbest HS-L_1 adaptive controller

the obtained results that, the fuzzy PDC-L_1 adaptive controller can efficiently nullifies the disturbances, uncertainties and produces satisfactory system performance by tackling nonlinearities.

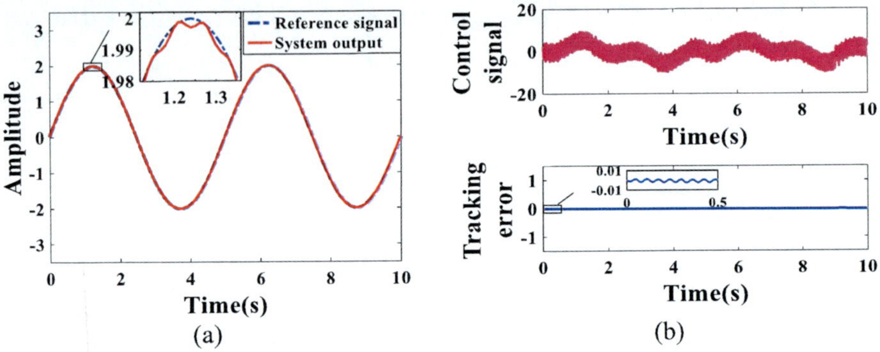

Fig. 4.5 Evaluation period **a** system response and corresponding **b** control signal with tracking error of case study-I (Duffing's oscillatory system) with disturbance for fuzzy PDC-L_1 adaptive controller

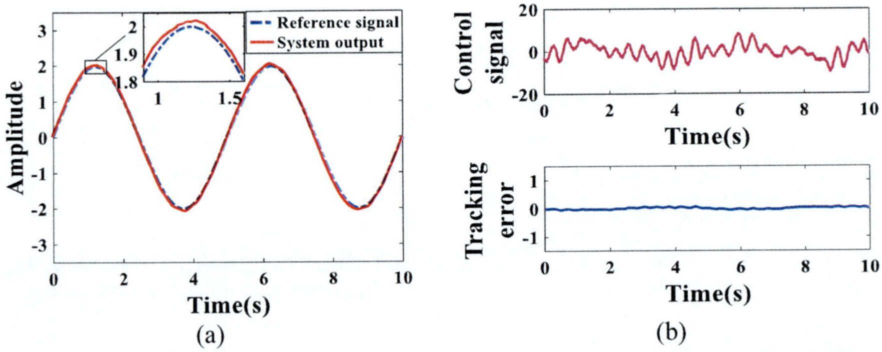

Fig. 4.6 Evaluation period **a** system response and corresponding **b** control signal with tracking error of case study-I (Duffing's oscillatory system) with white Gaussian noise for basic L_1 adaptive controller

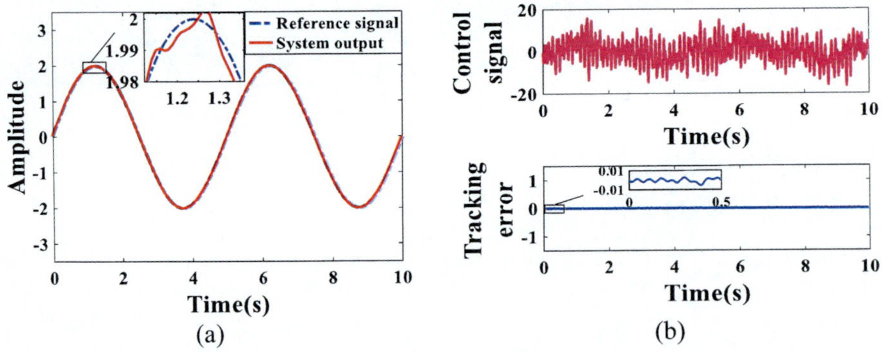

Fig. 4.7 Evaluation period **a** system response and corresponding **b** control signal with tracking error of case study-I (Duffing's oscillatory system) with white Gaussian noise for lbest HS-L_1 adaptive controller

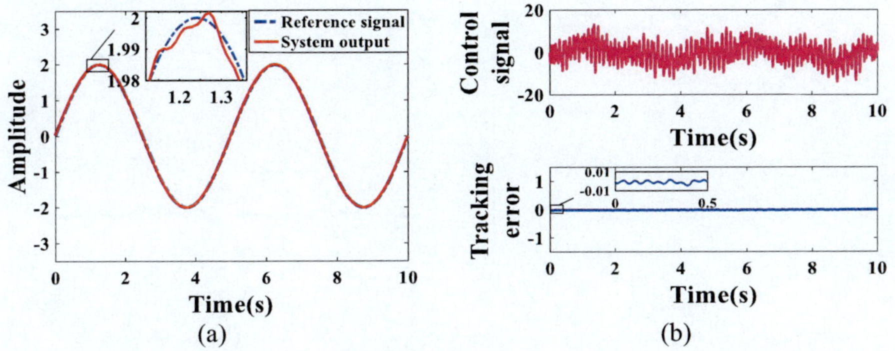

Fig. 4.8 Evaluation period **a** system response and corresponding **b** control signal with tracking error of case study-I (Duffing's oscillatory system) with white Gaussian noise for fuzzy PDC-L_1 adaptive controller

4.6.2 Case Study-II: Nonlinear Spring Mass Damper System

In this case study, the fuzzy PDC-L_1 adaptive controller is employed on a nonlinear spring mass damper system [7], described in Sect. 2.6.2, similar as (2.64):

$$\left.\begin{aligned} \dot{x}_1 &= x_2 \\ \dot{x}_2 &= -0.02x_1 - 0.67x_1^3 - 0.1x_2^3 + u(t) + d(t) \\ y &= x_1 \end{aligned}\right\} \tag{4.94}$$

The control objective is to track the desired trajectory when disturbance as shown in Fig. 2.2 is applied to the system. Then, the tuned controllers are subjected to the noise with higher frequency as provided in Fig. 2.3 without further tuning to observe robustness of the controllers. The integral absolute error (IAE) of the tracking responses are given in Table 4.2. The system responses with disturbance for basic L_1 adaptive, lbest HS-L_1 adaptive and fuzzy PDC-L_1 adaptive controller are depicted in Figs. 4.9(a), 4.10(a) and 4.11(a) respectively. The corresponding control effort with

Table 4.2 Comparative study of different control methodologies for case study-II (nonlinear spring mass damper)

Control methodologies	IAE value			
	For disturbance			For white Gaussian noise
	Best	Average	Std. Dev.	
Basic L_1 adaptive controller	0.0813	–	–	0.2694
lbest HS-L_1 adaptive controller	0.0140	0.0207	0.0034	0.0359
Fuzzy PDC-L_1 adaptive controller	0.0095	–	–	0.0116

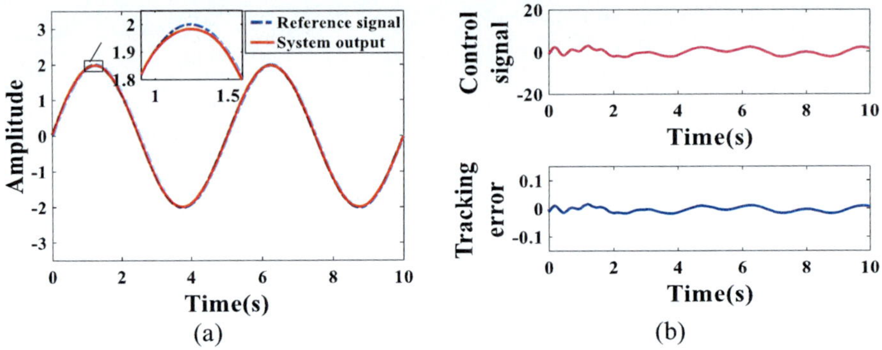

Fig. 4.9 Evaluation period **a** system response and corresponding **b** control signal with tracking error of case study-II (nonlinear spring mass damper) with disturbance for basic L_1 adaptive controller

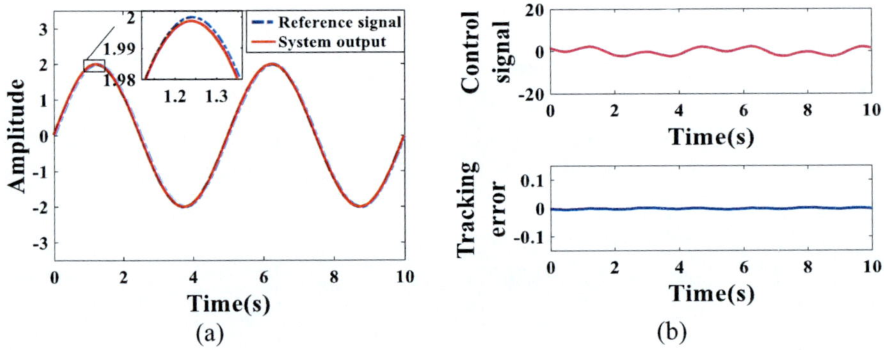

Fig. 4.10 Evaluation period **a** system response and corresponding **b** control signal with tracking error of case study-II (nonlinear spring mass damper) with disturbance for lbest HS-L_1 adaptive controller

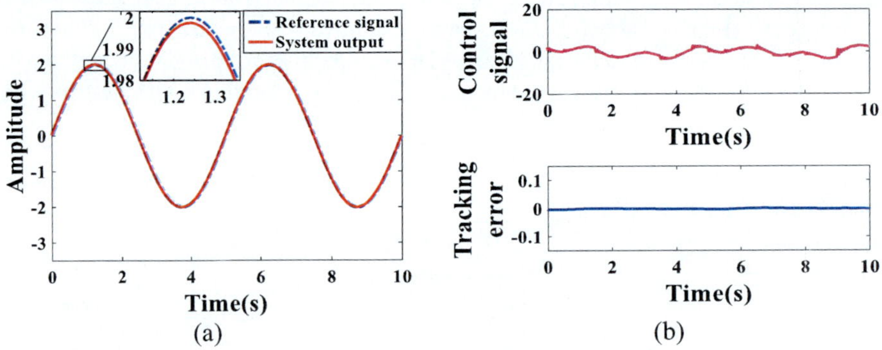

Fig. 4.11 Evaluation period **a** system response and corresponding **b** control signal with tracking error of case study-II (nonlinear spring mass damper) with disturbance for fuzzy PDC-L_1 adaptive controller

error signals are provided in Figs. 4.9(b), 4.10(b) and 4.11(b) respectively. From results it is evident that, the IAE value for the fuzzy PDC-L_1 adaptive controller is less than the other methods tested. When the tuned controllers are subjected to the 10 dB white Gaussian noise then, the performances of the system are portrayed in Figs. 4.12, 4.13 and 4.14 for above mentioned three controllers respectively. Results show that, when the tuned controllers are subjected to the unknown time varying noise suddenly then, the fuzzy PDC-L_1 adaptive controller performs well enough than the other controllers tested.

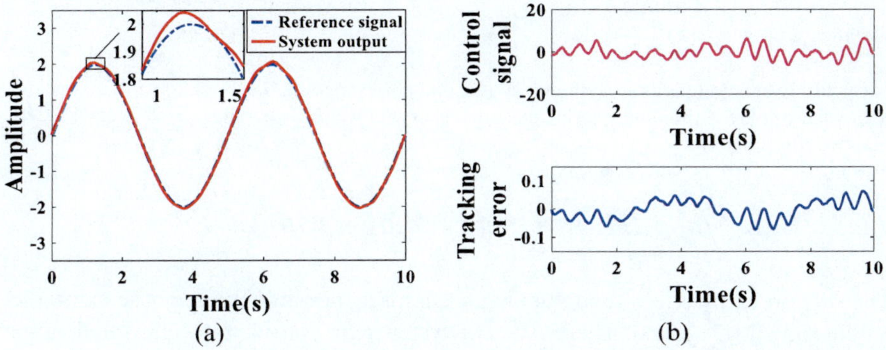

Fig. 4.12 Evaluation period **a** system response and corresponding **b** control signal with tracking error of case study-II (nonlinear spring mass damper) with white Gaussian noise for basic L_1 adaptive controller

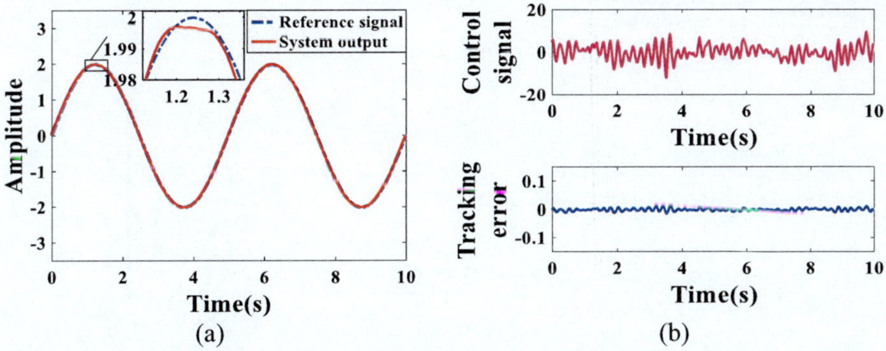

Fig. 4.13 Evaluation period **a** system response and corresponding **b** control signal with tracking error of case study-II (nonlinear spring mass damper) with white Gaussian noise for lbest HS-L_1 adaptive controller

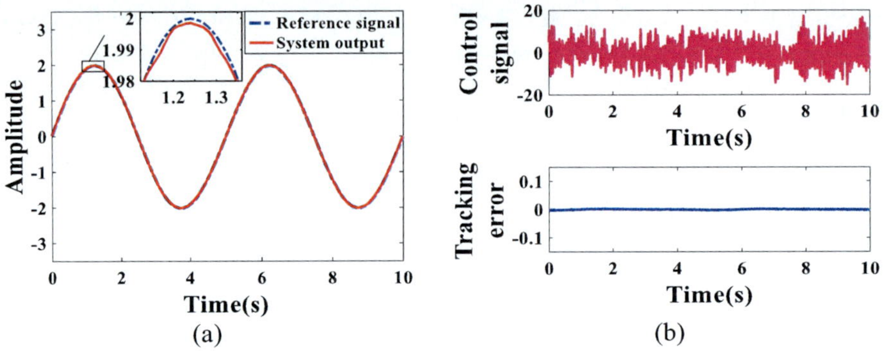

Fig. 4.14 Evaluation period **a** system response and corresponding **b** control signal with tracking error of case study-II (nonlinear spring mass damper) with white Gaussian noise for fuzzy PDC-L_1 adaptive controller

4.6.3 *Case Study-III: Inverted Pendulum with Cart*

The 4th order inverted pendulum with cart is a non-minimum phase, unstable, nonlinear, SIMO system. These types of systems are considered as the challenging problem from the control engineering point of view. Fuzzy logic is used in different literatures to control unstable non-minimum phase inverted pendulum with cart system [8, 9]. Fuzzy PDC controller is employed to tackle the nonlinearity present in the system [8]. However, at the same time to eliminate uncertainties and disturbances as well as nonlinearity present in this system, the fuzzy PDC-L_1 adaptive controller is utilized in this case study.

The dynamics of this system is provided in (2.65) as [10]:

$$\left.\begin{aligned} \dot{x}_1 &= x_2 \\ \dot{x}_2 &= \frac{-mg\,\cos x_3\,\sin x_3 + ml\sin x_3\, x_4^2 + u}{M + m\sin^2 x_3} + d_1(t) \\ \dot{x}_3 &= x_4 \\ \dot{x}_4 &= \frac{-ml\,\cos x_3\,\sin x_3\, x_4^2 + (M+m)g\,\sin x_3 - u\cos x_3}{Ml + ml\sin^2 x_3} + d_2(t) \end{aligned}\right\} \tag{4.95}$$

All units are provided in S.I. This system is elaborated in Sect. 2.6.3 therefore, omitted in this section. The control objective is to track the desired trajectory $y = 0.1$ *unit* for 10 s duration with 0.01 s time step and the integral absolute error (IAE) is calculated and tabulated in Table 4.3. At first, the system performance is evaluated without disturbances. The performances of the system are plotted in Figs. 4.15, 4.16, 4.17 for basic L_1 adaptive, lbest HS-L_1 adaptive and fuzzy PDC-L_1 adaptive controller respectively. Results show that, the fuzzy PDC-L_1 adaptive controller performs better than other controllers tested in this case study. As, this

Table 4.3 Comparative study of different control methodologies for case study-III (inverted pendulum with cart)

Control methodologies	IAE value			
	Best	Average	Std. Dev.	With disturbance
Basic L_1 adaptive controller	0.0465	–	–	0.0502
lbest HS-L_1 adaptive controller	0.0009	0.0033	0.0020	0.0078
Fuzzy PDC-L_1 adaptive controller	$2.13*10^{-05}$	–	–	0.0021

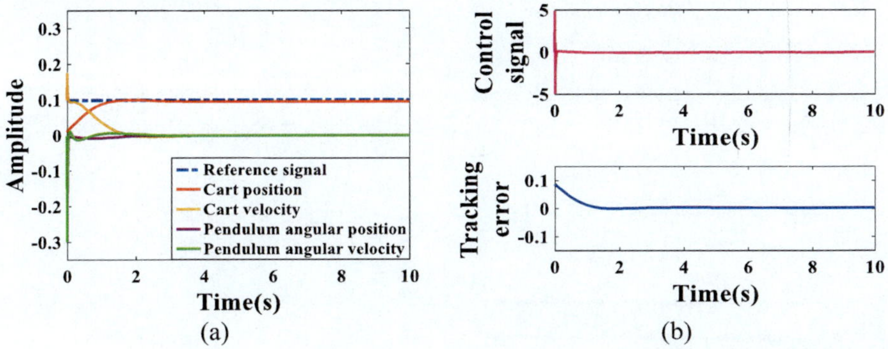

Fig. 4.15 Evaluation period **a** system response and corresponding **b** control signal with tracking error of case study-III (inverted pendulum with cart) without disturbance for basic L_1 adaptive controller

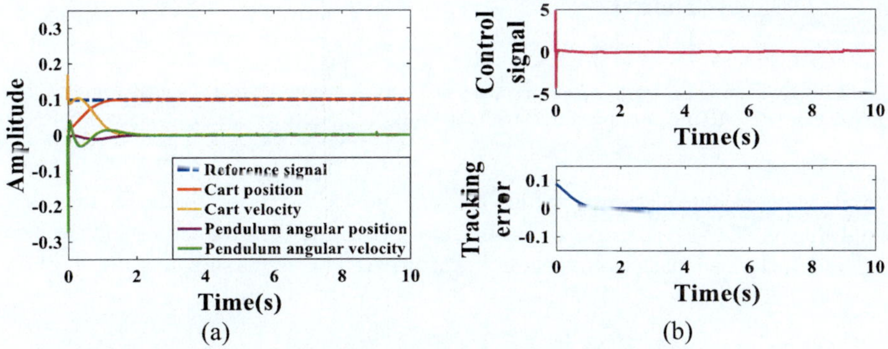

Fig. 4.16 Evaluation period **a** system response and corresponding **b** control signal with tracking error of case study-III (inverted pendulum with cart) without disturbance for lbest HS-L_1 adaptive controller

is a non-minimum phase system therefore it is shown that, it's all states are stabilized. Now, these tuned controllers are subjected to the disturbances as described in Sect. 2.6.3 and the respective results are portrayed in Figs. 4.18, 4.19, 4.20. In case of basic L_1 adaptive controller and lbest HS-L_1 adaptive controller the disturbances

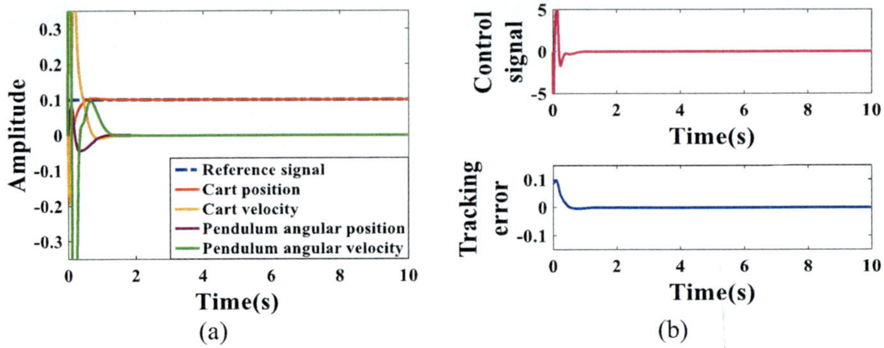

Fig. 4.17 Evaluation period **a** system response and corresponding **b** control signal with tracking error of case study-III (inverted pendulum with cart) without disturbance for fuzzy PDC-L_1 adaptive controller

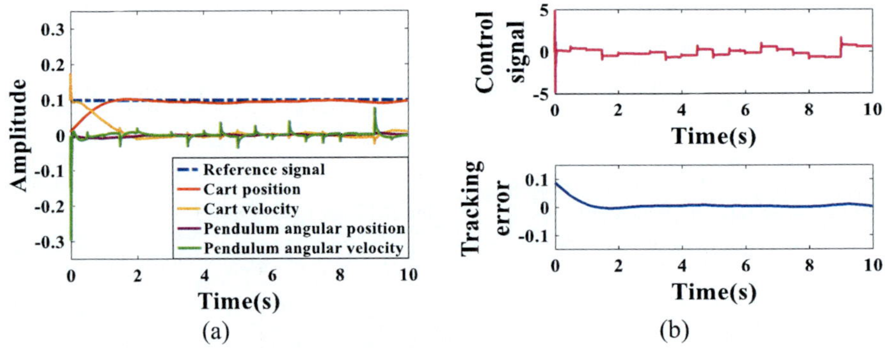

Fig. 4.18 Evaluation period **a** system response and corresponding **b** control signal with tracking error of case study-III (inverted pendulum with cart) with disturbance for basic L_1 adaptive controller

persist in the system as shown in Figs. 4.18(a) and 4.19(a) respectively. On the other hand, fuzzy PDC-L_1 adaptive controller eliminates the disturbance efficiently and provides fast transient as well as better tracking performance as shown in Fig. 4.20(a).

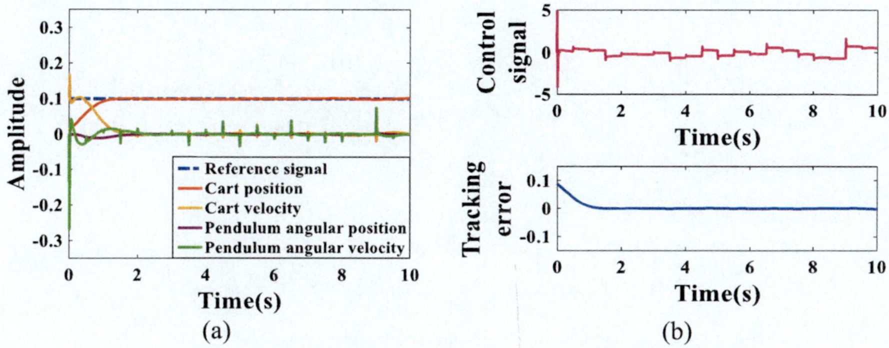

Fig. 4.19 Evaluation period **a** system response and corresponding **b** control signal with tracking error of case study-III (inverted pendulum with cart) with disturbance for lbest HS-L_1 adaptive controller

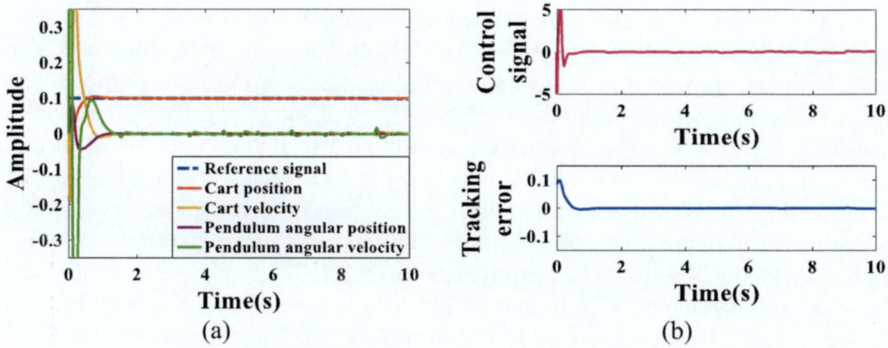

Fig. 4.20 Evaluation period **a** system response and corresponding **b** control signal with tracking error of case study-III (inverted pendulum with cart) with disturbance for fuzzy PDC-L_1 adaptive controller

4.6.4 Case Study-IV: Twin Rotor MIMO System

Different types of fuzzy controllers are designed in different literatures to control twin rotor MIMO system (TRMS) [11, 12]. Though, to tackle uncertainties and disturbances as well as nonlinearities, cross-coupling, in this case, fuzzy PDC-L_1 adaptive controller is employed to control TRMS.

The dynamics of the TRMS similar as (2.66) can be given as [13]:

$$\dot{x}_1 = x_2$$

$$\dot{x}_2 = \frac{1}{I_1}\begin{pmatrix} a_1 x_5^2 + b_1 x_5 - M_g \sin x_1 \\ -K_{gy} b_1 \cos x_1 x_4 x_5 + 0.016 \sin 2x_1 x_4^2 \\ -K_{gy} a_1 \cos x_1 x_4 x_5^2 - B_{1\phi} x_2 \end{pmatrix}$$

$$\dot{x}_3 = x_4$$

$$\dot{x}_4 = \frac{1}{I_2}\begin{pmatrix} a_2 x_6^2 + b_2 x_6 - 1.75(K_c a_1 x_5^2 + K_c b_1 x_5) \\ +0.375(K_c a_1 x_5^2 + K_c b_1 x_5) \exp^{-0.5t} - B_{1\varphi} x_4 \end{pmatrix} \tag{4.96}$$

$$\dot{x}_5 = -\frac{T_{10} x_5}{T_{11}} + \frac{K_{11} u_1}{T_{11}}$$

$$\dot{x}_6 = -\frac{T_{20} x_6}{T_{21}} + \frac{K_{22} u_2}{T_{21}}$$

Description of the system is elaborated in Sect. 2.6.4. The control objective is to track the desired variable step trajectory for 100 s duration with 0.01 s time step. The tracking error is provided in Table 4.4 in terms of IAE for comparison purpose. The system responses without disturbance for basic L_1 adaptive, lbest HS-L_1 adaptive and fuzzy PDC-L_1 adaptive controller are provided in Figs. 4.21(a), 4.22(a) and 4.23(a) respectively. Results show that, the fuzzy PDC-L_1 adaptive control method as shown in Fig. 4.23(a) provides fast transient responses as well as satisfactory tracking performances for both pitch and yaw rotor. The control efforts with error signals for basic L_1 adaptive, lbest HS-L_1 adaptive and fuzzy PDC-L_1 adaptive controller are depicted in Fig. 4.21(b), 4.22(b) and 4.23(b) respectively. Then, the pitch rotor is subjected to the Disturbance-I as described in Sect. 2.6.4 and the performances of the controllers without further tuning for both the rotors are provided in Table 4.5. In case of the fuzzy PDC-L_1 adaptive control method, by comparing the tabulated results of Tables 4.4 and 4.5, it is evident that, the yaw rotor performances do not vary when disturbance is subjected to the pitch rotor. Hence, it can be stated that, the fuzzy PDC-L_1 adaptive controller can provide decoupling performance for cross-coupled system efficiently. The system responses with some zoomed portion and control signal with error signal plots for basic L_1 adaptive, lbest HS-L_1 adaptive

Table 4.4 Comparative study of different control methodologies for case study-IV (twin rotor MIMO system) without disturbance

Control methodologies	IAE value without disturbance					
	Pitch angle			Yaw angle		
	Best	Average	Std. Dev.	Best	Average	Std. Dev.
Basic L_1 adaptive controller	1.4914	–	–	4.0749	–	–
lbest HS-L_1 adaptive controller	0.6037	1.3033	0.5577	2.6873	3.7344	0.5559
Fuzzy PDC-L_1 adaptive controller	0.4434	–	–	1.6714	–	–

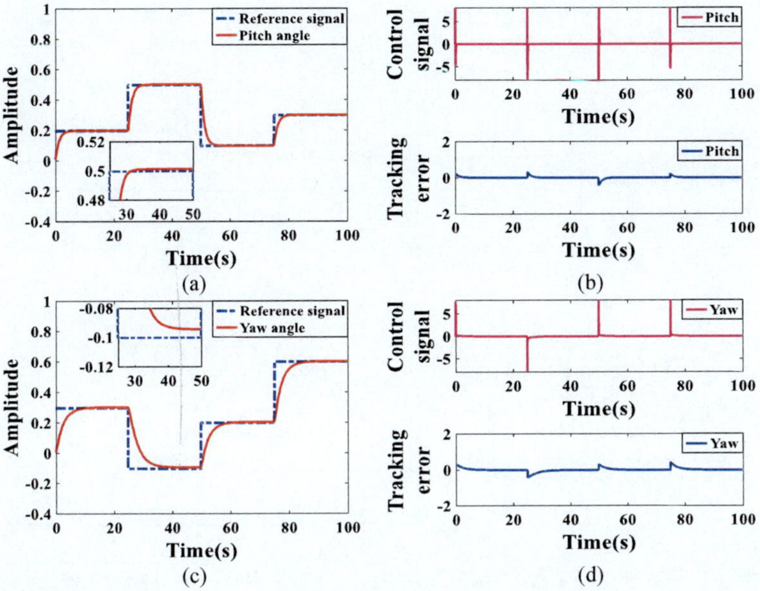

Fig. 4.21 Evaluation period **a** pitch angle and corresponding **b** control signal with tracking error, **c** yaw angle and corresponding **d** control signal with tracking error of case study-IV (twin rotor MIMO system) without disturbance for basic L_1 adaptive controller

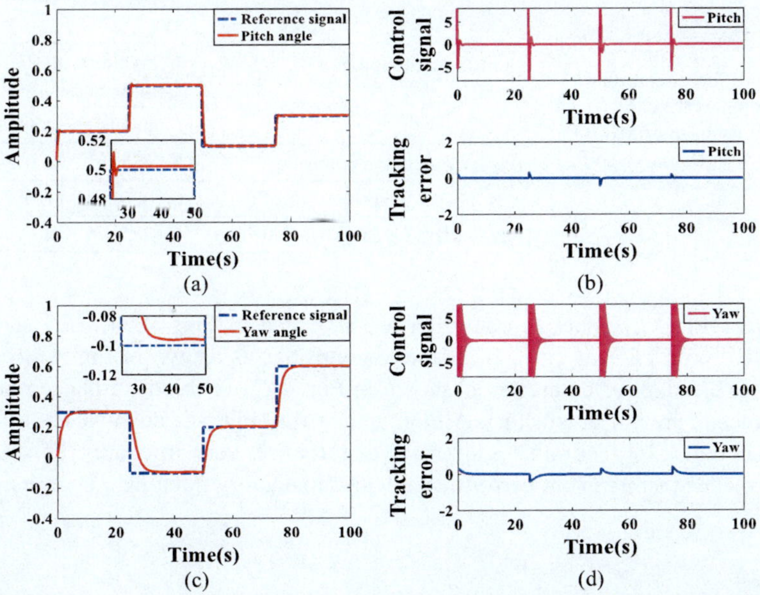

Fig. 4.22 Evaluation period **a** pitch angle and corresponding **b** control signal with tracking error, **c** yaw angle and corresponding **d** control signal with tracking error of case study-IV (twin rotor MIMO system) without disturbance for lbest HS-L_1 adaptive controller

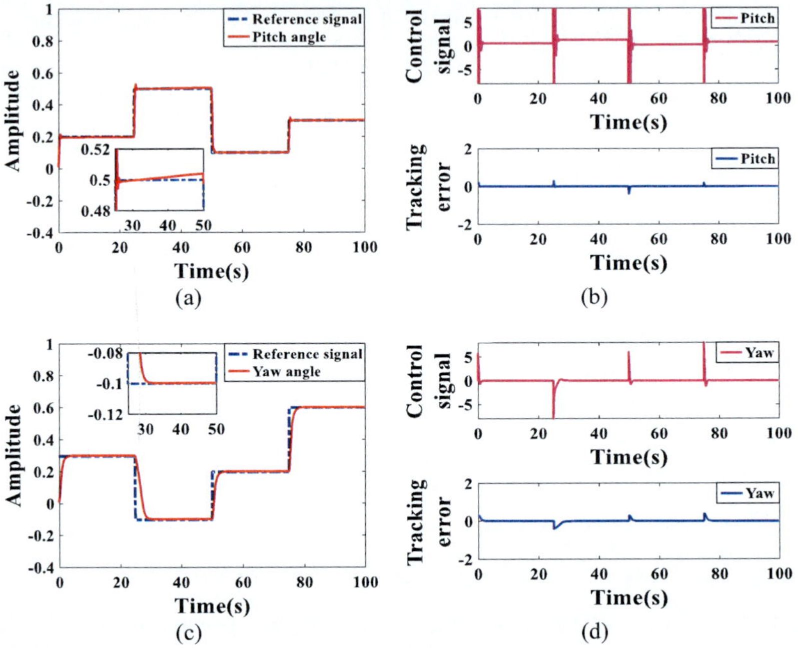

Fig. 4.23 Evaluation period **a** pitch angle and corresponding **b** control signal with tracking error, **c** yaw angle and corresponding **d** control signal with tracking error of case study-IV (twin rotor MIMO system) without disturbance for fuzzy PDC-L_1 adaptive controller

Table 4.5 Comparative study of different control methodologies for case study-IV (twin rotor MIMO system) with Disturbance-I

Control methodologies	IAE value with Disturbance-I	
	Pitch angle	Yaw angle
Basic L_1 adaptive controller	1.5655	4.0783
lbest HS-L_1 adaptive controller	0.7020	2.6873
Fuzzy PDC-L_1 adaptive controller	0.4585	1.6714

and fuzzy PDC-L_1 adaptive controller are provided in Figs. 4.24, 4.25 and 4.26 respectively. In case of basic L_1 adaptive controller as shown in Fig. 4.24(a) and lbest HS-L_1 adaptive controller as shown in Fig. 4.25(a), the disturbance rejection phenomenon are not satisfactory. Although the fuzzy PDC-L_1 adaptive controller as shown in Fig. 4.26(a) eliminates unknown disturbances very efficiently. This method also provides fast transient performance with satisfactory tracking responses.

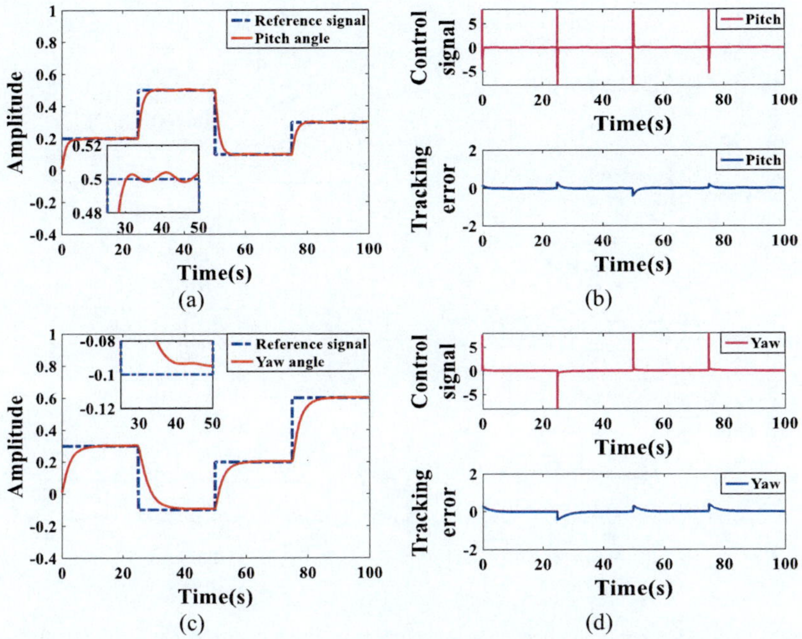

Fig. 4.24 Evaluation period **a** pitch angle and corresponding **b** control signal with tracking error, **c** yaw angle and corresponding **d** control signal with tracking error of case study-IV (twin rotor MIMO system) with Disturbance-I for basic L_1 adaptive controller

4.7 Summary

Basic L_1 adaptive controller can eliminate time varying uncertainties and disturbance well though fails to tackle system with nonlinearities. On the other hand, fuzzy PDC controller can provide good tracking performances for the nonlinear system but its performance degrades when uncertainties and disturbances include into the system. Again, fuzzy logic can be utilized to approximate different systems and controllers accurately. To ensure the advantages of both L_1 adaptive controller and fuzzy PDC controller, in this chapter, a fuzzy PDC-L_1 adaptive controller is designed. The nonlinear system is considered as the combination of linear consequence rule based fuzzy systems. Zone-wise L_1 adaptive controller components, viz., predictor, L_1 adaptation law, L_1 control law and low-pass filter are designed utilizing same premise part used to formulate the system. The conventional state feedback controller of the basic L_1 adaptive controller is replaced with fuzzy PDC controller. The linear consequence rule based fuzzy systems of respective components are combined to develop the overall models of those components. To design the components of fuzzy PDC-L_1 adaptive controller, the positions of singletons are formulated following the adaptation laws developed from the fuzzy Lyapunov

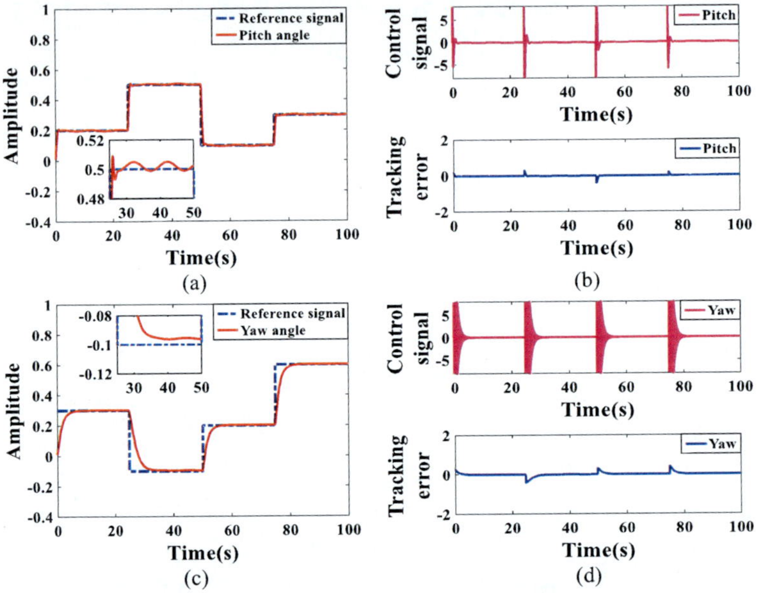

Fig. 4.25 Evaluation period **a** pitch angle and corresponding **b** control signal with tracking error, **c** yaw angle and corresponding **d** control signal with tracking error of case study-IV (twin rotor MIMO system) with Disturbance-I for lbest HS-L_1 adaptive controller

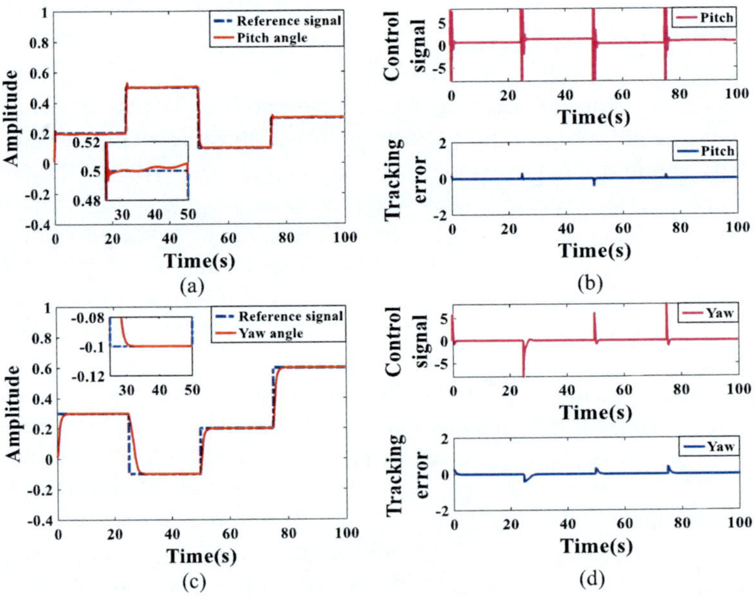

Fig. 4.26 Evaluation period **a** pitch angle and corresponding **b** control signal with tracking error, **c** yaw angle and corresponding **d** control signal with tracking error of case study-IV (twin rotor MIMO system) with Disturbance-I for fuzzy PDC-L_1 adaptive controller

stability criteria. Overall stability of this methodology is also guaranteed. The transient and steady state performances are analyzed. This controller is employed on four systems in simulation to show its effectiveness over basic L_1 adaptive controller and lbest HS-L_1 adaptive controller. The results show that, in case of the fuzzy PDC-L_1 adaptive control method, at the same time fuzzy L_1 adaptive controller efficiently nullifies the uncertainties, disturbances and fuzzy PDC controller handles nonlinearities to provide satisfactory transient and steady state tracking performance. Hence, the fuzzy PDC-L_1 adaptive controller outperforms the other control strategies.

After designing and evaluating the controllers in simulation case studies, their performances are validated on experimental case studies. Three experimental case studies are performed in this book where as in the next chapter the first experimental setup, viz., a single input single output (SISO) electrical actuator system is investigated.

References

1. Hovakimyan, N., Cao, C.: L1 Adaptive Control Theory Guaranteed Robustness with Fast Adaptation. SIAM, Philadelphia (2010)
2. Maiti, R., Das Sharma, K., Sarkar, G.: Linear consequence based fuzzy parallel distributed compensation type L1 adaptive controller for two link robot manipulator. IEEE Transactions on Circuits and Systems I: Regular Papers, vol. 66, no. 10, pp. 3978–3990, (2019)
3. Maiti, R., Das Sharma, K., Sarkar, G.: Design of hybrid L1 adaptive control methodologies, Doctoral Dissertation, University of Calcutta, Kolkata, India, (2021)
4. Maiti, R., Das Sharma, K., Sarkar, G.: Fuzzy predictor based L1 adaptive controller for nonlinear systems with disturbances. In: Proceedings of the 2017 IEEE Calcutta Conference (CALCON), Kolkata, pp. 253–257 (2017)
5. Pomet, J., Praly, L.: Adaptive nonlinear regulation: estimation from the Lyapunov equation. IEEE Trans. Autom. Control **37**(6), 729–740 (June 1992)
6. Maiti, R., Das Sharma, K., Sarkar, G.: Adaptive fuzzy low-pass filter based L1 adaptive controller design for twin rotor MIMO system. In: Proceedings of the IEEE Region 10 Symposium (TENSYMP 2019), Kolkata, India, pp. 744–749, (2019)
7. Maiti, R., Das Sharma, K., Sarkar, G.. lBest HS algorithm based concurrent L1 adaptive control for nonlinear systems. ISA .Trans. **96**, 390–414 (2020)
8. Roose, A.I., Yahya, S., Al-Rizzo, H.: Fuzzy-logic control of an inverted pendulum on a cart. Comput. Electr. Eng. **61**, 31–47 (2017)
9. Su, X., Xia, F., Liu, J., Wu, L.: Event-triggered fuzzy control of nonlinear systems with its application to inverted pendulum systems. Automatica **94**, 236–248 (2018)
10. Mishra, S.-K., Chandra, D.: Stabilization and tracking control of inverted pendulum using fractional order PID controllers. J. Eng. **2014**, 1–9 (2014)
11. Juang, J.-G., Liu, W.-K., Lin, R.-W.: A hybrid intelligent controller for a twin rotor MIMO system and its hardware implementation. ISA Trans. **50**(4), 609–619 (October 2011)
12. Roman, R.-C., Precup, R.-E., David, R.-C.: Second order intelligent proportional-integral fuzzy control of twin rotor aerodynamic systems. Procedia Comput Sci **139**, 372–380 (2018)
13. Shaik, F.A., Purwar, S., Pratap, B.: Real-time implementation of Chebyshev neural network observer for twin rotor control system. Expert Syst. Appl. **38**, 13043–13049 (2011)

Part IV
Applications

Chapter 5
Speed Control of Electrical Actuator

Abstract In this chapter, the controllers are employed on an electrical actuator, viz., DC motor experimental setup to validate their effectiveness. The control objective is to track a desired speed trajectory and to nullify the sudden occurrence of load disturbance without retuning the controller. Results show that, the lbest HS-L_1 adaptive controller provides better performance than the basic L_1 adaptive controller and the fuzzy PDC-L_1 adaptive controller performs well enough than all other controllers tested in this experimental case study.

5.1 Introduction

Electrical actuators have compact structure and easy interfacing phenomenon with computer controlled systems and they can provide immediate precise control in response to the input signals [1]. In different literatures, electrical actuators were employed to drive different systems, viz., robotic arms [2, 3], vehicle eco-pedal system [4], aircraft [5], electric vehicles [6], hard disk drive [7], etc. Consequently, control of electrical actuators gained attention over the last few decades. DC motors are considered as the continuous actuators in converting electrical energy to mechanical energy. Speed of these types of actuators can be controlled by armature voltage control and field flux control [8]. These control laws can be implemented by employing different techniques. Linear matrix inequality (LMI) based H_∞ optimal controller was utilized to control a DC motor with uncertainties and disturbances [9]. Beetle antennae search (BAS) and proportional integral derivative (PID) controller were brought together to obtain improved speed controlling of DC motor [10]. Pulse width modulation (PWM) technique was used extensively in speed control of DC motor [11, 12]. Speed of the DC motor was also controlled through fuzzy logic [13].

Owing to the enormous utilization of the electrical actuator, in this chapter, a DC shunt motor experimental setup, indigenously developed at the Systems Engineering Laboratory of Applied Physics Department, University of Calcutta, Kolkata, India, is controlled utilizing different controllers. Proportional integral derivative (PID), linear quadratic Gaussian (LQG), model reference adaptive controller (MRAC), basic L_1

R. Maiti et al., *Hybrid L_1 Adaptive Control*, Studies in Systems, Decision and Control 422, https://doi.org/10.1007/978-3-030-97102-1_5

adaptive, particle swarm optimization based L_1 (PSO-L_1) adaptive, basic harmony search based L_1 (HS-L_1) adaptive, local best HS based L_1 (lbest HS-L_1) adaptive and fuzzy parallel distributed compensation type L_1 (fuzzy PDC-L_1) adaptive controller are employed on this system with the aim to obtain precise and fast speed control and their performances are compared. At first, system identification of the DC motor is done in open loop condition with variable step trajectory. Then, controllers are developed for the identified model in simulation environment for another variable step trajectory. Ultimately, the tuned controllers are employed on the DC motor experimental setup to control its speed by armature voltage control method. These tuned controllers without further tuning are then subjected to 100% load disturbance at the time of tracking a fixed step trajectory. The results show that, the performance of lbest HS-L_1 adaptive controller is better than the basic L_1 adaptive controller and the performance of fuzzy PDC-L_1 adaptive controller is better than all other methods tested. The results also show that, the disturbance rejection phenomenon of the fuzzy PDC-L_1 adaptive controller is satisfactory than the other controllers compared.

5.2 Dynamical Model of Electrical Actuator

In this chapter, an electrical actuator, viz., a shunt wound DC motor, as provided in Fig. 5.1, is considered for experimental case study. The schematic diagram of the DC motor model is provided in Fig. 5.2. The armature of the DC motor contains resistance R_a in series with reactance L_a. The armature is connected in parallel with field which consists of resistance R_{sh} and reactance L_{sh} in series. The angular speed of the DC motor is given as ω_m.

The voltage equation of the armature circuit of the DC motor is considered as:

$$V_{in} = R_a I_a + L_a \frac{dI_a}{dt} + K_b \omega_m \tag{5.1}$$

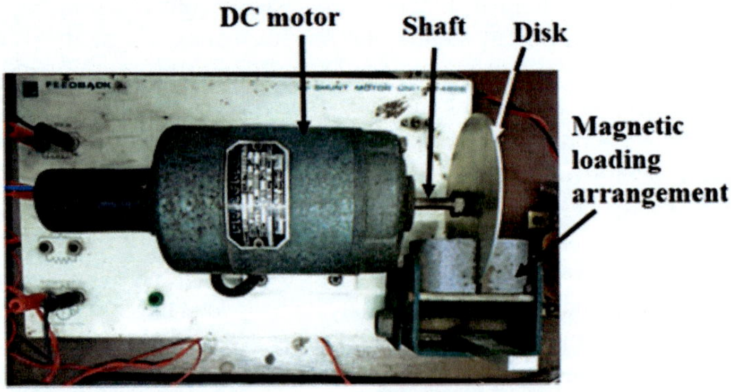

Fig. 5.1 Experimental model of electrical actuator

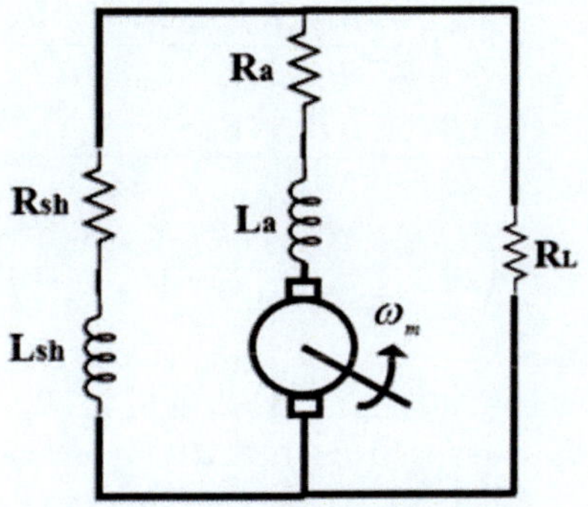

Fig. 5.2 Schematic diagram of electrical actuator

where V_{in} is the input voltage, I_a is the armature current, ω_m is the angular speed of the motor. R_a and L_a are the armature resistance and armature reactance respectively. K_b is the back e.m.f. coefficient. All units are considered in S.I.

Now, the torque equation of the DC motor can be stated as:

$$T_L + J_m \frac{d\omega_m}{dt} + B_m \omega_m = K_T I_a \tag{5.2}$$

where T_L is the load torque, K_T is the motor torque coefficient, B_m is the damping ratio and J_m is the moment of inertia of the motor, all units are in S.I.

Considering the load torque $T_L = 0$, (5.2) yields:

$$J_m \frac{d\omega_m}{dt} + B_m \omega_m = K_T I_a \tag{5.3}$$

Now, utilizing (5.1) and (5.3), it is obtained:

$$\frac{d^2\omega_m}{dt^2} + \left(\frac{R_a}{L_a} + \frac{B_m}{J_m}\right)\frac{d\omega_m}{dt} + \left(\frac{B_m R_a}{J_m L_a} + \frac{K_b K_T}{J_m L_a}\right)\omega_m = \frac{K_d K_T}{J_m L_a} V_{in} \tag{5.4}$$

where K_d is the driver circuit coefficient.

As, V_{in} is the input to the DC motor therefore, it can be written as control signal $u(t)$. Equation (5.4) again can be written as:

$$\ddot{\omega}_m + \left(\frac{R_a}{L_a} + \frac{B_m}{J_m}\right)\dot{\omega}_m + \left(\frac{B_m R_a}{J_m L_a} + \frac{K_b K_T}{J_m L_a}\right)\omega_m = \frac{K_d K_T}{J_m L_a} u(t) \tag{5.5}$$

In experimental setup of the DC motor, parametric uncertainties and measurement noise include into the system model. The load is also considered as the disturbance to the system. Hence, the DC motor model is written as [14]:

$$
\begin{aligned}
&\ddot{\omega}_m + \left(\frac{R_a + \Delta R_a}{L_a + \Delta L_a} + \frac{B_m + \Delta B_m}{J_m + \Delta J_m}\right)\dot{\omega}_m \\
&+\left(\frac{(B_m + \Delta B_m)\,(R_a + \Delta R_a)}{(J_m + \Delta J_m)\,(L_a + \Delta L_a)} + \frac{(K_b + \Delta K_b)\,(K_T + \Delta K_T)}{(J_m + \Delta J_m)\,(L_a + \Delta L_a)}\right)\omega_m + d(t) \\
&= \left(\frac{(K_d + \Delta K_d)\,(K_T + \Delta K_T)}{(J_m + \Delta J_m)\,(L_a + \Delta L_a)}\right)u(t)
\end{aligned}
\tag{5.6}
$$

where ΔR_a, ΔL_a,ΔJ_m, ΔB_m, ΔK_b, ΔK_T and ΔK_d are the uncertainties related to those respective terms, $d(t)$ represents the time varying disturbances.

Again, (5.6) can be written as:

$$
\ddot{\omega}_m + U_1\dot{\omega}_m + U_2\omega_m + d(t) = U_3 u(t) \tag{5.7}
$$

where

$$
U_1 = \left(\frac{R_a + \Delta R_a}{L_a + \Delta L_a} + \frac{B_m + \Delta B_m}{J_m + \Delta J_m}\right)
$$

or

$$
U_1 = \left(\frac{R_a}{L_a} + \frac{B_m}{J_m} + \frac{L_a\Delta R_a - R_a\Delta L_a}{L_a(L_a + \Delta L_a)} + \frac{J_m\Delta B_m - B_m\Delta J_m}{J_m(J_m + \Delta J_m)}\right),
$$

$$
U_2 = \left(\frac{(B_m + \Delta B_m)\,(R_a + \Delta R_a)}{(J_m + \Delta J_m)\,(L_a + \Delta L_a)} + \frac{(K_b + \Delta K_b)\,(K_T + \Delta K_T)}{(J_m + \Delta J_m)\,(L_a + \Delta L_a)}\right)
$$

or

$$
U_2 = \left(\begin{aligned}
&\frac{B_m R_a}{J_m L_a} + \frac{K_b K_T}{J_m L_a} \\
&+\frac{R_a\Delta B_m + B_m\Delta R_a + \Delta B_m\Delta R_a - B_m R_a(L_a\Delta J_m + J_m\Delta L_a + \Delta J_m\Delta L_a)}{J_m L_a(J_m + \Delta J_m)\,(L_a + \Delta L_a)} \\
&+\frac{K_T\Delta K_b + K_b\Delta K_T + \Delta K_b\Delta K_T - K_b K_T(L_a\Delta J_m + J_m\Delta L_a + \Delta J_m\Delta L_a)}{J_m L_a(J_m + \Delta J_m)\,(L_a + \Delta L_a)}
\end{aligned}\right)
$$

and

$$
U_3 = \left(\frac{(K_d + \Delta K_d)\,(K_T + \Delta K_T)}{(J_m + \Delta J_m)\,(L_a + \Delta L_a)}\right)
$$

or

$$U_3 = \left(\frac{K_d K_T}{J_m L_a} + \frac{\begin{matrix} K_T \Delta K_d + K_d \Delta K_T + \Delta K_d \Delta K_T \\ - K_d K_T (L_a \Delta J_m + J_m \Delta L_a + \Delta J_m \Delta L_a) \end{matrix}}{J_m L_a (J_m + \Delta J_m)(L_a + \Delta L_a)} \right),$$

Therefore, (5.7) can be given as:

$$\begin{aligned} &\ddot{\omega}_m + \left(\frac{R_a}{L_a} + \frac{B_m}{J_m} + \Delta U_1' \right) \dot{\omega}_m + \left(\frac{B_m R_a}{J_m L_a} + \frac{K_b K_T}{J_m L_a} + \Delta U_2' \right) \omega_m + d(t) \\ &= \left(\frac{K_d K_T}{J_m L_a} + \Delta U_3' \right) u(t) \end{aligned} \tag{5.8}$$

where

$$\Delta U_1' = \frac{L_a \Delta R_a - R_a \Delta L_a}{L_a (L_a + \Delta L_a)} + \frac{J_m \Delta B_m - B_m \Delta J_m}{J_m (J_m + \Delta J_m)},$$

$$\begin{aligned} \Delta U_2' = & \frac{R_a \Delta B_m + B_m \Delta R_a + \Delta B_m \Delta R_a - B_m R_a (L_a \Delta J_m + J_m \Delta L_a + \Delta J_m \Delta L_a)}{J_m L_a (J_m + \Delta J_m)(L_a + \Delta L_a)} \\ & + \frac{K_T \Delta K_b + K_b \Delta K_T + \Delta K_b \Delta K_T - K_b K_T (L_a \Delta J_m + J_m \Delta L_a + \Delta J_m \Delta L_a)}{J_m L_a (J_m + \Delta J_m)(L_a + \Delta L_a)} \end{aligned}$$

and

$$\Delta U_3' = \frac{K_T \Delta K + K_d \Delta K_T + \Delta K_d \Delta K_T - K_d K_T (L_a \Delta J_m + J_m \Delta L_a + \Delta J_m \Delta L_a)}{J_m L_a (J_m + \Delta J_m)(L_a + \Delta L_a)}$$

are respectively the uncertainties related with angular speed, angular acceleration and input voltage of the DC motor.

In matrix form, the DC motor model can be expressed as [15]:

$$\begin{aligned} \begin{bmatrix} \dot{x}_1 \\ \dot{x}_2 \end{bmatrix} = & \begin{bmatrix} 0 & 1 \\ -\left(\frac{R_a}{L_a} + \frac{B_m}{J_m} + \Delta U_1' \right) & -\left(\frac{B_m R_a}{J_m L_a} + \frac{K_b K_T}{J_m L_a} + \Delta U_2' \right) \end{bmatrix} \begin{bmatrix} x_1 \\ x_2 \end{bmatrix} \\ & + \begin{bmatrix} 0 \\ \left(\frac{K_d K_T}{J_m L_a} + \Delta U_3' \right) \end{bmatrix} u + \begin{bmatrix} 0 \\ d(t) \end{bmatrix} \end{aligned}$$

or

$$\begin{bmatrix} \dot{x}_1 \\ \dot{x}_2 \end{bmatrix} = \begin{bmatrix} 0 & 1 \\ -\left(\frac{R_a}{L_a} + \frac{B_m}{J_m}\right) & -\left(\frac{B_m R_a}{J_m L_a} + \frac{K_b K_T}{J_m L_a}\right) \end{bmatrix} \begin{bmatrix} x_1 \\ x_2 \end{bmatrix}$$
$$+ \begin{bmatrix} 0 \\ 1 \end{bmatrix} \left(\left(\frac{K_d K_T}{J_m L_a} + \Delta U_3' \right) u + \begin{bmatrix} -\Delta U_1' & -\Delta U_2' \end{bmatrix} \begin{bmatrix} x_1 \\ x_2 \end{bmatrix} + d(t) \right)$$

where $\omega = \left(\frac{K_d K_T}{J_m L_a} + \Delta U_3'\right)$ represents unknown constant,

$\underline{\theta} = \begin{bmatrix} \theta_1 & \theta_2 \end{bmatrix} = \begin{bmatrix} -\Delta U_1' & -\Delta U_2' \end{bmatrix}$ represents time varying uncertainties and $\sigma = d(t)$ represents time varying disturbances.

5.3 Description of Experimental Setup of Electrical Actuator

The electrical actuator, used in this study, is a shunt wound DC motor of 25 W, 50 V, 3000 rpm. The control signal is generated in the sophisticated software MATLAB® in PC and provided to the DC motor as voltage input through an ARDUINO® microcontroller board and driver circuit as shown in Figs. 5.3 and 5.4. A taco-generator is attached to the shaft of the DC motor to measure the speed in rpm. The rpm is converted into voltage within the system setup with a conversion ratio of 2 V/1000 rpm. The output voltage is fed back to the MATLAB® through the ARDUINO® microcontroller board. An aluminium disk is attached to the shaft of the DC motor where a magnet load may be subjected as per the requirement by observing the scale marked on the load. This load, attached beside the DC motor, is capable of producing up to 100% load disturbance to the DC motor experimental setup.

5.4 System Identification of Electrical Actuator

After formulating the dynamic model of the DC motor, the parameter values of it have to be known to design the controllers. To perform the system identification, the DC motor is run in open loop condition for 10 min with time step $\Delta t = 0.1$ s with variable step input as follows:

$$v_{oc}(t) = \begin{cases} 14.4\,u(t)\,, & \text{for } 0 \text{ s} \le t \le 300 \text{ s} \\ 25.92\,u(t), & \text{for } 300 \text{ s} < t \le 600 \text{ s} \end{cases}$$

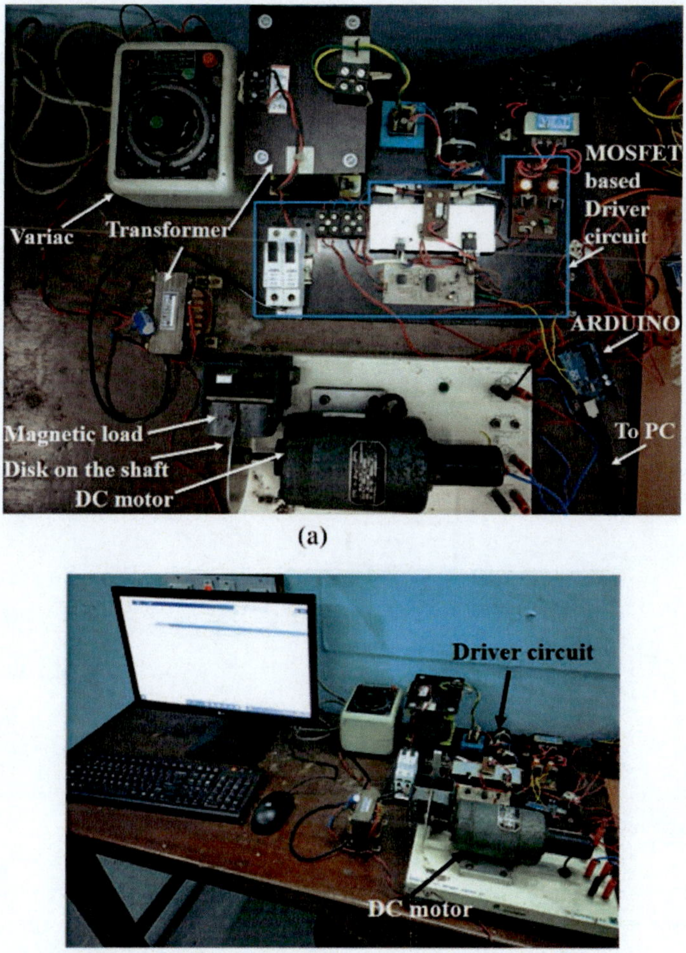

Fig. 5.3 Experimental setup of electrical actuator for speed control

This variable input voltage as shown in Fig. 5.5(a) is provided to the DC motor experimental setup. The output of the DC motor in terms of voltage is obtained in the MATLAB® where it is converted into speed in terms of rpm by a conversion ratio of 2 V/1000 rpm. The output speed of the DC motor is depicted in Fig. 5.5(b) with blue line. Now, with the same variable step input, the same output of the experimental setup is to be generated by the simulation model of the DC motor as shown in Fig. 5.5(b) with red line. The parameters of the DC motor are tuned employing a representative stochastic optimization technique, viz., particle swarm optimization (PSO) with 30 particles for 200 iterations. The whole process run for 10 times to obtain the best optimal parameter setting of the DC motor which are provided in Table 5.1.

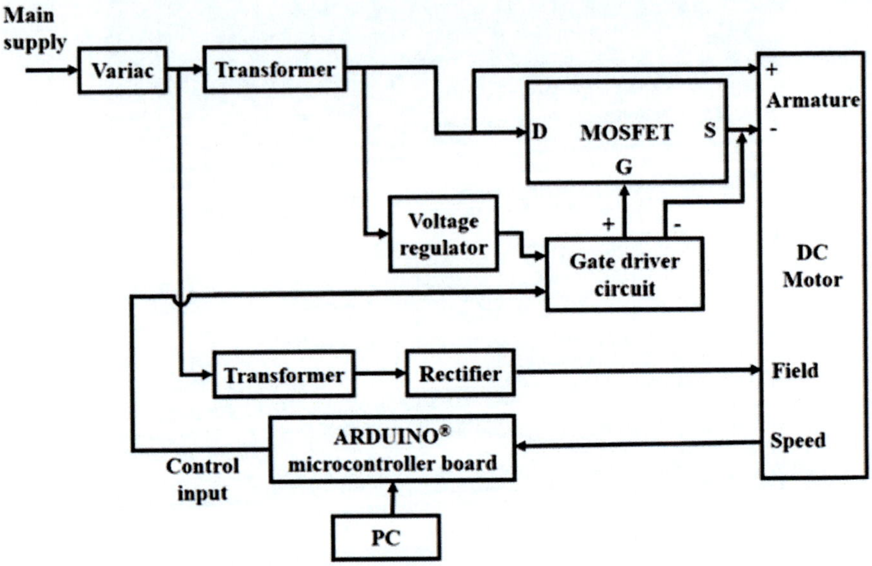

Fig. 5.4 Circuit layout of electrical actuator experimental setup

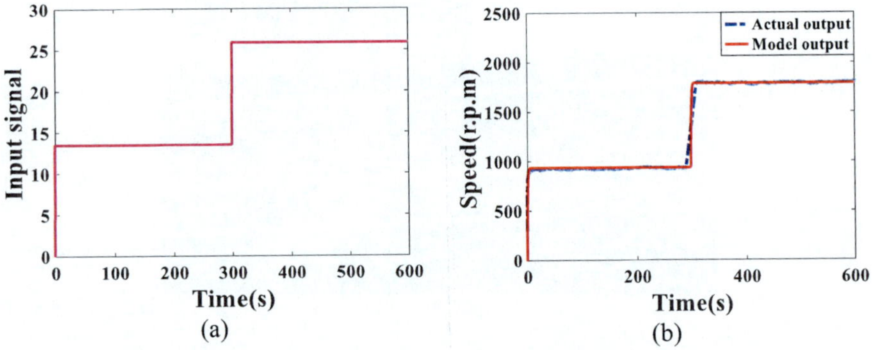

Fig. 5.5 **a** Input signal and corresponding **b** system output plot in open loop configuration for parameter estimation of electrical actuator experimental setup

Table 5.1 Estimated nominal parameter values of DC motor

R_a	L_a	J	B	K	K_b	K_T
0.44102 Ω	0.00074 H	0.00744 kg m^2	0.00098 N·m/(rad/s)	0.60990	0.00500 V/(rad/s)	0.00500 V/(rad/s)

This obtained parameters are utilized to design the DC motor model in simulation environment and the controllers are developed. Then, the controllers are implemented in real time environment as discussed in the next section.

5.5 Experimental Case Study of Electrical Actuator

The DC motor experimental setup with time varying uncertainties and disturbances is used to validate the effectiveness of the lbest HS-L_1 adaptive and fuzzy PDC-L_1 adaptive controllers, designed in the previous chapters, over other controllers. At first, three conventional controllers, viz., PID controller, LQG and MRAC are tested. Then, basic L_1 adaptive controller, stochastic optimization based L_1 adaptive controller, viz., PSO-L_1 adaptive, metaheuristic HS-L_1 adaptive and metaheuristic lbest HS-L_1 adaptive controller are examined. Finally, fuzzy PDC-L_1 adaptive controller is employed on the DC motor. The DC motor experimental setup is run for 10 min duration with time step 0.1 s with the control objective to track the desired speed trajectories. The two reference speed trajectories are given as follows.

Reference-I: A four step variable trajectory:

$$r(t) = \begin{cases} 1000 \text{ rpm} & 0 \text{ s} \le t \le 150 \text{ s} \\ 1500 \text{ rpm} & 150 \text{ s} < t \le 300 \text{ s} \\ 800 \text{ rpm} & 300 \text{ s} < t \le 450 \text{ s} \\ 1200 \text{ rpm} & 450 \text{ s} < t \le 600 \text{ s} \end{cases}$$

Reference-II: A single step trajectory:

$r(t) = 800\,u(t)\;\; 0 \text{ s} \le t \le 600$ s with 100% loading as disturbance.

The controllers are designed and/ or tuned with reference-I in simulation environment and then the best optimized parameter setting is provided to design the controller in real time environment. The designed controllers without further tuning are subjected to the reference-II in real time environment. The results in term of integral absolute error (IAE) are provided in Table 5.2. The responses and the control effort, error signal plots of DC motor experimental setup with reference-I are depicted in Figs. 5.6, 5.7, 5.8, 5.9, 5.10, 5.11, 5.12, 5.13 for above mentioned controllers. The system responses and control effort with error signals for reference-II are depicted in Figs. 5.14, 5.15, 5.16, 5.17, 5.18, 5.19, 5.20, 5.21.

From Fig. 5.6(a), Fig. 5.7(a) and Fig. 5.8(a) respectively it can be stated that, the conventional controllers, viz., PID, LQG and MRAC cannot eliminate uncertainties and disturbances properly and their tracking performances are not satisfactory. The performance of basic L_1 adaptive controller as shown in Fig. 5.9(a) demonstrates better performance than the conventional controllers in case of disturbance rejection though its tracking performance is not satisfactory. The stochastically and metaheuristically optimized L_1 adaptive controllers perform better than the basic

Table 5.2 Comparative study of different control methodologies for electrical actuator experimental setup

Control methodologies	IAE value	
	Reference-I	Reference-II
PID controller	13,751.69	7248.74
LQG	33,978.17	82,687.15
MRAC	32,844.20	74,633.07
Basic L_1 adaptive controller	10,002.30	6038.56
PSO-L_1 adaptive controller	9989.54	1766.42
HS-L_1 adaptive controller	9994.71	1838.63
lbest HS-L_1 adaptive controller	9328.89	1631.86
Fuzzy PDC-L_1 adaptive controller	7455.49	1518.68

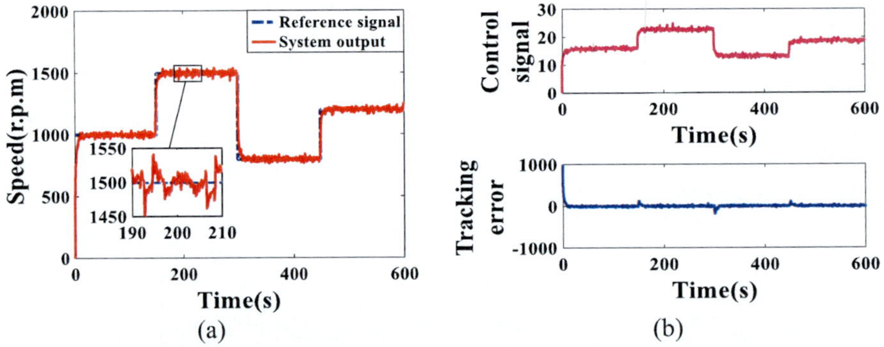

Fig. 5.6 Evaluation period **a** system response and corresponding **b** control signal with tracking error of electrical actuator experimental setup with reference-I for PID controller

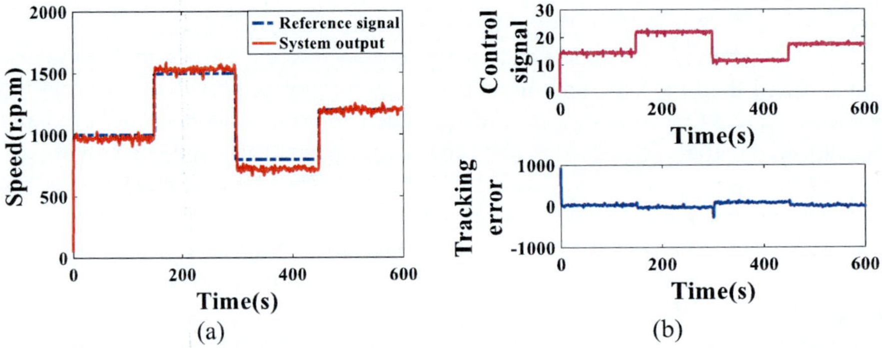

Fig. 5.7 Evaluation period **a** system response and corresponding **b** control signal with tracking error of electrical actuator experimental setup with reference-I for LQG

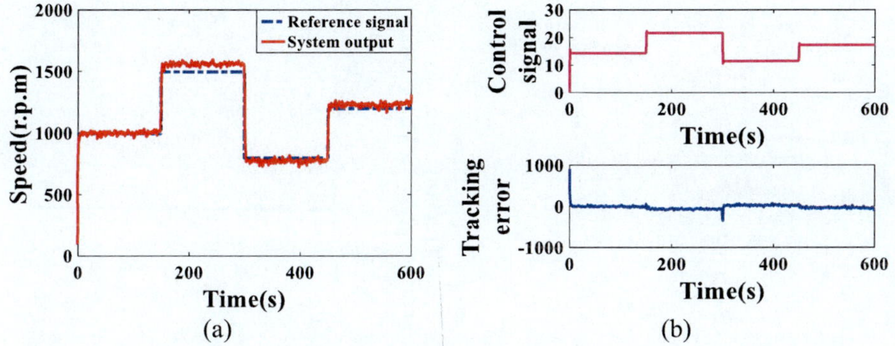

Fig. 5.8 Evaluation period **a** system response and corresponding **b** control signal with tracking error of electrical actuator experimental setup with reference-I for MRAC

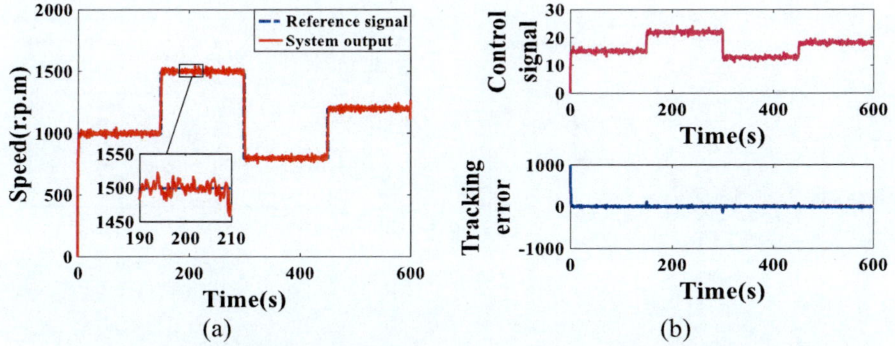

Fig. 5.9 Evaluation period **a** system response and corresponding **b** control signal with tracking error of electrical actuator experimental setup with reference-I for basic L_1 adaptive controller

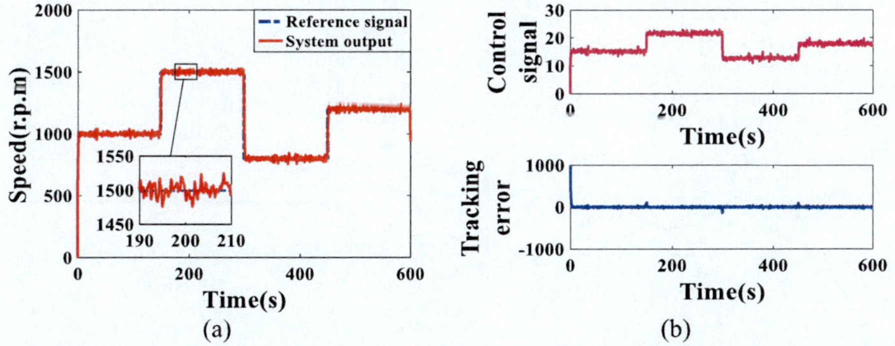

Fig. 5.10 Evaluation period **a** system response and corresponding **b** control signal with tracking error of electrical actuator experimental setup with reference-I for PSO-L_1 adaptive controller

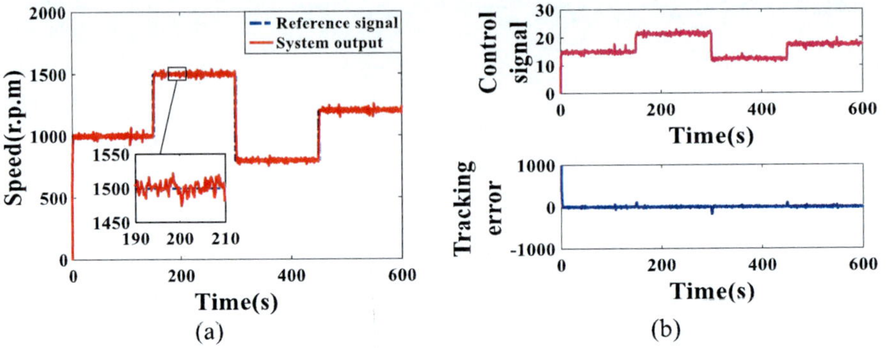

Fig. 5.11 Evaluation period **a** system response and corresponding **b** control signal with tracking error of electrical actuator experimental setup with reference-I for HS-L_1 adaptive controller

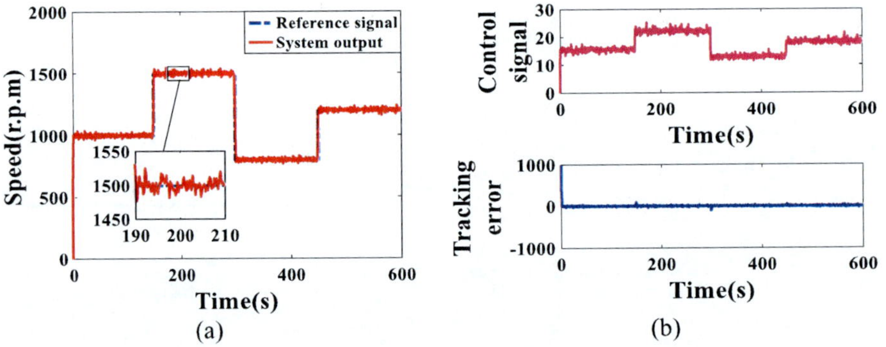

Fig. 5.12 Evaluation period **a** system response and corresponding **b** control signal with tracking error of electrical actuator experimental setup with reference-I for lbest HS-L_1 adaptive controller

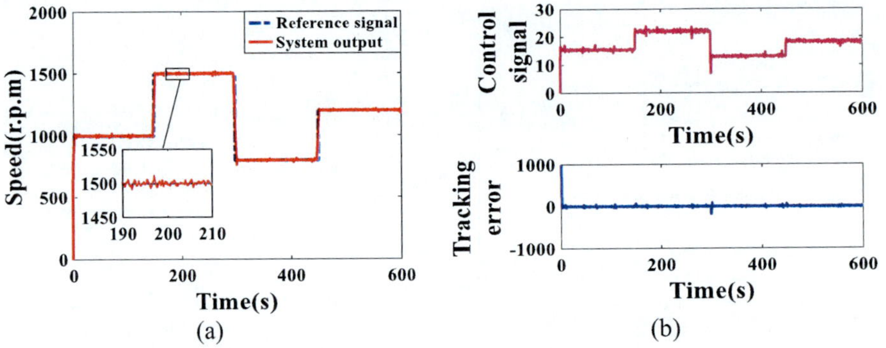

Fig. 5.13 Evaluation period **a** system response and corresponding **b** control signal with tracking error of electrical actuator experimental setup with reference-I for fuzzy PDC-L_1 adaptive controller

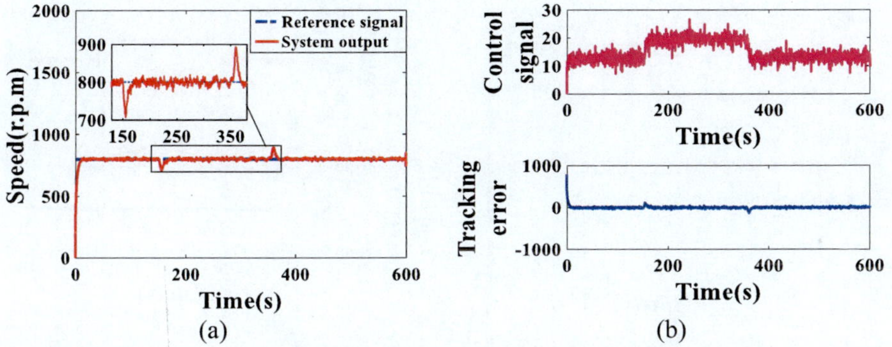

Fig. 5.14 Evaluation period **a** system response and corresponding **b** control signal with tracking error of electrical actuator experimental setup with reference-II for PID controller

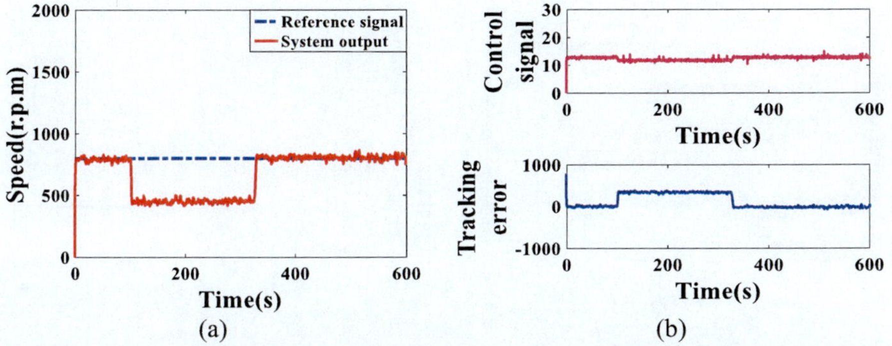

Fig. 5.15 Evaluation period **a** system response and corresponding **b** control signal with tracking error of electrical actuator experimental setup with reference-II for LQG

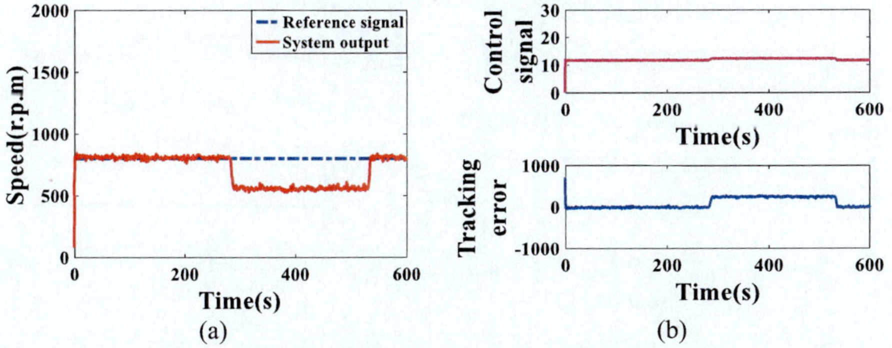

Fig. 5.16 Evaluation period **a** system response and corresponding **b** control signal with tracking error of electrical actuator experimental setup with reference-II for MRAC

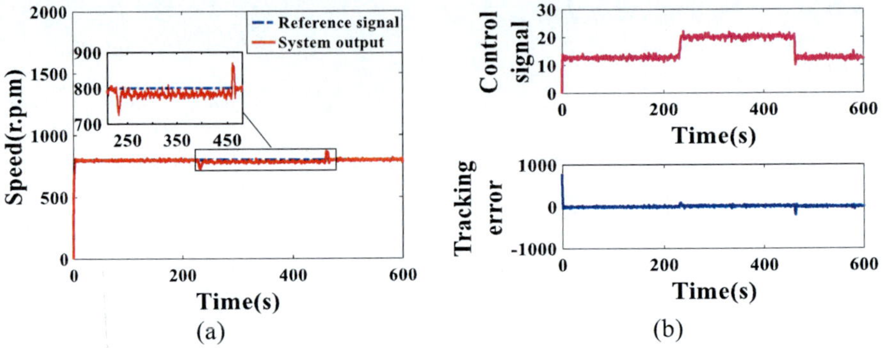

Fig. 5.17 Evaluation period **a** system response and corresponding **b** control signal with tracking error of electrical actuator experimental setup with reference-II for basic L_1 adaptive controller

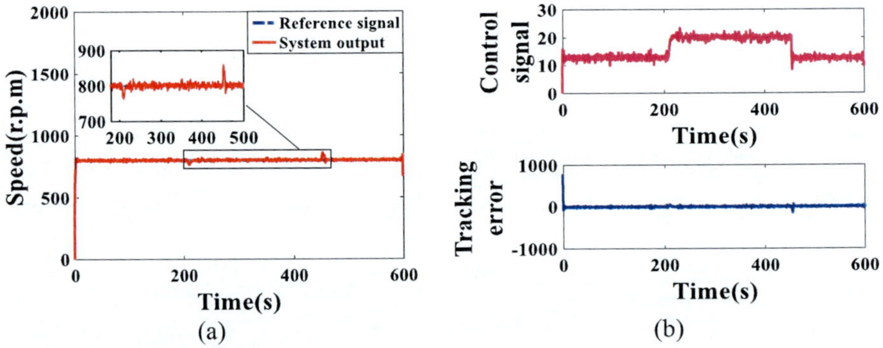

Fig. 5.18 Evaluation period **a** system response and corresponding **b** control signal with tracking error of electrical actuator experimental setup with reference-II for PSO-L_1 adaptive controller

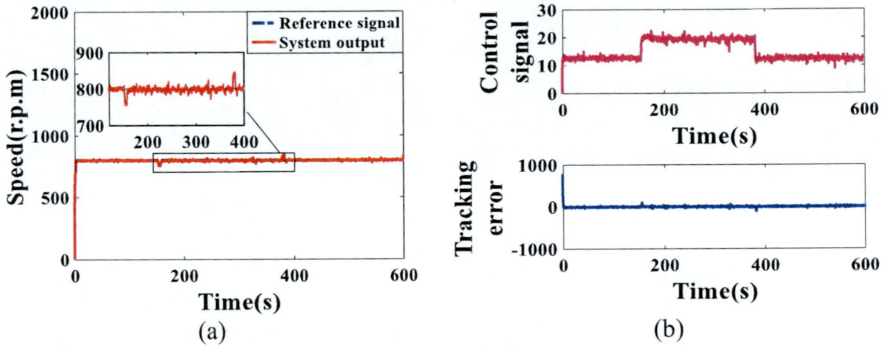

Fig. 5.19 Evaluation period **a** system response and corresponding **b** control signal with tracking error of electrical actuator experimental setup with reference-II for HS-L_1 adaptive controller

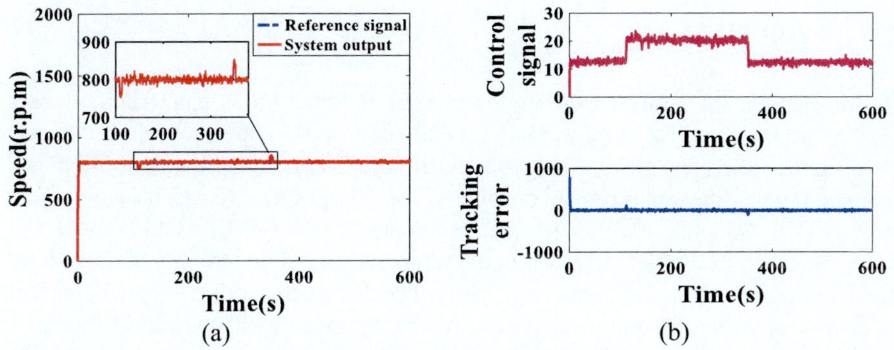

Fig. 5.20 Evaluation period **a** system response and corresponding **b** control signal with tracking error of electrical actuator experimental setup with reference-II for lbest HS-L_1 adaptive controller

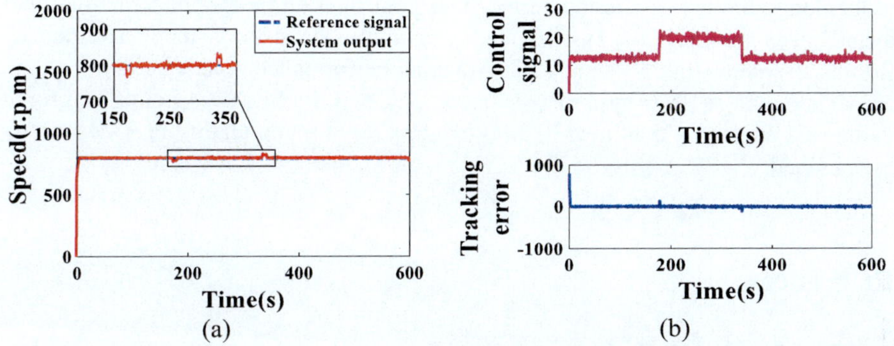

Fig. 5.21 Evaluation period **a** system response and corresponding **b** control signal with tracking error of electrical actuator experimental setup with reference-II for fuzzy PDC-L_1 adaptive controller

L_1 adaptive controller as shown in Table 5.2 and Figs. 5.10(a), 5.11(a) and 5.12(a). Among them, the metaheuristic lbest HS-L_1 adaptive controller performs satisfactory due to the appropriate selection of optimal parameters. The fuzzy PDC-L_1 adaptive controller performs best among all other controllers tested in this case study. It provides fast transient performance as well as satisfactory tracking performance as provided in Table 5.2 and Fig. 5.13(a). From results it can be stated that, when 100% load disturbance is subjected to the tuned controllers without further tuning then, the fuzzy PDC-L_1 adaptive controller performs better than the other controllers examined.

5.6 Summary

In this chapter, the control methodologies are employed on an electrical actuator, viz., a shunt wound DC motor experimental setup to validate their effectiveness. Performances of lbest HS-L_1 adaptive and fuzzy PDC-L_1 adaptive controllers are compared with different types of controllers, viz., PID, LQG, MRAC, basic L_1 adaptive, PSO-L_1 adaptive and HS-L_1 adaptive controller. The PID, LQG and MRAC fail to track desired trajectories when the references are time varying and also they cannot eliminate load disturbances properly. The disturbance rejection phenomenon of the basic L_1 adaptive controller is better than the conventional controllers though it is not satisfactory enough. The performances of stochastically and metaheuristically optimised L_1 adaptive controllers are better than the basic L_1 adaptive controller. The lbest HS-L_1 adaptive controller performs better than the PSO-L_1 adaptive and HS-L_1 adaptive controller. Finally, the fuzzy PDC-L_1 adaptive controller performs best among all other controllers tested. This controller eliminates uncertainties and disturbances present in the system quickly and provides fast transient performance. Results also show that, its steady state tracking performance is satisfactory.

After tackling a SISO system, the concern is to tackle a MIMO system. Therefore, in the next chapter, a nonlinear MIMO two link robot manipulator (TLRM) system is examined.

References

1. Bajpai, P.: Hydraulics. In: Biermann's Handbook of Pulp and Paper, vol. 2, 3rd edn, pp. 455–482. Paper and Board Making, Elsevier (2018)
2. Rouzbeh, B., Bone, G.M., Ashby, G., Li, E.: Design, implementation and control of an improved hybrid pneumatic-electric actuator for robot arms. IEEE Access **7**, 14699–14713 (2019)
3. Sakaino, S., Furuya, T., Tsuji, T.: Bilateral control between electric and hydraulic actuators using linearization of hydraulic actuators. IEEE Trans. Industr. Electron. **64**(6), 4631–4641 (2017)
4. Kim, Y., Gu, B., Jung, I., Won, S., Hur, J.: Analysis and design of slotted tubular linear actuator for the eco-pedal system of a vehicle. IEEE Trans. Magn. **48**(2), 939–942 (2012)
5. Barater, D., et al.: Multistress characterization of fault mechanisms in aerospace electric actuators. IEEE Trans. Ind. Appl. **53**(2), 1106–1115 (2017)
6. Kim, S.K.H.L.T.: Quad-bus motor drive system for electrified vehicles based on a dual-output–single-inductor structure. IET Electr. Power Appl. **13**(12), 1985–1992 (2019)
7. Lee, H.-S., Kim, Y.-H., Hwang, T.-Y., Kim, C.-S.: VCM design to improve dynamic performance of an actuator in a disk drive. IEEE Trans. Magn. **41**(2), 774–778 (2005)
8. Chapman, S.J.: Electric Machinery Fundamentals, 5th edn. McGraw-Hill, New York (2012)
9. Lu, R., Li, S., Xue, L.: Robust H∞ optimal speed control of DC motor using LMI approach. In: Proceedings of the Chinese Control and Decision Conference, Yantai, Shandong, pp. 4350–4354 (2008)
10. Lin, X., Liu, Y., Wang, Y.: Design and research of DC motor speed control system based on improved bas. In: Proceedings of the 2018 Chinese Automation Congress (CAC), Xi'an, China, pp. 3701–3705 (2018)
11. Yurkevich, V.D., Stepanov, N.A.: PWM speed control of DC motor based on singular perturbation technique. In: Proceedings of the 2014 6th International Congress on Ultra

Modern Telecommunications and Control Systems and Workshops (ICUMT), St. Petersburg, pp. 434–440 (2014)
12. Agung, I.G.A.P.R., Huda, S., Wijaya, I.W.A: Speed control for DC motor with pulse width modulation (PWM) method using infrared remote control based on ATmega16 microcontroller. In: Proceedings of the 2014 International Conference on Smart Green Technology in Electrical and Information Systems (ICSGTEIS), Kuta, 2014, pp. 108–112
13. Sadiq, A.A., Bakare, G.A., Anene, E.C., Mamman, H.B.: A fuzzy-based speed control of DC motor using combined armature voltage and field current. In: Proceedings of the 3rd IFAC International Conference on Intelligent Control and Automation Science, Chengdu, China, pp. 387–392 (2013)
14. Maiti, R., Das Sharma, K., Sarkar, G.: Modelling of parameter uncertainties and disturbance rejection of a real-life system employing L1 adaptive control scheme. In: Lectures on Modelling and Simulation; A selection from AMSE # 2017, vol. 2017, no. 2, pp. 98–108
15. Maiti, R., Das Sharma, K., Sarkar, G.: Adaptive fuzzy filter based L1 adaptive controller design for electrical actuator containing uncertainties and disturbances. Communicated (2020)

Chapter 6
Angular Position Control of Two Link Robot Manipulator

Abstract In this chapter, the lbest HS-L_1 adaptive controller and fuzzy PDC-L_1 adaptive controller are evaluated on a two link robot manipulator (TLRM) experimental setup to control its two joint angles. TLRM epitomizes a MIMO nonlinear system with uncertainties and disturbances. The lbest HS-L_1 adaptive controller provides better tracking phenomenon than the basic L_1 adaptive controller. The fuzzy PDC-L_1 adaptive controller nullifies uncertainties, disturbances and handles nonlinearities better than all other methods tested.

6.1 Introduction

Two link robot manipulator (TLRM) is a nonlinear system with unknown dynamics, time varying uncertainties and disturbances which exemplifies a challenging system from control engineering point of view [1–4]. Different attempts were made in different literatures [5] to reduce the control complexity of such under actuated system. Trajectory tracking control of TLRM was performed in [6, 7]. Back stepping controllers were employed for trajectory tracking of robot manipulator system [8]. Closed loop nonlinear predictive controller was developed to tackle variable stiffness actuated two link robot [9]. A self-adaptive neural network with fast transient performance was employed to control TLRM [10]. An observer-controller was proposed to avoid problem in measuring velocity of the joint [11]. However, control of TLRM system considering all the complexities, viz., nonlinearities, structured and unstructured dynamics, parametric uncertainties and disturbances concurrently is absent in the literatures as per author's knowledge and belief.

To investigate the performance of controllers in controlling nonlinear system with unmodelled dynamics, a TLRM experimental setup, indigenously developed at Systems Engineering Laboratory of Applied Physics Department, University of Calcutta, Kolkata, India, is considered in this case study. At first, the TLRM has to be modelled by its dynamics and kinematics to design controller for it. This TLRM is an emulated model of large scale industrial manipulator system. In this work, the modelling and controller designing of the laboratory scale TLRM system is done as a way that it can be imitated in large scale industrial manipulator system. Modelling of

R. Maiti et al., *Hybrid L_1 Adaptive Control*, Studies in Systems, Decision and Control 422, https://doi.org/10.1007/978-3-030-97102-1_6

TLRM via Euler–Lagrange system dynamics [12] requires the exact parameter values of TLRM which are difficult to measure accurately for industrial manipulator system. To avoid such problem, in this chapter, the TLRM is modelled through the universal approximator fuzzy logic system. The two links of this system are rotated with the help of two servo motors mounted at the bottom of each link. These two motors are excited with the control input required to follow the reference trajectories. At first, different types of controllers, viz., proportional integral derivative (PID), linear quadratic Gaussian (LQG), model reference adaptive controller (MRAC), basic L_1 adaptive, particle swarm optimization based L_1 (PSO-L_1) adaptive, basic harmony search based L_1 (HS-L_1) adaptive, local best HS based L_1 (lbest HS-L_1) adaptive and fuzzy parallel distributed compensation type L_1 (fuzzy PDC-L_1) adaptive controller are at first developed in simulation environment for that identified model. Then, those tuned controllers are utilized to control TLRM experimental setup. The obtained results portray that, the lbest HS-L_1 adaptive and fuzzy PDC-L_1 adaptive controllers are efficient in tackling such nonlinear system with uncertainties, measurement noise, etc. Fuzzy PDC-L_1 adaptive controller appropriately models unstructured dynamics of the system and produces effective joint angle control of TLRM. This controller outperforms other control strategies tested in this case study.

6.2 Dynamical Model of TLRM

Two link robot manipulator (TLRM) consists of two links of length l_1 and l_2. The first link is of mass m_1 and second link is of mass m_2. All parameters are in S.I. unit. The experimental model of TLRM is depicted in Fig. 6.1 and the schematic diagram is shown in Fig. 6.2. Let, the point at which the first link is fixed with the base, is considered as (0, 0) and at any instant it creates an angle of θ_1 rad from the abscissa x axis. The second link is attached with the first link at point (x_1, y_1) and rotates θ_2

Fig. 6.1 Experimental model of TLRM

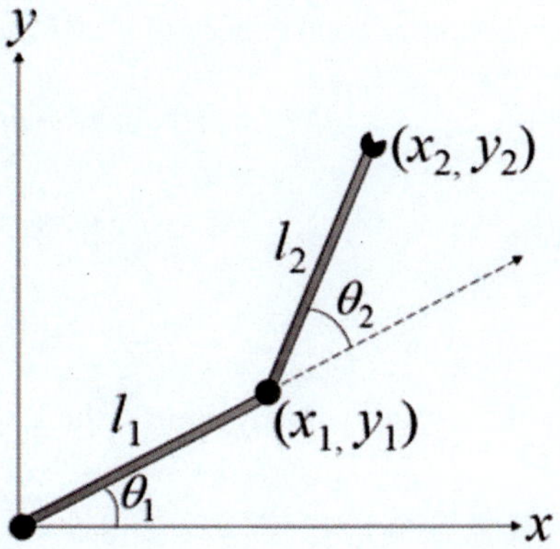

Fig. 6.2 Schematic diagram of TLRM

rad from the axis of the first link. The position of the end effector, connected at the tips of the second link, is (x_2, y_2).

Now, from Fig. 6.2, the dynamics of TLRM is formulated as follows.

For first link, it can be written as:

$$x_1 = l_1 \cos \theta_1 \tag{6.1}$$

and

$$y_1 = l_1 \sin \theta_1 \tag{6.2}$$

Similarly, for second link:

$$x_2 = l_1 \cos \theta_1 + l_2 \cos(\theta_1 + \theta_2) \tag{6.3}$$

and

$$y_2 = l_1 \sin \theta_1 + l_2 \sin(\theta_1 + \theta_2) \tag{6.4}$$

Now, total kinetic energy of TLRM can be formulated as:

$$K = \frac{1}{2}m_1\dot{x}_1^2 + \frac{1}{2}m_1\dot{y}_1^2 + \frac{1}{2}m_2\dot{x}_2^2 + \frac{1}{2}m_2\dot{y}_2^2$$

or

$$\begin{aligned} K = {} & \frac{1}{2}m_1 l_1^2\dot{\theta}_1^2 + \frac{1}{2}m_2 l_1^2\dot{\theta}_1^2 + \frac{1}{2}m_2 l_2^2(\dot{\theta}_1 + \dot{\theta}_2)^2 \\ & + m_2 l_1 l_2 \cos\theta_2(\dot{\theta}_1^2 + \dot{\theta}_1\dot{\theta}_2) \end{aligned} \tag{6.5}$$

Total potential energy of TLRM is considered as:

$$P = m_1 g l_1 \sin\theta_1 + m_2 g l_1 \sin\theta_1 + m_2 g l_2 \sin(\theta_1 + \theta_2)$$

or

$$P = (m_1 + m_2) g l_1 \sin\theta_1 + m_2 g l_2 \sin(\theta_1 + \theta_2) \tag{6.6}$$

From Lagrange dynamics, the Lagrangian for the system becomes:

$$L = K - P$$

or

$$\begin{aligned} L = &\frac{1}{2}(m_1 + m_2) l_1^2 \dot{\theta}_1^2 + \frac{1}{2} m_2 l_2^2 (\dot{\theta}_1 + \dot{\theta}_2)^2 + m_2 l_1 l_2 \cos\theta_2 (\dot{\theta}_1^2 + \dot{\theta}_1 \dot{\theta}_2) \\ &-(m_1 + m_2) g l_1 \sin\theta_1 - m_2 g l_2 \sin(\theta_1 + \theta_2) \end{aligned} \tag{6.7}$$

The dynamics of the system for ith link is given by:

$$\tau_i = \frac{d}{dt}\left(\frac{\partial L}{\partial \dot{\theta}_i}\right) - \left(\frac{\partial L}{\partial \theta_i}\right) \tag{6.8}$$

where τ_i is the torque at ith joint.

Therefore, the torques for first and second joint become τ_1 and τ_2 respectively. The partial derivative of the Lagrangian (6.7) with respect to $\dot{\theta}_1$ is provided as:

$$\frac{\partial L}{\partial \dot{\theta}_1} = (m_1 + m_2) l_1^2 \dot{\theta}_1 + m_2 l_2^2 (\dot{\theta}_1 + \dot{\theta}_2) + m_2 l_1 l_2 \cos\theta_2 (2\dot{\theta}_1 + \dot{\theta}_2) \tag{6.9}$$

Taking time derivative of (6.9) yields:

$$\begin{aligned} \frac{d}{dt}\left(\frac{\partial L}{\partial \dot{\theta}_1}\right) = &(m_1 + m_2) l_1^2 \ddot{\theta}_1 + m_2 l_2^2 (\ddot{\theta}_1 + \ddot{\theta}_2) \\ &+ m_2 l_1 l_2 \cos\theta_2 (2\ddot{\theta}_1 + \ddot{\theta}_2) - m_2 l_1 l_2 \sin\theta_2 (2\dot{\theta}_1 + \dot{\theta}_2)\dot{\theta}_2 \end{aligned}$$

or

$$\begin{aligned} \frac{d}{dt}\left(\frac{\partial L}{\partial \dot{\theta}_1}\right) = &\left((m_1 + m_2) l_1^2 + m_2 l_2^2 + 2 m_2 l_1 l_2 \cos\theta_2\right)\ddot{\theta}_1 \\ &+ \left(m_2 l_2^2 + m_2 l_1 l_2 \cos\theta_2\right)\ddot{\theta}_2 - m_2 l_1 l_2 \sin\theta_2 (2\dot{\theta}_1 + \dot{\theta}_2)\dot{\theta}_2 \end{aligned} \tag{6.10}$$

The partial derivative of the Lagrangian (6.7) with respect to θ_1 is provided as:

$$\frac{\partial L}{\partial \theta_1} = -(m_1 + m_2)gl_1 \cos\theta_1 - m_2 g l_2 \cos(\theta_1 + \theta_2) \tag{6.11}$$

The partial derivative of the Lagrangian (6.7) with respect to $\dot{\theta}_2$ is provided as:

$$\frac{\partial L}{\partial \dot{\theta}_2} = m_2 l_2^2 (\dot{\theta}_1 + \dot{\theta}_2) + m_2 l_1 l_2 \cos\theta_2 \dot{\theta}_1 \tag{6.12}$$

Taking time derivative of (6.12) yields:

$$\frac{d}{dt}\left(\frac{\partial L}{\partial \dot{\theta}_2}\right) = m_2 l_2^2 (\ddot{\theta}_1 + \ddot{\theta}_2) - m_2 l_1 l_2 \sin\theta_2 \dot{\theta}_1 \dot{\theta}_2 + m_2 l_1 l_2 \cos\theta_2 \ddot{\theta}_1$$

or

$$\frac{d}{dt}\left(\frac{\partial L}{\partial \dot{\theta}_2}\right) = \left(m_2 l_2^2 + m_2 l_1 l_2 \cos\theta_2\right)\ddot{\theta}_1 + m_2 l_2^2 \ddot{\theta}_2 - m_2 l_1 l_2 \sin\theta_2 \dot{\theta}_1 \dot{\theta}_2 \tag{6.13}$$

The partial derivative of the Lagrangian (6.7) with respect to θ_2 is provided as:

$$\frac{\partial L}{\partial \theta_2} = -m_2 l_1 l_2 \sin\theta_2 (\dot{\theta}_1^2 + \dot{\theta}_1 \dot{\theta}_2) - m_2 g l_2 \cos(\theta_1 + \theta_2) \tag{6.14}$$

Therefore, the torque of first joint becomes:

$$\begin{aligned} \tau_1 &= \left((m_1 + m_2) l_1^2 + m_2 l_2^2 + 2 m_2 l_1 l_2 \cos\theta_2\right) \ddot{\theta}_1 \\ &+ \left(m_2 l_2^2 + m_2 l_1 l_2 \cos\theta_2\right) \ddot{\theta}_2 - m_2 l_1 l_2 \sin\theta_2 (2\dot{\theta}_1 + \dot{\theta}_2)\, \dot{\theta}_2 \\ &+ (m_1 + m_2) g l_1 \cos\theta_1 + m_2 g l_2 \cos(\theta_1 + \theta_2) \end{aligned} \tag{6.15}$$

The torque of second joint becomes:

$$\begin{aligned} \tau_2 &= \left(m_2 l_2^2 + m_2 l_1 l_2 \cos\theta_2\right) \ddot{\theta}_1 + m_2 l_2^2 \ddot{\theta}_2 - m_2 l_1 l_2 \sin\theta_2 \dot{\theta}_1 \dot{\theta}_2 \\ &+ m_2 l_1 l_2 \sin\theta_2 (\dot{\theta}_1 + \dot{\theta}_2)\, \dot{\theta}_1 + m_2 g l_2 \cos(\theta_1 + \theta_2) \end{aligned} \tag{6.16}$$

Now, the overall dynamic model of TLRM in matrix form can be expressed as:

$$\begin{bmatrix} \tau_1 \\ \tau_2 \end{bmatrix} = \begin{bmatrix} \begin{pmatrix} (m_1 + m_2)l_1^2 + m_2 l_2^2 \\ +2m_2 l_1 l_2 \cos\theta_2 \end{pmatrix} & m_2 l_2^2 + m_2 l_1 l_2 \cos\theta_2 \\ m_2 l_2^2 + m_2 l_1 l_2 \cos\theta_2 & m_2 l_2^2 \end{bmatrix} \begin{bmatrix} \ddot{\theta}_1 \\ \ddot{\theta}_2 \end{bmatrix}$$
$$+ \begin{bmatrix} 0 & -m_2 l_1 l_2 \sin\theta_2 \\ m_2 l_1 l_2 \sin\theta_2 & 0 \end{bmatrix} \begin{bmatrix} \dot{\theta}_1^2 \\ \dot{\theta}_2^2 \end{bmatrix} + \begin{bmatrix} -2m_2 l_1 l_2 \dot{\theta}_1 \dot{\theta}_2 \sin\theta_2 \\ 0 \end{bmatrix}$$
$$+ \begin{bmatrix} (m_1 + m_2) g l_1 \cos\theta_1 + m_2 g l_2 \cos(\theta_1 + \theta_2) \\ m_2 g l_2 \cos(\theta_1 + \theta_2) \end{bmatrix} \tag{6.17}$$

Equation (6.17) again can be written as:

$$\tau = J(\theta)\ddot{\theta} + B(\theta, \dot{\theta}) + G(\theta) \tag{6.18}$$

where

$$J(\theta) = \begin{bmatrix} (m_1 + m_2)l_1^2 + m_2 l_2^2 + 2m_2 l_1 l_2 \cos\theta_2 & m_2 l_2^2 + m_2 l_1 l_2 \cos\theta_2 \\ m_2 l_2^2 + m_2 l_1 l_2 \cos\theta_2 & m_2 l_2^2 \end{bmatrix}$$

is the coupled inertia matrix,

$$B(\theta, \dot{\theta}) = \begin{bmatrix} 0 & -m_2 l_1 l_2 \sin\theta_2 \\ m_2 l_1 l_2 \sin\theta_2 & 0 \end{bmatrix} \begin{bmatrix} \dot{\theta}_1^2 \\ \dot{\theta}_2^2 \end{bmatrix} + \begin{bmatrix} -2m_2 l_1 l_2 \dot{\theta}_1 \dot{\theta}_2 \sin\theta_2 \\ 0 \end{bmatrix}$$

is the Coriolis and centrifugal force matrix and

$$G(\theta) = \begin{bmatrix} (m_1 + m_2) g l_1 \cos\theta_1 + m_2 g l_2 \cos(\theta_1 + \theta_2) \\ m_2 g l_2 \cos(\theta_1 + \theta_2) \end{bmatrix}$$

is the gravity matrix.

After formulating the mathematical model of TLRM, description of its experimental setup is provided in the next section.

6.3 Description of Experimental Setup of TLRM

In experimental setup of the two link robot manipulator, the first link is attached with a servo motor, mounted on a disk. The disk is positioned at a certain height from the ground that the manipulator can move freely in the downward direction. The second link is attached with another servo motor, mounted at the tip of the first link. End effector may be attached directly to the tip of the second link. The servo motors, utilized to drive the links, are of rating 5 V, 2 A. The experimental setup of the TLRM is connected to the PC where two control signals are generated in a sophisticated software MATLAB®. These two control signals are given to the two joint angles of the system in terms of voltages. Two joint angles of the TLRM in radian as output are obtained from the rotary encoders provided within the servo motors and are given to the MATLAB® for feedback. This shows that, TLRM is a multi-input multi-output (MIMO) system. The experimental setup of the TLRM is shown in Fig. 6.3 whereas, the circuit layout is provided in Fig. 6.4.

After the mathematical modelling and description of the experimental setup, system identification is performed to formulate the dynamical model of TLRM as follows.

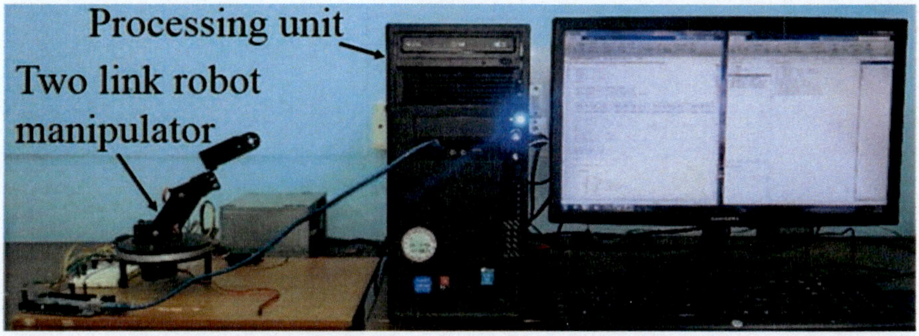

Fig. 6.3 Experimental setup of TLRM

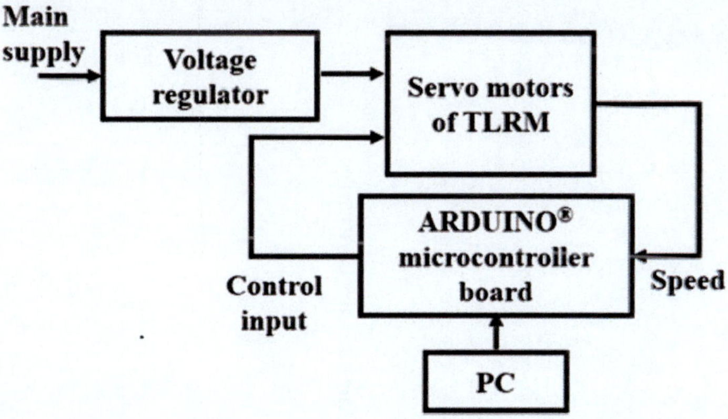

Fig. 6.4 Circuit layout of TLRM

6.4 System Identification of TLRM

The TLRM, utilized in this work, is an emulated model of large scale industrial manipulator system. System identification and controller designing of this laboratory scale TLRM system are performed as a way that it can be replicated on large scale industrial TLRM. Euler–Lagrange system dynamics of TLRM needs the exact parameter values of the system. In case of industrial TLRM system, it may be very difficult to measure the parameter values, required to formulate the system dynamics. As well as proper dynamic model cannot be formulated for system with high nonlinearities and complex relation between parameters. Considering such problem of industrial TLRM, in this chapter, the model of the TLRM is obtained utilizing Takagi–Sugeno (T-S) fuzzy logic.

To perform system identification, the TLRM experimental system is run in open loop condition with two inputs given to the two joint angles. The first joint angle is subjected to the input as shown in Fig. 6.5a and given as:

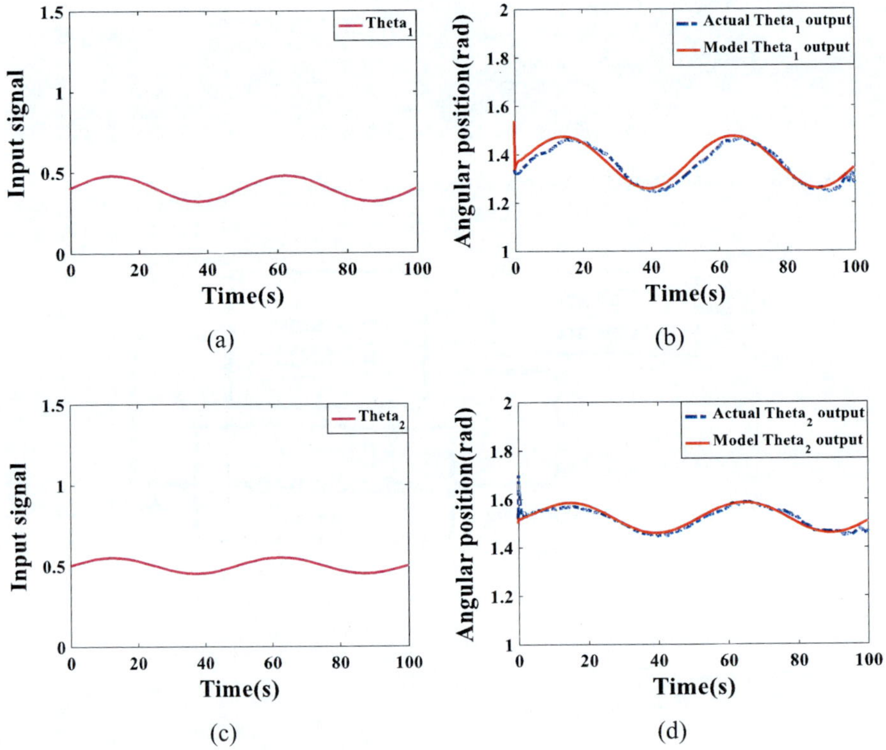

Fig. 6.5 **a** Input signal and corresponding **b** system output plot of first joint angle, **c** input signal and corresponding **d** system output plot of second joint angle in open loop configuration for parameter estimation of TLRM experimental setup

$$u_1 = 0.4 + 0.08 \sin(2\pi \times 0.02 \times t) \tag{6.19}$$

The second joint angle is given the input as depicted in Fig. 6.5c and given as:

$$u_2 = 0.5 + 0.05 \sin(2\pi \times 0.02 \times t) \tag{6.20}$$

The responses of the first joint angle and second joint angle for two corresponding inputs (6.19) and (6.20) are provided in Fig. 6.5b and Fig. 6.5d respectively with blue lines. The aim is to design a dynamical model of TLRM in simulation environment that, the system model can produce same outputs of the experimental system in open loop condition with the same inputs. The nonlinear two link robot manipulator model is designed by combining twenty five number of linear consequence rule based fuzzy systems. The errors between the reference and corresponding joint angle output of two joints are considered as the inputs to the fuzzy logic system. The linear consequence outputs of the fuzzy logic system are augmented together to produce the overall nonlinear TLRM system model with parameter uncertainties and measurement noise. The open loop responses of the TLRM experimental setup with blue lines and the open loop responses of the TLRM model with red lines are depicted in Fig. 6.5b for first joint angle and in Fig. 6.5d for second joint angle.

After modelling the TLRM system in simulation, different control methodologies are employed on it to evaluate their performances.

6.5 Experimental Case Study of TLRM

In this chapter, the effectiveness of the lbest HS-L_1 adaptive controller and fuzzy PDC-L_1 adaptive controller over PID, LQG, MRAC, basic L_1 adaptive, PSO-L_1 adaptive, HS-L_1 adaptive controller are validated by employing on a nonlinear, uncertain, disturbed MIMO two link robot manipulator (TLRM) experimental setup. The control objective is to track the desired trajectories as follows.

For first joint angle the reference is:

$$\theta_1 = 1.54 + 0.088 \sin(2\pi \times 0.02 \times t) \tag{6.21}$$

For second joint angle the reference is:

$$\theta_2 = 1.5 + 0.04 \sin(2\pi \times 0.02 \times t) \tag{6.22}$$

To design the controller for TLRM experimental setup, it is essential to design and/or tune the above mentioned controllers in simulation environment and then, those tuned controllers are employed on the experimental setup. The controller produces two control inputs for two motors in terms of voltage and two joint angle output are feedback from the TLRM experimental setup in terms of radian. The integral absolute error (IAE) of θ_1 and θ_2 for different control methodologies are tabulated in Table 6.1 and the performances are depicted in Figs. 6.6, 6.7, 6.8, 6.9, 6.10, 6.11, 6.12 and 6.13. Figure 6.6 depicts the performances of PID controller, Fig. 6.7 portrays the performances of LQG and Fig. 6.8 shows the performances

Table 6.1 Comparative study of different control methodologies for TLRM experimental setup

Control methodologies	IAE value		
	Theta$_1$	Theta$_2$	Total
PID controller	13.7201	6.7839	20.5041
LQG	4.2243	13.6230	17.8473
MRAC	7.4416	3.5960	11.0376
Basic L_1 adaptive controller	3.6545	3.6352	7.2897
PSO-L_1 adaptive controller	2.5189	3.2798	5.7987
HS-L_1 adaptive controller	2.2192	3.2855	5.5047
lbest HS-L_1 adaptive controller	2.0047	3.1775	5.1822
Fuzzy PDC-L_1 adaptive controller	1.4897	1.4864	2.9761

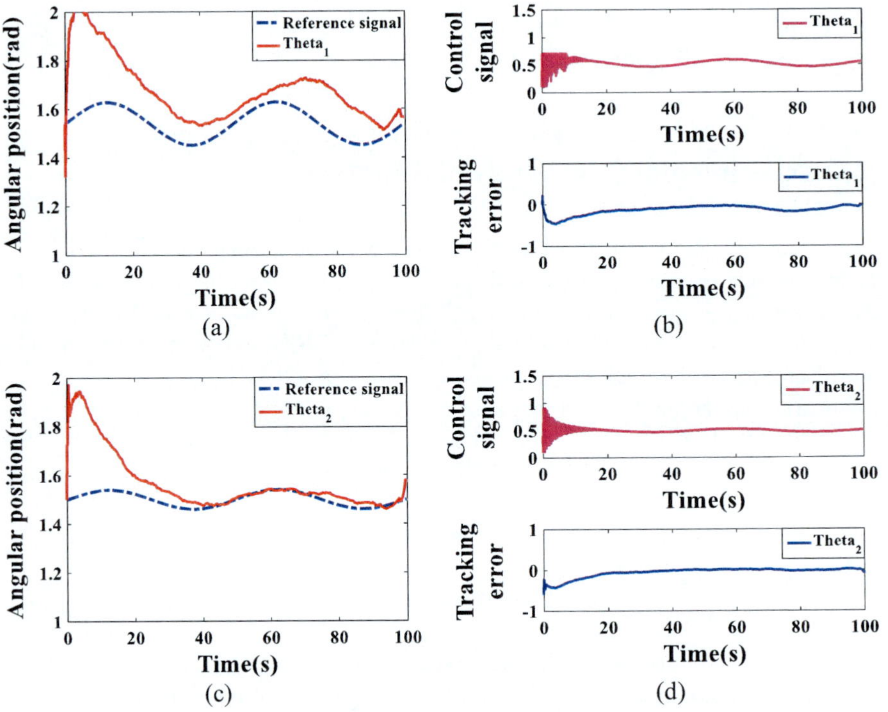

Fig. 6.6 Evaluation period **a** first joint angle response and corresponding **b** control signal with tracking error, **c** second joint angle response and corresponding **d** control signal with tracking error of TLRM experimental setup for PID controller

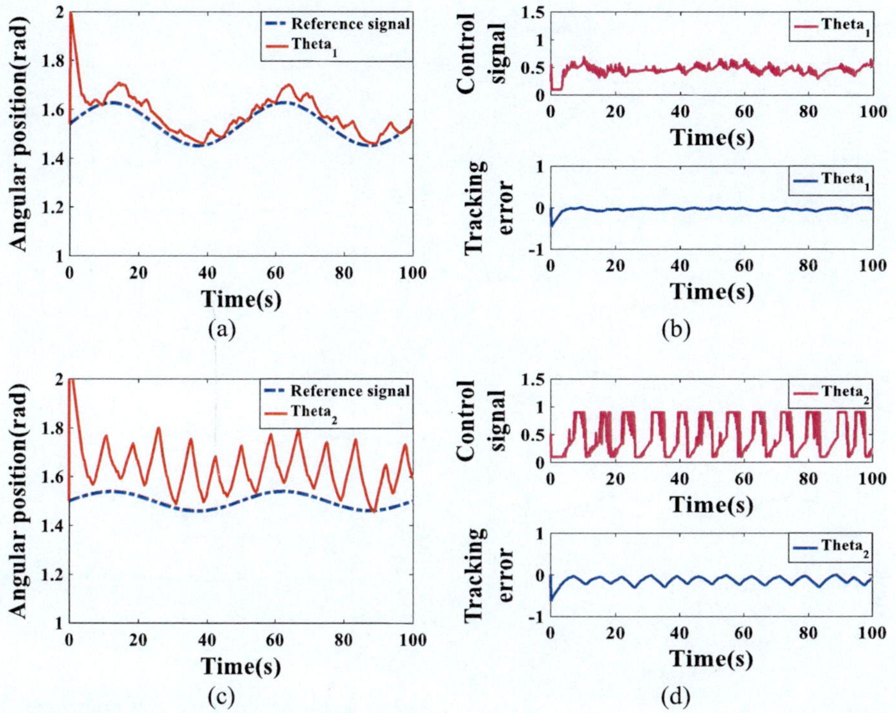

Fig. 6.7 Evaluation period **a** first joint angle response and corresponding **b** control signal with tracking error, **c** second joint angle response and corresponding **d** control signal with tracking error of TLRM experimental setup for LQG

of MRAC for both the links. The responses show that, their transient performances are sluggish and tracking performances are very poor. They cannot properly nullify the disturbances also. The performances of basic L_1 adaptive controller as shown in Fig. 6.9 are better than the classical controllers and MRAC.

The performances of lbest HS-L_1 adaptive controller as shown in Fig. 6.12 are better than the basic L_1 adaptive controller as well as PSO-L_1 adaptive and HS-L_1 adaptive controller. In case of lbest HS-L_1 adaptive controller, the disturbance rejection phenomenon and tracking performance for θ_1 is good though, tracking performance for θ_2 is not satisfactory. The performances of the fuzzy PDC-L_1 adaptive controller as shown in Fig. 6.13 are better than all other controllers tested. Its disturbance rejection phenomenon is satisfactory as well as tracking performances for both θ_1 and θ_2 are well enough than the other methods tested. In this method, the zone-wise linear design of the nonlinear system and corresponding zone-wise controller provides fast transient performance without any delay, satisfactory disturbances, uncertainties rejection and good tracking performance with high robustness.

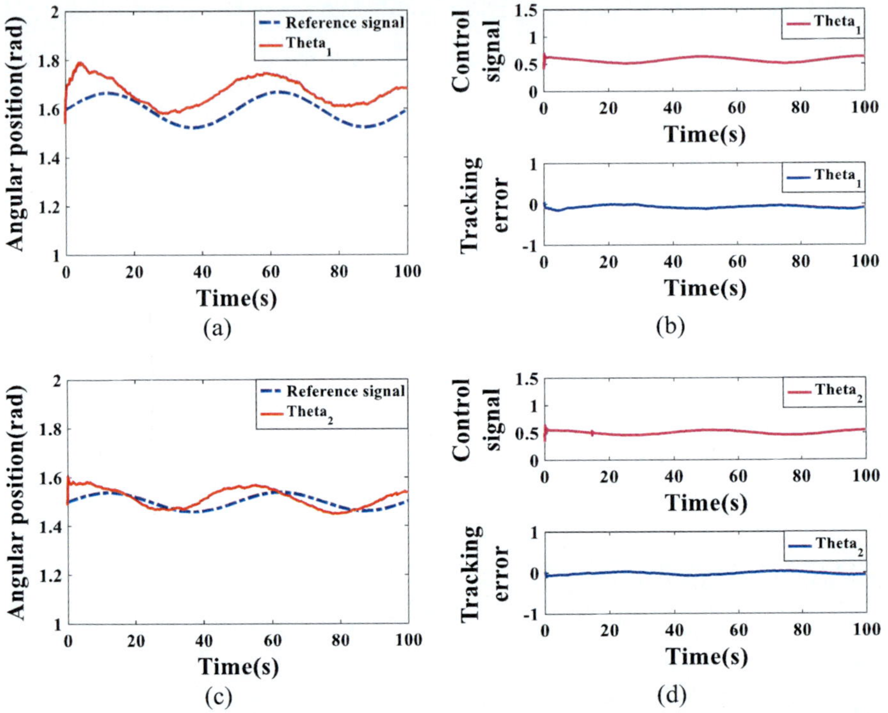

Fig. 6.8 Evaluation period **a** first joint angle response and corresponding **b** control signal with tracking error, **c** second joint angle response and corresponding **d** control signal with tracking error of TLRM experimental setup for MRAC

6.6 Summary

In this chapter, the lbest HS-L_1 adaptive and fuzzy PDC-L_1 adaptive control methodologies are employed on a two link robot manipulator (TLRM) experimental setup which exemplifies a nonlinear system with unstructured dynamics, uncertainties and disturbances. The TLRM, used in this work, is the emulated version of large scale industrial TLRM system. TLRM is a multi-input multi-output (MIMO) system where the angular positions of its two joints have to be controlled. At first, the system model is formulated by means of the fuzzy PDC logic. Then, different types of controllers, viz., PID, LQG, MRAC, basic L_1 adaptive, PSO-L_1 adaptive, HS-L_1 adaptive, lbest HS-L_1 adaptive and fuzzy PDC-L_1 adaptive controller are employed on this system. The results show that, the lbest HS-L_1 adaptive controller performs better than the basic L_1 adaptive controller whereas, the fuzzy PDC-L_1 adaptive controller performs better than all of the controllers tested in this work.

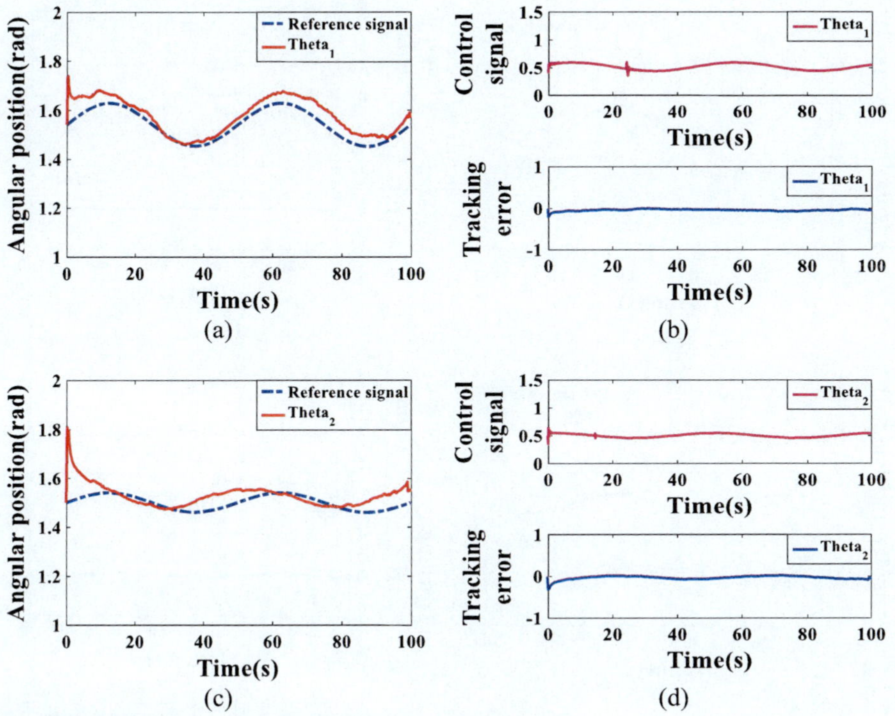

Fig. 6.9 Evaluation period **a** first joint angle response and corresponding **b** control signal with tracking error, **c** second joint angle response and corresponding **d** control signal with tracking error of TLRM experimental setup for basic L_1 adaptive controller

To investigate the effectiveness of the lbest HS-L_1 adaptive and fuzzy PDC-L_1 adaptive controllers in controlling system containing large delay along with nonlinearities, uncertainties, disturbances, etc., a delay dependent air heater system is considered in the next chapter.

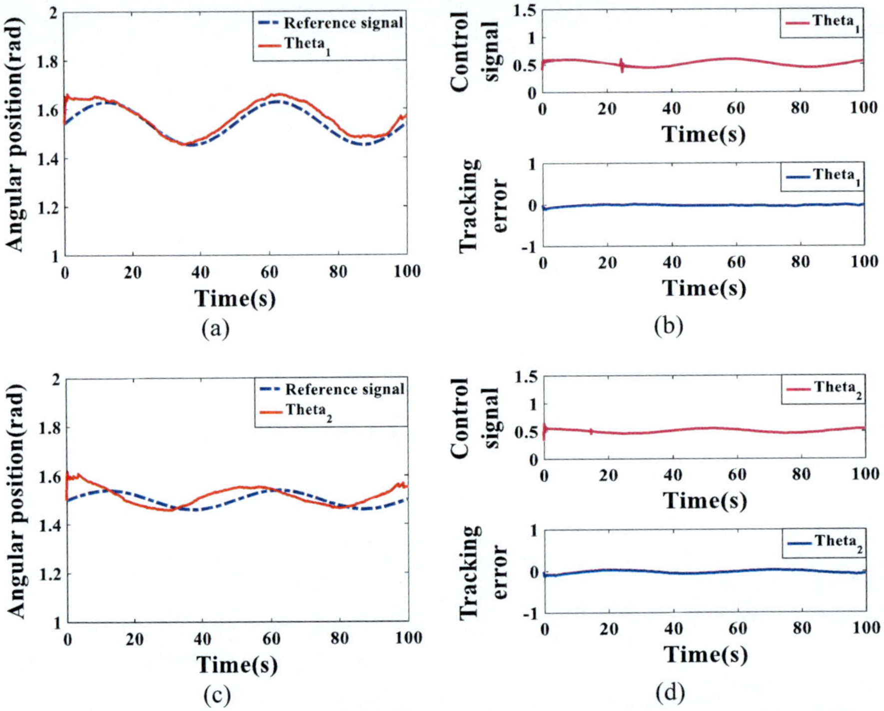

Fig. 6.10 Evaluation period **a** first joint angle response and corresponding **b** control signal with tracking error, **c** second joint angle response and corresponding **d** control signal with tracking error of TLRM experimental setup for PSO-L_1 adaptive controller

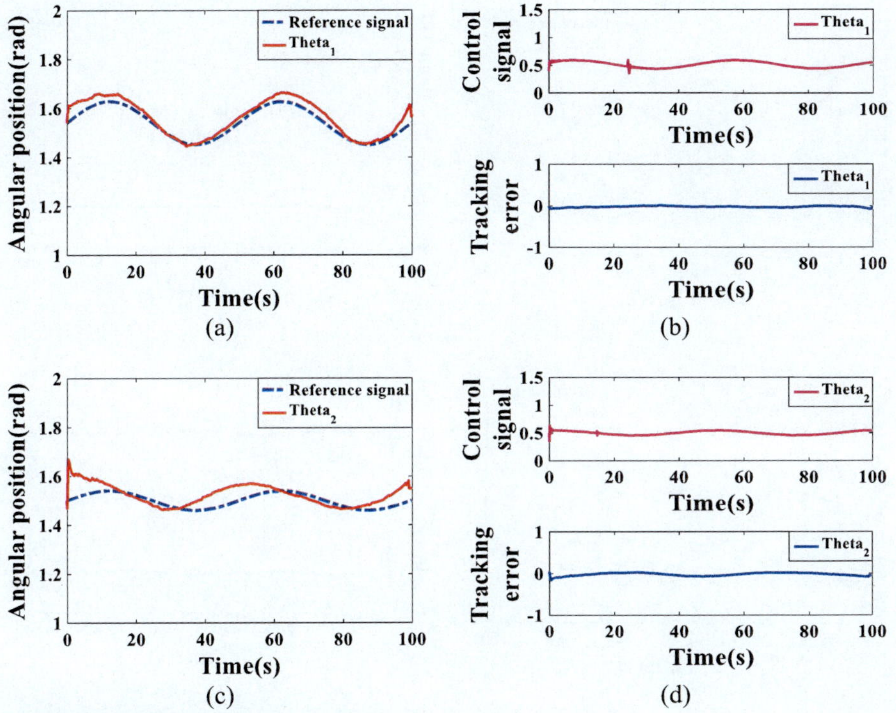

Fig. 6.11 Evaluation period **a** first joint angle response and corresponding **b** control signal with tracking error, **c** second joint angle response and corresponding **d** control signal with tracking error of TLRM experimental setup for HS-L_1 adaptive controller

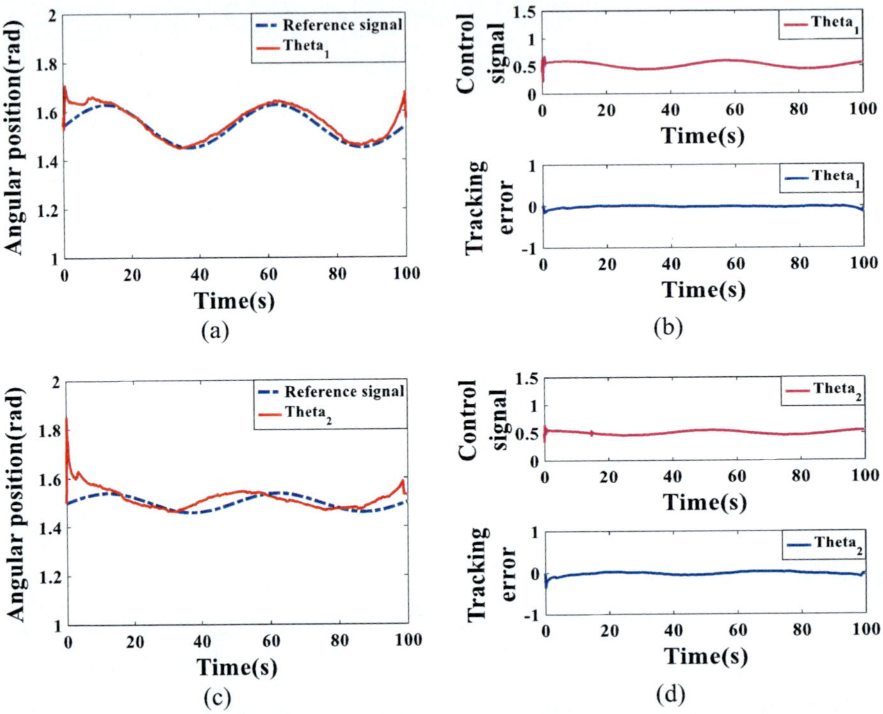

Fig. 6.12 Evaluation period **a** first joint angle response and corresponding **b** control signal with tracking error, **c** second joint angle response and corresponding **d** control signal with tracking error of TLRM experimental setup for lbest HS-L_1 adaptive controller

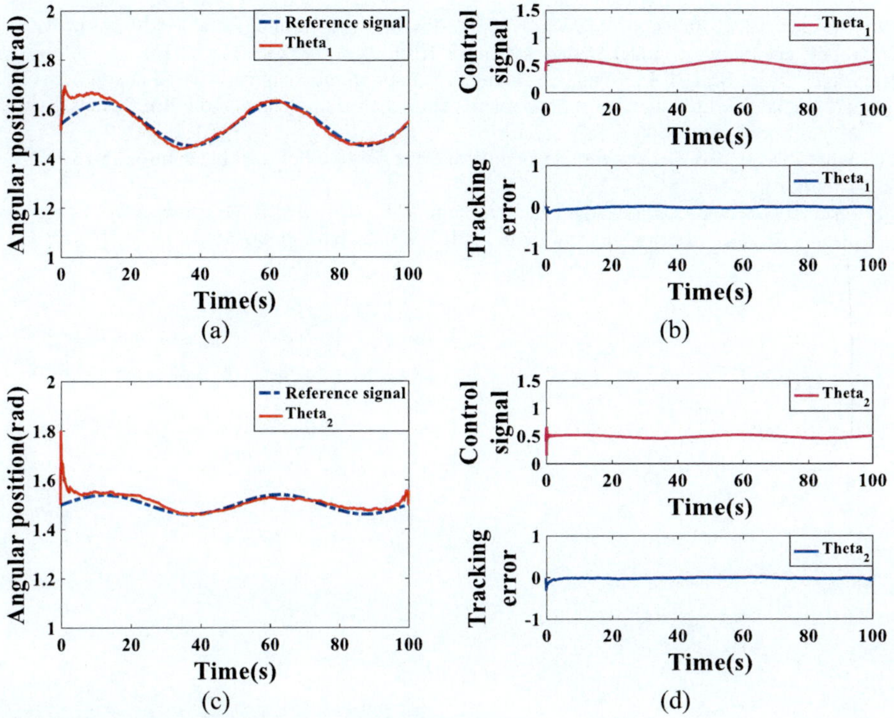

Fig. 6.13 Evaluation period **a** first joint angle response and corresponding **b** control signal with tracking error, **c** second joint angle response and corresponding **d** control signal with tracking error of TLRM experimental setup for fuzzy PDC-L_1 adaptive controller

References

1. Oh, S., Kong, K.: Two-degree-of-freedom control of a two-link manipulator in the rotating coordinate system. IEEE Trans. Industr. Electron. **62**(9), 5598–5607 (2015)
2. Wai, R., Muthusamy, R.: Fuzzy-neural-network inherited sliding-mode control for robot manipulator including actuator dynamics. IEEE Trans. Neural Netw. Learn. Syst. **24**(2), 274–287 (2013)
3. Wai, R., Muthusamy, R.: Design of fuzzy-neural-network-inherited backstepping control for robot manipulator including actuator dynamics. IEEE Trans. Fuzzy Syst. **22**(4), 709–722 (2014)
4. Chen, Y., Ma, G., Lin, S., Ning, S., Gao, J.: Computed-torque plus robust adaptive compensation control for robot manipulator with structured and unstructured uncertainties. IMA J. Math. Control. Inf. **33**(1), 37–52 (2016)
5. Pucci, D., Romano, F., Nori, F.: Collocated adaptive control of underactuated mechanical systems. IEEE Trans. Rob. **31**(6), 1527–1536 (2015)
6. Galicki, M.: Finite-time trajectory tracking control in a task space of robotic manipulators. Automatica **67**, 165–170 (2016)
7. Galicki, M.: Finite-time control of robotic manipulators. Automatica **51**, 49–54 (2015)
8. Smaeilzadeh, S.M., Golestani, M.: Finite-time fault-tolerant adaptive robust control for a class of uncertain non-linear systems with saturation constraints using integral backstepping approach. IET Control Theory Appl. **12**(15), 2109–2117 (2018)

9. Zhakatayev, A., Rubagotti, M., Varol, H.A.: Closed-loop control of variable stiffness actuated robots via nonlinear model predictive control. IEEE Access **3**, 235–248 (2015)
10. Li, T., Duan, S., Liu, J., Wang, L., Huang, T.: A spintronic memristor-based neural network with radial basis function for robotic manipulator control implementation. IEEE Trans. Syst., Man, Cybern.: Syst. **46**(4), 582–588 (2016)
11. Yih, C.: Extended Nicosia-Tomei velocity observer-based robot-tracking control. IET Control Theory Appl. **6**(1), 51–61 (2012)
12. Fischer, N., Kan, Z., Kamalapurkar, R., Dixon, W.E.: Saturated RISE feedback control for a class of second-order nonlinear systems. IEEE Trans. Autom. Control **59**(4), 1094–1099 (2014)

Chapter 7
Temperature Control of Air Heater System

Abstract In this chapter, temperature control of a delay dependent, nonlinear air heater system is investigated. The results show that, the lbest HS-L_1 adaptive controller tracks the reference temperature trajectory better than the basic L_1 adaptive controller. The fuzzy PDC-L_1 adaptive controller outperforms the other controllers by nullifying time varying uncertainties and disturbances utilizing fuzzy L_1 adaptive controller and handling nonlinearities, delays with the help of fuzzy PDC controller.

7.1 Introduction

In this chapter, a nonlinear, delay dependent air heater system containing uncertainties and disturbances is investigated owing to its enormous applications in different industrial processes, viz., variable air volume (VAV) system [1], heating, ventilating, air-conditioning (HVAC) systems [2], air pre-heater [3], boiler arrangement [4], etc. The hot air mass flow through a duct following conduction, convection and radiation introduces nonlinearities and uncertainties into the heat transfer model [5, 6]. The temperature sensors, utilized at different positions of the system, incorporate measurement noise in the system output. Change in the blower fan speed creates variation in the air mass flow rate which predominantly introduces disturbances and uncertainties in the air heater system. Model predictive controllers (MPC) [2, 7], iterative learning control (ILC) scheme [8] gain enormous attention to control air-conditioning systems. Though they require accurate modelling of the thermal system which is difficult due to the difference in heater material, duct material, ambient temperature, air mass flow rate, etc., which makes these controllers unsuitable for large scale systems.

In large scale industrial process, controllers have to maintain desired temperatures at various points far from the heater which introduces transport delay in the heat transfer model. A harmony search (HS) based adaptive fuzzy controller was developed to control a delayed air heater system [9] though the system model was considered as linear and no explicit measures were taken to cope up with delay exist in the system. Presence of large delay in any system causes unsatisfactory transient performance and even instability of the system [10]. Hence, explicit measures

R. Maiti et al., *Hybrid L_1 Adaptive Control*, Studies in Systems, Decision and Control 422, https://doi.org/10.1007/978-3-030-97102-1_7

have to be taken to tackle large delay present in the system. Pure time delay system was controlled through Smith predictor and proportional integral derivative (PID) control [11]. Though, the dependency on accurate mathematical model of the Smith predictor degrades its performance in case of time varying parameters. Random time delay compensation was proposed in [12], where the network-induced delay was estimated by employing a Kalman filter and then modelled utilizing Padé approximation. However, linearization of the system model in [12] could not provide appropriate control action. In another work, an adaptive fuzzy predictive control scheme was employed to control nonlinear systems with time-varying delay [13]. It utilized low adaptation gain which provided sluggish transient performance and no separate means were taken to nullify disturbances present in the system.

Fuzzy parallel distributed compensation (PDC) strategy can be utilized efficiently to model nonlinearities present in the system and to produce efficient control action by generating zone wise state feedback control law [14]. Though, it fails to nullify time varying uncertainties and disturbances properly. L_1 adaptive controller achieves much attention [15, 16] to tackle systems with time varying uncertainties and disturbances providing fast transient performance with high robustness. However, the performance of L_1 adaptive controller degrades due to the use of fixed state feedback control gain when nonlinearities, delays enter into the system.

In this chapter, an experimental model of air heater system, indigenously developed at the Systems Engineering Laboratory of Applied Physics Department, University of Calcutta, Kolkata, India, is investigated. Controllers are implemented for the laboratory scale system as a representative way that it can readily be implemented in large scale HVAC systems. In this work, at first, the overall air heater system model is effectively formulated by conjugating heat transfer model quantities, viz., nonlinearities, time varying uncertainties, and disturbances as well as transport delay. Then, the air heater is controlled through different types of controllers, viz., proportional integral derivative (PID), linear quadratic Gaussian (LQG), model reference adaptive controller (MRAC), basic L_1 adaptive, particle swarm optimization based L_1 (PSO-L_1) adaptive, basic harmony search based L_1 (HS-L_1) adaptive, local best HS based L_1 (lbest HS-L_1) adaptive and fuzzy parallel distributed compensation type L_1 (fuzzy PDC-L_1) adaptive controller. In case of fuzzy PDC-L_1 adaptive controller, the unknown constant, time varying uncertainties, time varying disturbances are tackled utilizing fuzzy L_1 adaptive controller and the time delays, nonlinearities are coped up with fuzzy PDC logic. Hence, this controller provides superior performance to maintain desired temperature of the air heater system.

7.2 Dynamical Model of Air Heater System

The air heater experimental model as provided in Fig. 7.1 has a hollow 'U' shaped duct of polyvinyl chloride (PVC) through which hot air flows. A heating element with a temperature sensor named as Sensor 1 is mounted at the inner side of a heat sink, placed at the inlet of air flow duct. The heat sink is utilized to clamp the heating

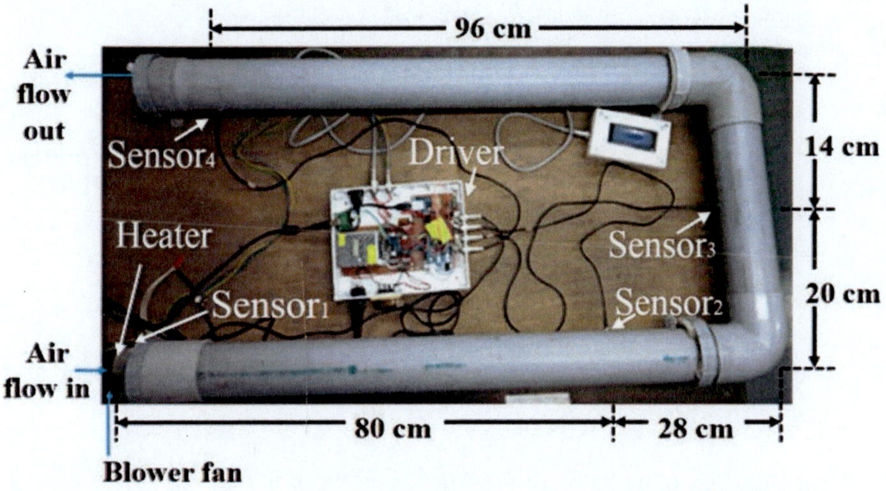

Fig. 7.1 Experimental model of air heater system

material by providing safety to the other materials of the air heater system. At the outer side of that heat sink a blower fan with maximum speed 2500 rpm [17] is mounted as shown in Fig. 7.2a. This forces the hot air to blow through the duct and it also acts as the non-uniform disturbance to the system. At a straight distance of 80 cm from the first sensor, the second sensor, viz., Sensor 2 is mounted. After 28 cm of that, the pipe bents in 90° angle. Sensor 3 is mounted after 20 cm of that bent. 14 cm after that sensor, the pipe bents again at 90° angle. From that bent at a distance of 96 cm, the Sensor 4 is mounted and after that the hot air flows out from the duct.

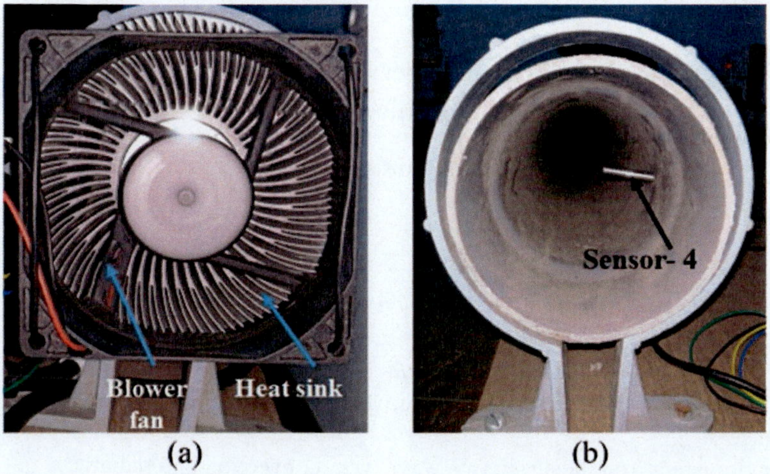

Fig. 7.2 Front side view of **a** inlet, and **b** outlet of the air flow duct

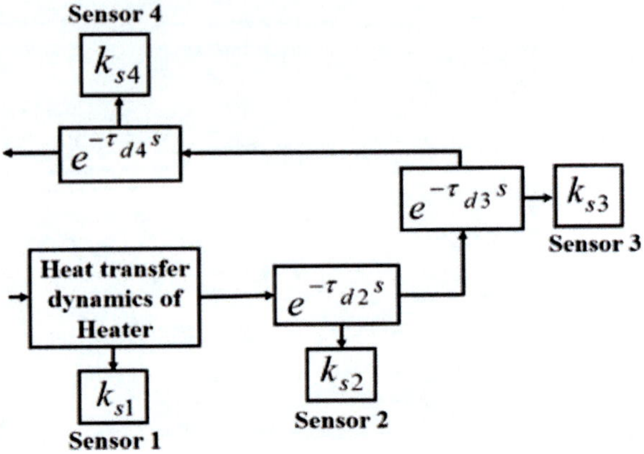

Fig. 7.3 Schematic diagram of air heater system

The front side view of outlet of air flow duct is provided in Fig. 7.2b. Therefore, at Sensor 4 large transport delay is present along with nonlinearities, uncertainties and disturbances.

The air heater system model consists of heat transfer dynamics and transport delay at temperature sensors as shown in Fig. 7.3.

The overall air heater system dynamics is formulated for Sensor 4 due to the maximum transport delay inclusion as elaborated follows.

Let, $\dot{W}_h$ represents the rate of electric energy input to the heater and $\dot{Q}_h$ is the rate of heat transfer from the heater. Then, assuming the heater temperate remains uniform, the energy equation is considered as:

$$\dot{W}_h - \dot{Q}_h = m_h c_{vh} \frac{dT}{dt} \tag{7.1}$$

where m_h and c_{vh} are the mass and specific heat of the heater respectively. Therefore, $m_h c_{vh}$ is the thermal capacity of the heater.

In this air heater system, the heater experiences considerable amount of temperature change (from ambient temperature $T_a \approx 30$ °C to maximum temperature ~100 °C) which causes change in the resistance of the heating material which is given as:

$$\dot{W}_h = \frac{E^2}{R} = \frac{E^2}{R_0(1 + \alpha_r(T - T_0))} \tag{7.2}$$

where E is the applied voltage, R and R_0 are the resistance at operating temperature (T) and 0 °C temperate (T_0) respectively. α_r is the temperature coefficient of resistance.

Heat transfers from heater in two ways, viz., convection and radiation.

Convection heat transfer dynamics is considered as:

$$\dot{Q}_{h\_conv} = h_h A_h (T - T_a) \tag{7.3}$$

where A_h is the surface area of heating element, h_h is the convection heat transfer coefficient and T_a is the ambient temperature.

Radiation heat transfer dynamics is considered as:

$$\dot{Q}_{h\_rad} = \varepsilon_h \sigma_h A_h (T^4 - T_a^4) \tag{7.4}$$

where σ_h is the Stefan-Boltzmann constant, ε_h is the emissivity of the heater surface.

Therefore, the complete thermal model of heater becomes:

$$\dot{T} = \frac{1}{m_h c_{vh}} \left[\frac{E^2}{R_0(1 + \alpha_r (T - T_0))} - h_h A_h (T - T_a) - \varepsilon_h \sigma_h A_h \left(T^4 - T_a^4 \right) \right] \tag{7.5}$$

Let, the heater model (7.5) is represented as:

$$\dot{\underline{x}}_h = A_1 \underline{x}_h + b_1 u_h \tag{7.6}$$

$$y_h = c_1 \underline{x}_h \tag{7.7}$$

where $\underline{x}_h$ is the state of the heater, y_h is the output of the heater, A_1, b_1, c_1 are the system matrix, input matrix and output matrix of the heater respectively.

The heat flows from the heater through the long air duct where at different positions temperature sensors are attached. The complete air heater system model is formulated considering the heat transfer dynamics and maximum transport delay $e^{-\tau_{d4} s}$ of Sensor 4 as depicted in Fig. 7.3. Figure 7.3 also depicts that, the output of the heater model is considered as the input of the delay model. The delay present in the air heater system model can be realized utilizing Padé approximation method. Though the Padé approximation provides linear time invariant (LTI) systems equivalent to the delayed system. Therefore, the uncertainties present in the time varying delay due to the change in ambient temperature and natural disturbances cannot be modelled properly. Instead, the delay can be efficiently modelled by combining linear fuzzy systems models. These two types of fuzzy modelling are discussed in the following sections.

7.2.1 *Padé Approximation Based Delay Modelling*

The delay $e^{-\tau_{d4} s}$ is modelled utilizing 3rd order Padé approximation. The output of the heater model is considered as the input of the delay model, viz., $u_{dl} = y_h = c_1 \underline{x}_h$.

Therefore, the delay model is represented as:

$$\dot{\underline{x}}_{dl} = A_2\underline{x}_{dl} + b_2u_{dl} = A_2\underline{x}_{dl} + b_2c_1\underline{x}_h \tag{7.8}$$

$$y_{dl} = c_2\underline{x}_{dl} \tag{7.9}$$

where $\underline{x}_{dl}$ is the state and y_{dl} is the output of the delayed system, A_2, b_2 and c_2 are the system matrix, input matrix and output matrix of the delayed system respectively.

As, the delay model is formulated utilizing 3rd order Padé approximation, therefore, it has three states, viz., $\underline{x}_{dl} = [x_1, x_2, x_3]$. $E = u$ is the input to the system.

Let, $x_{dl_1} = x_1$.

Then, the overall delay dependent air heater system model becomes [18]:

$$\dot{x}_1 = x_2 \tag{7.10}$$

$$\dot{x}_2 = x_3 \tag{7.11}$$

$$\dot{x}_3 = A_{2_{31}}x_1 + A_{2_{32}}x_2 + A_{2_{33}}x_3 + b_{2_3}c_1x_4 \tag{7.12}$$

$$\dot{x}_4 = \frac{1}{m_h c_{vh}}\left[\begin{array}{l}\dfrac{u^2}{R_0(1+\alpha_r(x_4 - T_0))} \\ -h_h A_h(x_4 - T_a) - \varepsilon_h\sigma_h A_h\left(x_4^4 - T_a^4\right)\end{array}\right] \tag{7.13}$$

7.2.2 *Fuzzy Linear Consequence Rule Based Delay Modelling*

The delay can be represented as the summation of w (finite) number of linear systems as follows:

$$\dot{\underline{x}}_{dl} = \sum_{i=0}^{w} A_{2_i}\delta^i\underline{x}_{dl} + \sum_{i=0}^{w} b_{2_i}\delta^i u_{dl}$$

or

$$\dot{\underline{x}}_{dl} = \sum_{i=0}^{w} A_{2_i}\delta^i\underline{x}_{dl} + \sum_{i=0}^{w} b_{2_i}\delta^i c_1\underline{x}_h \tag{7.14}$$

$$y_{dl} = \sum_{i=0}^{w} c_{2_i}\delta^i\underline{x}_{dl} \tag{7.15}$$

where $\underline{x}_{dl}$ is the state and y_{dl} is the output of the delayed system, A_2, b_2 and c_2 are the system matrix, input matrix and output matrix of the delayed system respectively. δ represents the delay.

Let, $x_{dl} = x_1$ and $x_h = x_2$.

Then, the overall air heater system model with delay becomes:

$$\dot{x}_1 = \sum_{i=0}^{w} A_{2_i}\delta^i x_1 + \sum_{i=0}^{w} b_{2_i}\delta^i c_1 x_2 \tag{7.16}$$

$$\dot{x}_2 = \frac{1}{m_h c_{vh}}\begin{bmatrix} \dfrac{E^2}{R_0(1+\alpha_r(x_2 - T_0))} \\ -h_h A_h(x_2 - T_a) - \varepsilon_h \sigma_h A_h\left(x_2^4 - T_a^4\right) \end{bmatrix} \tag{7.17}$$

In matrix form, it can be written as:

$$\begin{bmatrix} \dot{x}_1 \\ \dot{x}_2 \end{bmatrix} = \begin{bmatrix} \sum\limits_{i=0}^{q} A_{2_i}\delta^i & \sum\limits_{i=0}^{q} b_{2_i}\delta^i c_1 \\ 0 & -\varepsilon_h\sigma_h A_h x_2^3 \end{bmatrix}\begin{bmatrix} x_1 \\ x_2 \end{bmatrix} + \begin{bmatrix} 0 \\ \frac{1}{m_h c_{vh}} \end{bmatrix}\begin{bmatrix} \dfrac{u^2}{R_0(1+\alpha_r(x_2 - T_0))} + [0 - h_h A_h]\begin{bmatrix} x_1 \\ x_2 \end{bmatrix} \\ +\left[h_h A_h T_a + \varepsilon_h\sigma_h A_h T_a^4\right] \end{bmatrix} \tag{7.18}$$

Again it can be written as:

$$\underline{\dot{x}}(t) = A_0\underline{x}(t) + b\left(f\underline{u}(t) + \underline{\theta}(t)\,\underline{x} + \underline{\sigma}(t)\right) \tag{7.19}$$

where $\underline{x}(t)$ is the system state, $\underline{u}(t)$ is the control input.

$$A_0 = \begin{bmatrix} \sum\limits_{i=0}^{w} A_{2_i}\delta^i & \sum\limits_{i=0}^{w} b_{2_i}\delta^i c_1 \\ 0 & -\varepsilon_h\sigma_h A_h x_2^3 \end{bmatrix}$$

is the open loop system matrix, contains nonlinearities and delays.

$b = \left[0\ 1/m_h c_{vh}\right]^T$ is the input matrix. $\frac{u}{R_0(1+\alpha_r(x_2-T_0))} = f$ represents nonlinearities related with the control input of the system, $\left[0\ -h_h A_h\right] = \underline{\theta}(t)$ represents parametric uncertainties vector, $\left[h_h A_h T_a + \varepsilon_h\sigma_h A_h T_a^4\right] = \sigma(t)$ represents disturbances present in the system.

Now, the delays, viz., $\sum_{i=0}^{w} A_{2_i}\delta^i$ and $\sum_{i=0}^{w} b_{2_i}\delta^i$ and nonlinearities present in the air heater system are represented as the combinations of W number of linear consequence rule based fuzzy systems. Therefore, the overall air heater system model is given as:

$$\underline{\dot{x}}(t) = \sum_{w=1}^{W} \left(A_{0_w} \underline{\xi}_w \underline{x}(t) + b_w \underline{\xi}_w \left(f_w \underline{\xi}_w \underline{u}(t) + \underline{\theta}(t)\underline{x} + \underline{\sigma}(t) \right) \right) \tag{7.20}$$

$$\underline{y}(t) = \sum_{w=1}^{W} \underline{\xi}_w c_w \underline{x}(t) \tag{7.21}$$

where $\underline{\xi}_w$ is the fuzzy basis function as described in the Chap. 4.

After developing the mathematical model, description of experimental setup of the air heater system is provided in the next section.

7.3 Description of Experimental Setup of Air Heater System

The experimental setup of the air heater system as shown in Fig. 7.4 consists of the air heater system, driver circuit and PC. The control signal is generated in PC within a sophisticated software MATLAB® and provided in terms of voltage to the system through the ARDUINO® microcontroller board and driver circuit. The blower fan also gets input in the similar way. Consequently, the heater element heated up and the blower fan, which also acts as the non-uniform, time varying disturbance to the system, forces the hot air to blow through the duct. The temperate of different places of the duct are sensed by the temperature sensors (DS18B20) and fed back to the

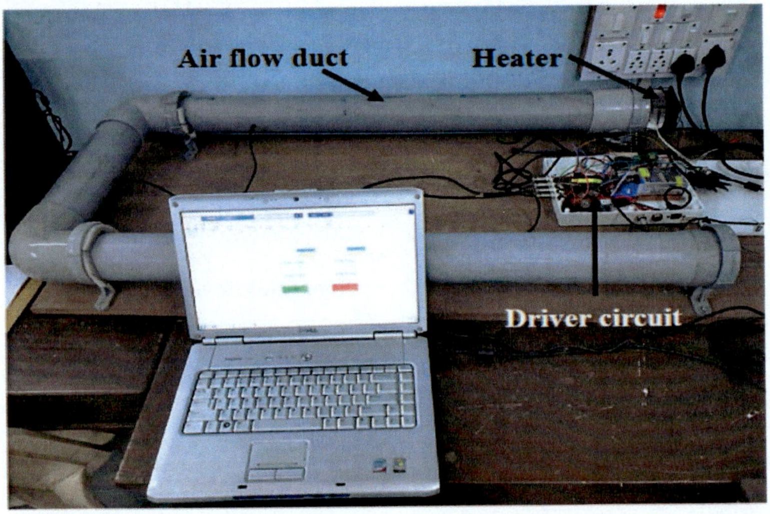

Fig. 7.4 Experimental setup of air heater system

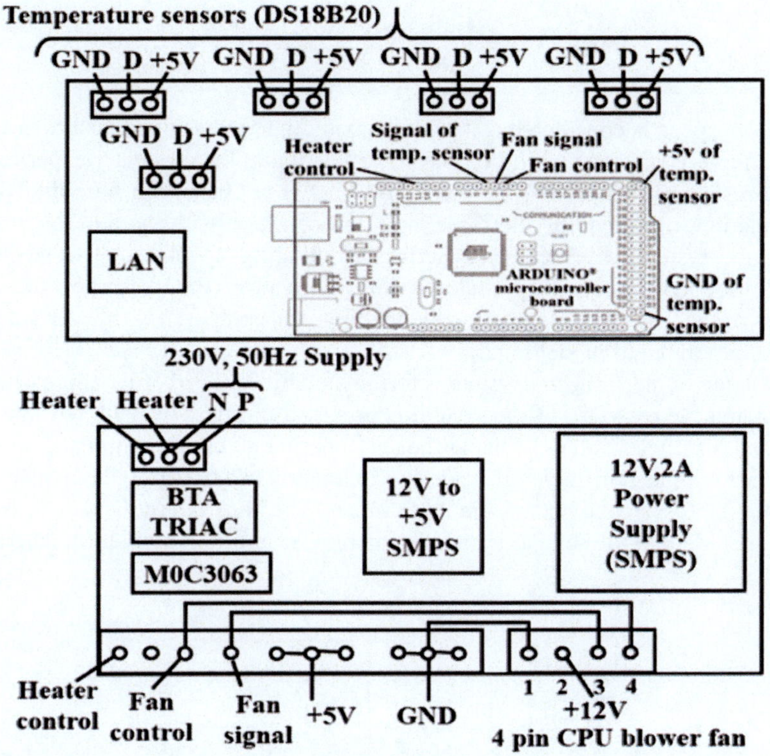

Fig. 7.5 Layout of the driver circuit of air heater system

MATLAB® through the ARDUINO® microcontroller board and driver circuit. The driver circuit layout is provided in Fig. 7.5.

Remarks 7.1 The DS18B20 can sense temperature up to 125 °C [19]. For safety purpose, the temperature of Sensor 1, attached beside the heater, is limited to 100 °C. It has been seen from the experiment that, in Sensor 4 temperature reduces near about 2.5 times that of the Sensor 1. In turn, temperature of Sensor 4 is limited to 40 °C.

At first, the system identification is performed and then the controllers are developed for the air heater system.

7.4 System Identification of Air Heater System

To identify the parameter values of the air heater system, a variable step input is provided to the experimental setup of the air heater system in open loop condition which is given as:

$$\text{Input: } u(t) = \begin{cases} 30\ \text{V} & 0\ \text{s} \le t \le 1800\ \text{s} \\ 90\ \text{V} & 1800\ \text{s} < t \le 3600\ \text{s} \end{cases}$$

The fan speed is considered as 70% of maximum speed as fixed disturbance.

This input is also shown in Fig. 7.6a. For this input, the air heater experimental system response at Sensor 4 is shown both in Fig. 7.6b, c with blue line. Then, in simulation, by providing the same input in open loop with the aim to get same output, the system identification is performed utilizing a representative stochastic optimization technique, viz., particle swarm optimization (PSO). To show the effectiveness of the fuzzy PDC logic based modelling, it is compared with Padé 3rd order approximation based modelling.

The delay of the air heater system is formulated by Padé 3rd order approximation method and the response of the system is provided in Fig. 7.6b with red line. The estimated parameter values of the air heater experimental setup with Padé approximation are tabulated in Table 7.1. Padé approximation of delay provides a linear time invariant (LTI) system of exact parameter values. The uncertainties and disturbances which include due to the heat transfer dynamics, changes in ambient temperature,

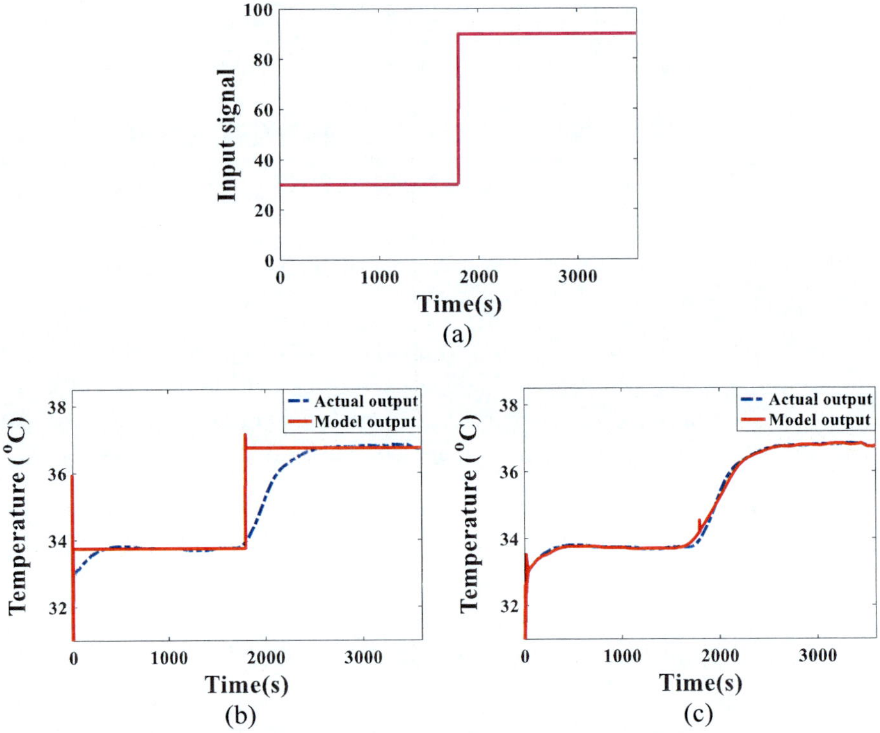

Fig. 7.6 **a** Input signal and corresponding open loop system output plot of **b** Padé and **c** fuzzy PDC logic based delay modelling for parameter estimation of air heater experimental setup

Table 7.1 Estimated nominal parameter values of air heater system with Padé approximation of delay

h_h	A_h	m_h	c_{vh}
0.4389 W/m^2 K	0.7659 m^2	2.8994 kg	4.5743 J/kg K

ε_h	σ_h	R_0	α_r	τ
0.0099	0.0330	9.6700 Ω	0.3169	5.0983 s

etc., influence on the delay which cannot be identified by Padé approximation and hence it yields improper modelling of the system.

To obtain accurate model, the delayed, nonlinear system is formulated utilizing fuzzy PDC logic also considering the uncertainties and disturbances. The nonlinearities and delays present in the system are realized by combining 25 linear fuzzy systems and the system response is shown in Fig. 7.6c with red line. The integral absolute error of Padé approximation and fuzzy PDC logic based approximation are 92.8367 and 18.9369 respectively. The results show that, the fuzzy PDC logic can model the air heater system with nonlinearities, transport delay, uncertainties and disturbances better than the Padé approximation method.

In the next section, different types of controllers are employed on air heater system.

7.5 Experimental Case Study of Air Heater System

7.5.1 Experiment-I

In this case, the temperature control of Sensor 4 is considered due to the presence of maximum transport delay. The control objective is to maintain a desired temperature at Sensor 4 by tackling nonlinearities, transport delay, uncertainties and blower fan disturbance present in the system. The desired trajectory is set as follows:

$$Reference\ signal\text{: Desired temperature } (T_d) = 36\,^{\circ}\text{C},\ 0\,\text{s} \leq t \leq 1200\,\text{s}$$

with fixed fan speed (disturbance) 100% of maximum speed.

Hence, as of Remarks 7.1, the Sensor 1 temperature will be ~90 °C.

The ambient temperature is considered as $T_a = 31$ °C.

The controllers are designed for Sensor 4 and then, the temperature tracking performance of Sensor 1 without delay and Sensor 4 with delay are tabulated in Table 7.2 in terms of IAE. Different types of controllers, viz., PID, LQG, MRAC, basic L_1 adaptive, PSO-L_1 adaptive, HS-L_1 adaptive, lbest HS-L_1 adaptive and fuzzy PDC-L_1 adaptive controllers are employed on the air heater experimental setup. The system responses of Sensor 1 and respective control signal, error signal plots for above mentioned controllers are provided in Figs. 7.7, 7.8, 7.9, 7.10, 7.11, 7.12, 7.13 and

Table 7.2 Comparative study of different control methodologies in experiment-I for delay dependent air heater experimental setup

Control methodologies	IAE value	
	Sensor 1	Sensor 4
PID controller	623.7500	175.6250
LQG	617.9000	154.6750
MRAC	713.7000	154.3000
Basic L_1 adaptive controller	568.0000	118.2700
PSO-L_1 adaptive controller	456.0000	110.2500
HS-L_1 adaptive controller	525.0500	103.5000
lbest HS-L_1 adaptive controller	505.6850	80.3700
Fuzzy PDC-L_1 adaptive controller	352.8500	74.1150

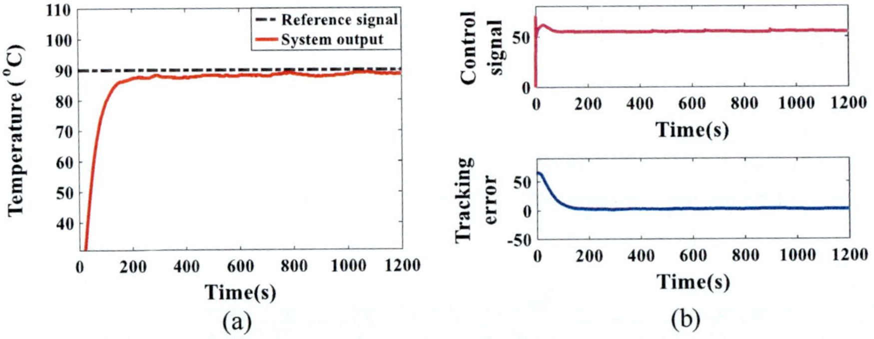

Fig. 7.7 Evaluation period **a** system response and corresponding **b** control signal with tracking error at Sensor 1 of air heater experimental setup for PID controller

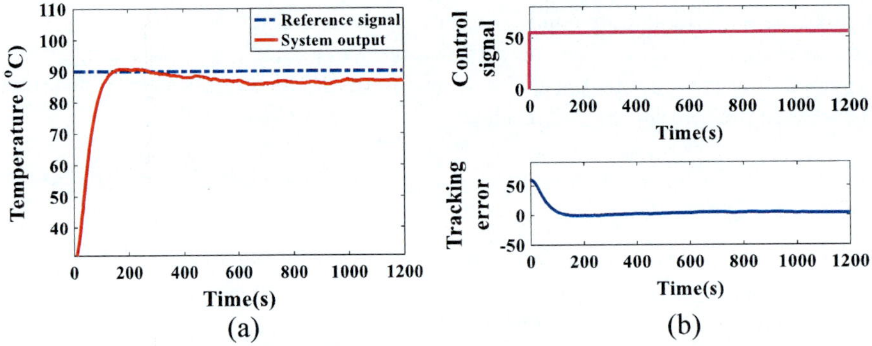

Fig. 7.8 Evaluation period **a** system response and corresponding **b** control signal with tracking error at Sensor 1 of air heater experimental setup for LQG

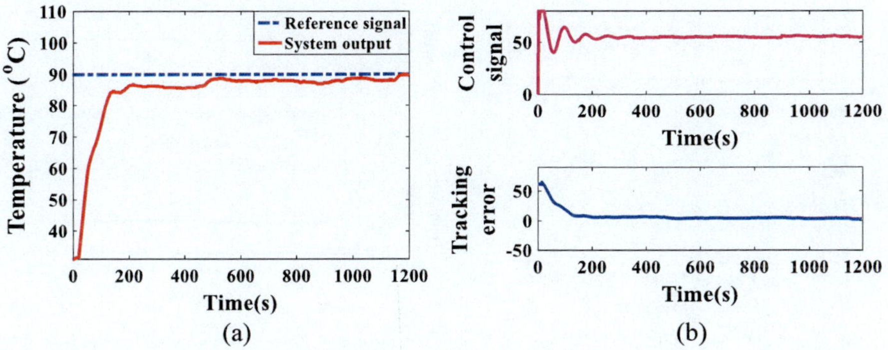

Fig. 7.9 Evaluation period **a** system response and corresponding **b** control signal with tracking error at Sensor 1 of air heater experimental setup for MRAC

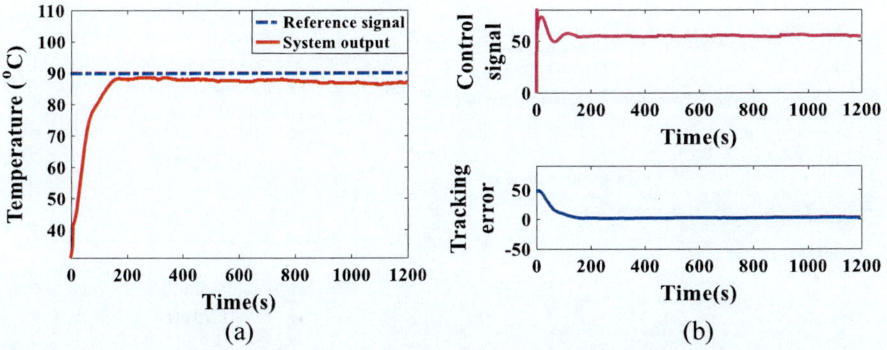

Fig. 7.10 Evaluation period **a** system response and corresponding **b** control signal with tracking error at Sensor 1 of air heater experimental setup for basic L_1 adaptive controller

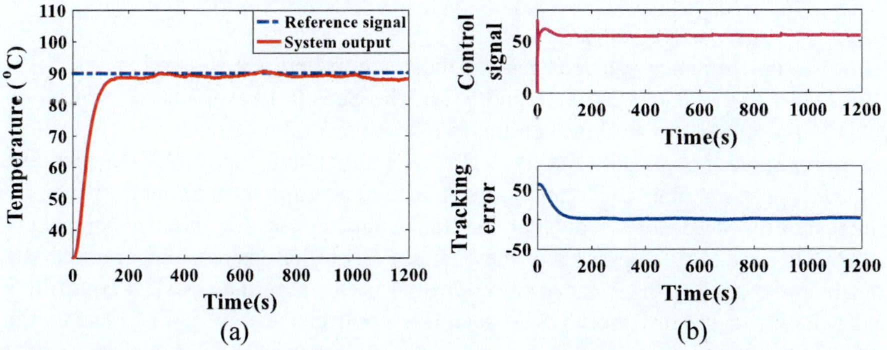

Fig. 7.11 Evaluation period **a** system response and corresponding **b** control signal with tracking error at Sensor 1 of air heater experimental setup for PSO-L_1 adaptive controller

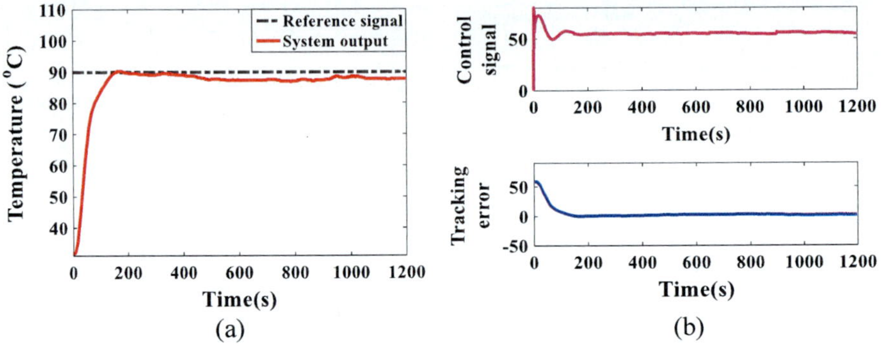

Fig. 7.12 Evaluation period **a** system response and corresponding **b** control signal with tracking error at Sensor 1 of air heater experimental setup for HS-L_1 adaptive controller

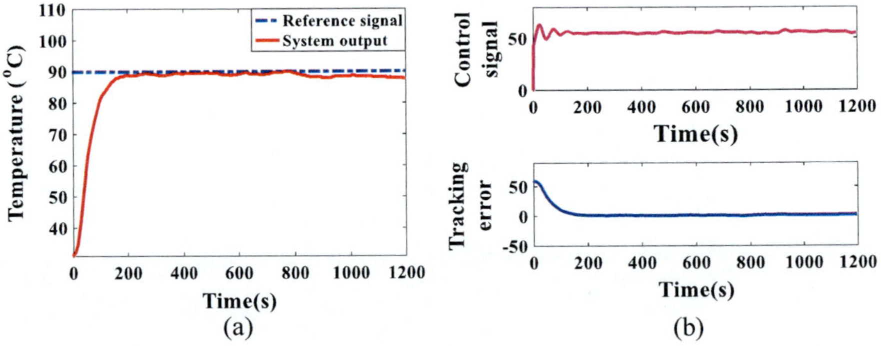

Fig. 7.13 Evaluation period **a** system response and corresponding **b** control signal with tracking error at Sensor 1 of air heater experimental setup for lbest HS-L_1 adaptive controller

7.14. The performances of Sensor 4 for those controllers are shown in Figs. 7.15, 7.16, 7.17, 7.18, 7.19, 7.20, 7.21 and 7.22. The Sensor 1 results show that, if no delay is present in the system then the transient performance of all the controllers are comparable though only the lbest HS-L_1 adaptive and fuzzy PDC-L_1 adaptive controllers can maintain the temperature at desired position throughout the run time. The responses of Sensor 4 show that, the transient responses for classical controllers and MRAC are very sluggish and they cannot reach the desired temperature. The transient response of the L_1 adaptive controller is fast than the classical controllers and MRAC. In case of lbest HS-L_1 adaptive controller, the response reaches the desired temperature though cannot maintain the temperature throughout the run time by nullifying time varying uncertainties and external disturbances. The transient and tracking response of the fuzzy PDC-L_1 adaptive controller is best among all other controllers tested.

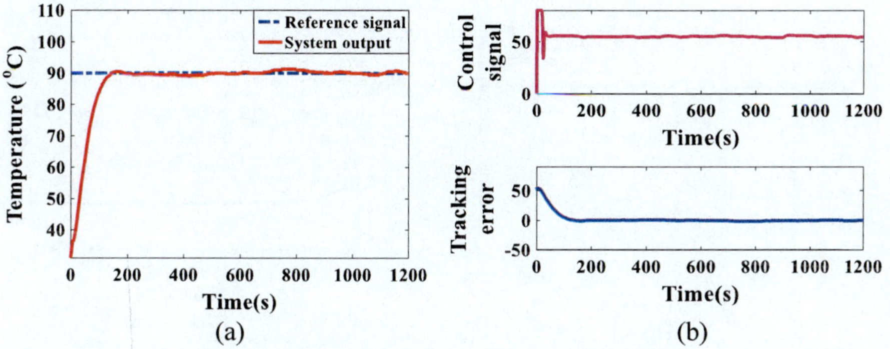

Fig. 7.14 Evaluation period **a** system response and corresponding **b** control signal with tracking error at Sensor 1 of air heater experimental setup for fuzzy PDC-L_1 adaptive controller

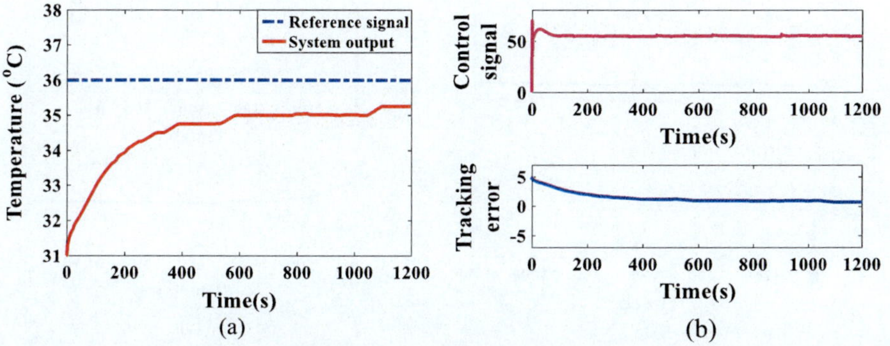

Fig. 7.15 Evaluation period **a** system response and corresponding **b** control signal with tracking error at Sensor 4 of air heater experimental setup for PID controller

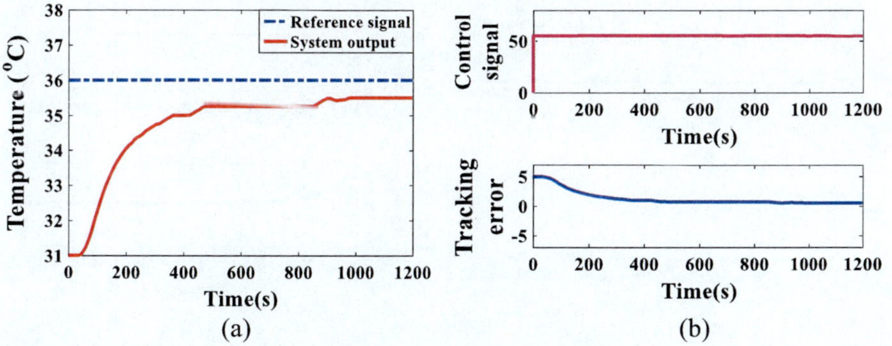

Fig. 7.16 Evaluation period **a** system response and corresponding **b** control signal with tracking error at Sensor 4 of air heater experimental setup for LQG

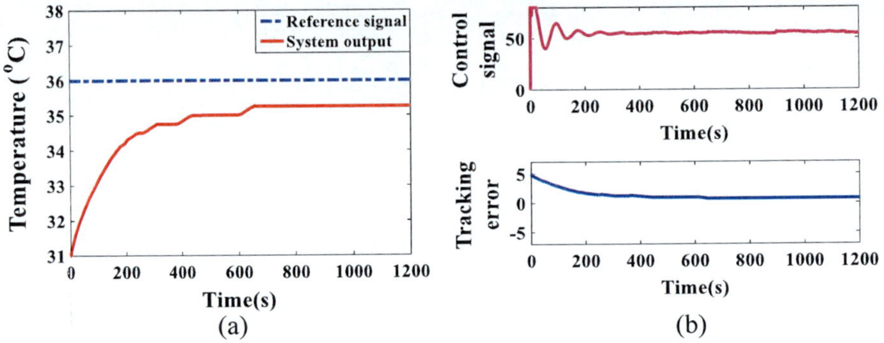

Fig. 7.17 Evaluation period **a** system response and corresponding **b** control signal with tracking error at Sensor 4 of air heater experimental setup for MRAC

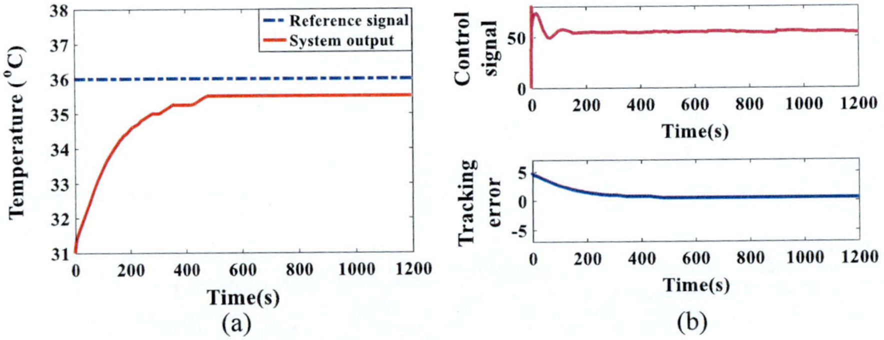

Fig. 7.18 Evaluation period **a** system response and corresponding **b** control signal with tracking error at Sensor 4 of air heater experimental setup for basic L_1 adaptive controller

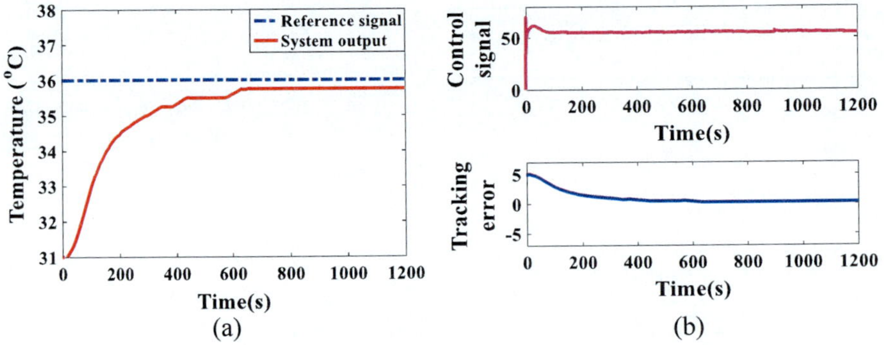

Fig. 7.19 Evaluation period **a** system response and corresponding **b** control signal with tracking error at Sensor 4 of air heater experimental setup for PSO-L_1 adaptive controller

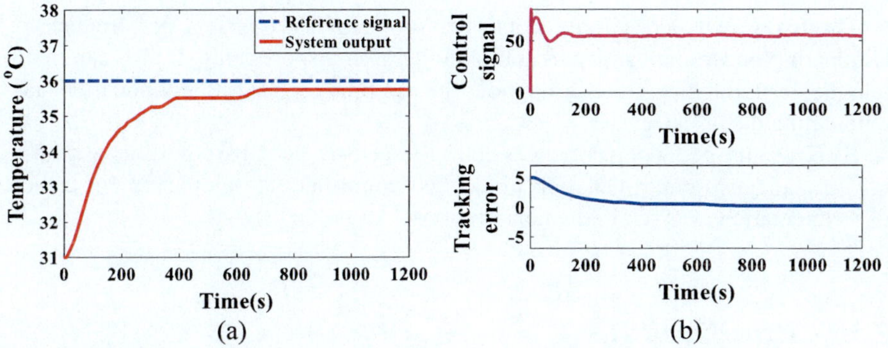

Fig. 7.20 Evaluation period **a** system response and corresponding **b** control signal with tracking error at Sensor 4 of air heater experimental setup for HS-L_1 adaptive controller

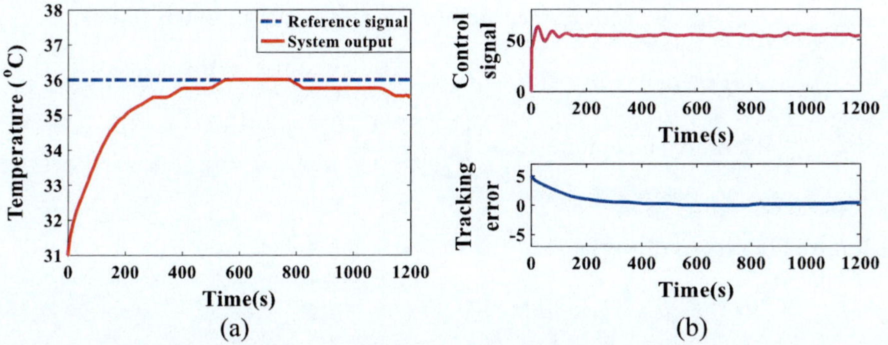

Fig. 7.21 Evaluation period **a** system response and corresponding **b** control signal with tracking error at Sensor 4 of air heater experimental setup for lbest HS-L_1 adaptive controller

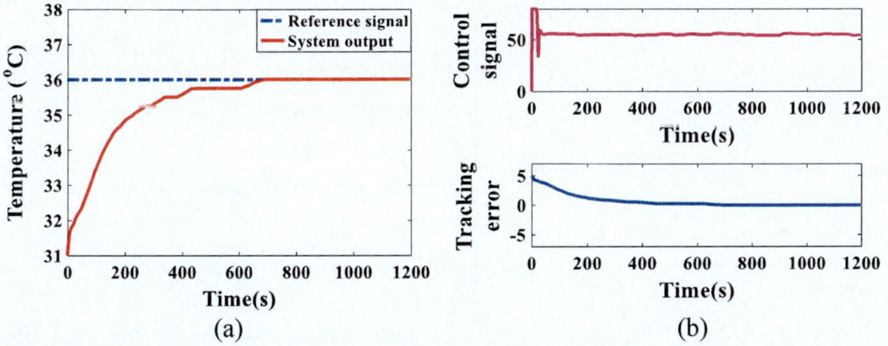

Fig. 7.22 Evaluation period **a** system response and corresponding **b** control signal with tracking error at Sensor 4 of air heater experimental setup for fuzzy PDC-L_1 adaptive controller

The fuzzy PDC-L_1 adaptive controller provides fast transient performance by eliminating uncertainties, disturbances and large transport delay. It also can maintain the desired temperature throughout the run time by tackling nonlinearities and eliminating disturbances.

By observing better performances of the lbest HS-L_1 adaptive controller and fuzzy PDC-L_1 adaptive controller, only these two controllers are employed for critical experimentation as given in the next section.

7.5.2 Experiment-II

In this experiment, the control objective is that, the Sensor 4 has to track a variable step trajectory with a fixed disturbance. Then, those controllers are subjected to the variable disturbance without further tuning. The references are set as follows:

$$Reference\ signal\text{-}I:$$

$$\text{Desired temperature } T_d = \begin{cases} 33.75\,^{\circ}\text{C}, & 0 \leq t \leq 1800\,\text{s} \\ 36.75\,^{\circ}\text{C}, & 1800\,\text{s} < t \leq 3600\,\text{s} \end{cases}$$

with fixed fan speed (disturbance) 70% of maximum speed.

$$Reference\ signal\text{-}II:$$

$$\text{Desired temperature } T_d = \begin{cases} 33.75\,^{\circ}\text{C}, & 0 \leq t \leq 1800\,\text{s} \\ 36.75\,^{\circ}\text{C}, & 1800\,\text{s} < t \leq 3600\,\text{s} \end{cases}$$

$$\text{with variable fan speed (disturbance)} \begin{cases} 60\% \text{ of maximum speed,} \\ 0\,\text{s} \leq t \leq 900\,\text{s} \\ 100\% \text{ of maximum speed,} \\ 900\,\text{s} < t \leq 1800\,\text{s} \\ 80\% \text{ of maximum speed,} \\ 1800\,\text{s} < t \leq 2700\,\text{s} \\ 90\% \text{ of maximum speed,} \\ 2700\,\text{s} < t \leq 3600\,\text{s} \end{cases}$$

The IAE values with fixed disturbance and variable disturbance for lbest HS-L_1 adaptive and fuzzy PDC-L_1 adaptive controllers are tabulated in Table 7.3. The system responses with fixed fan disturbance and corresponding control signal with error signal plots are provided in Figs. 7.23 and 7.24 respectively for lbest HS-L_1 adaptive and fuzzy PDC-L_1 adaptive controller. Then, the variable step disturbances are imposed on the system and the performances of the respective controllers without

Table 7.3 Comparative study of different control methodologies in experiment-II for delay dependent air heater experimental setup

Control methodologies	IAE value for Sensor 4	
	With fixed disturbance	With variable disturbance
lbest HS-L_1 adaptive controller	176.2950	178.4500
Fuzzy PDC-L_1 adaptive controller	143.3400	144.4650

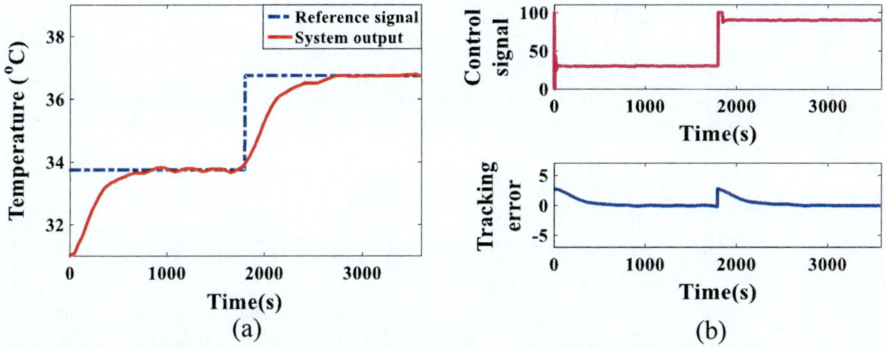

Fig. 7.23 Evaluation period **a** system response and corresponding **b** control signal with tracking error at Sensor 4 of air heater experimental setup with fixed disturbance for lbest HS-L_1 adaptive controller

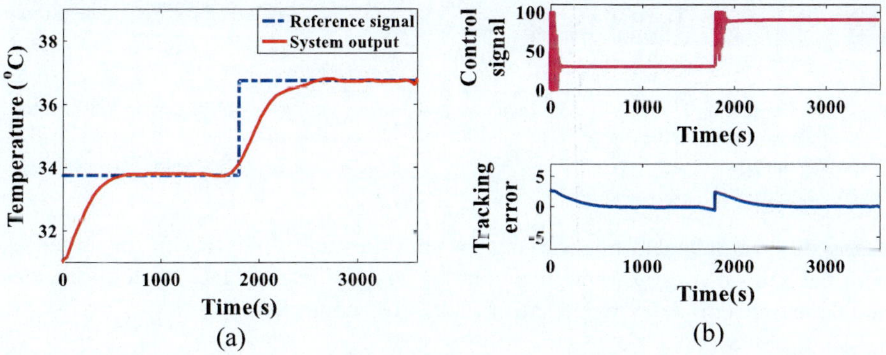

Fig. 7.24 Evaluation period **a** system response and corresponding **b** control signal with tracking error at Sensor 4 of air heater experimental setup with fixed disturbance for fuzzy PDC-L_1 adaptive controller

further tuning are investigated and depicted in Figs. 7.25 and 7.26. When time varying disturbances are imposed on the system, then the performance of the lbest HS-L_1 adaptive controller degrades whereas, the response of the fuzzy PDC-L_1 adaptive controller remains very satisfactory. Hence, the fuzzy PDC-L_1 adaptive controller

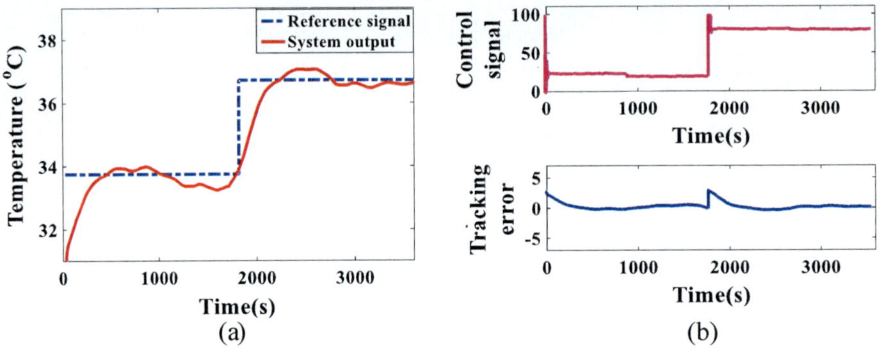

Fig. 7.25 Evaluation period **a** system response and corresponding **b** control signal with tracking error at Sensor 4 of air heater experimental setup with variable disturbance for lbest HS-L_1 adaptive controller

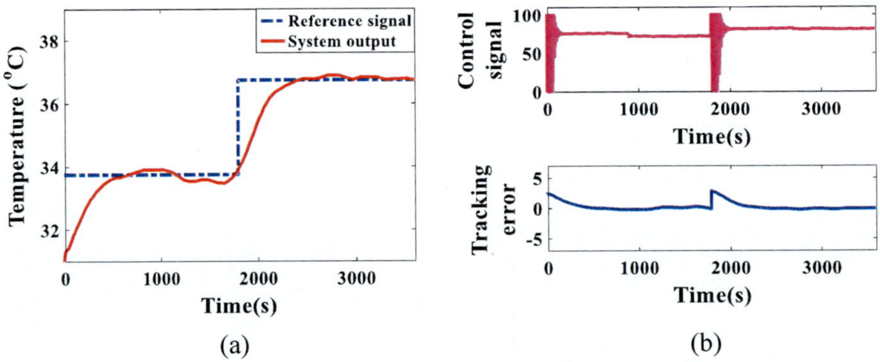

Fig. 7.26 Evaluation period **a** system response and corresponding **b** control signal with tracking error at Sensor 4 of air heater experimental setup with variable disturbance for fuzzy PDC-L_1 adaptive controller

successfully eliminates time varying uncertainties and time varying disturbances with the help of fuzzy L_1 adaptive controller and efficiently tackles nonlinearities and large transport delay by means of fuzzy PDC controller.

7.6 Summary

In this chapter, an air heater experimental setup is investigated which contains nonlinearities, time varying uncertainties, disturbances as well as large delay. At first, the air heater system is modelled and system identification is done. Then, the performances of different types of controllers, viz., PID, LQG, MRAC, basic L_1 adaptive,

PSO-L_1 adaptive, HS-L_1 adaptive controller are compared with lbest HS-L_1 adaptive and fuzzy PDC-L_1 adaptive controller. The experiments portray that, the transient performance of basic L_1 adaptive controller is fast though it's tracking performance for systems with nonlinearities and delays is not satisfactory. The lbest HS-L_1 adaptive controller performs better than the basic L_1 adaptive controller utilizing optimal parameter setting of L_1 adaptive controller. Though it cannot maintain the desired temperature by nullifying time varying disturbances and uncertainties properly. The fuzzy PDC-L_1 adaptive controller outperforms the other controllers tested in this work. To perform critical performance analysis of lbest HS-L_1 adaptive and fuzzy PDC-L_1 adaptive control methodologies, they are subjected to the variable disturbance without further tuning in case of tracking a variable step trajectory. The results show that, the fuzzy PDC-L_1 adaptive controller performs superior by tackling nonlinearities and delays with the help of fuzzy PDC logic as well as successfully handles time varying uncertainties and disturbances by utilizing fuzzy L_1 adaptive controller.

The conclusion of the book with future prospective of the current works are discussed in the next chapter.

References

1. Allen, W.H., Rubaai, A., Chawla, R.: Fuzzy neural network-based health monitoring for HVAC system variable-air-volume unit. IEEE Trans. Ind. Appl. **52**(3), 2513–2524 (2016)
2. Ma, Y., Borrelli, F., Hencey, B., Coffey, B., Bengea, S., Haves, P.: Model predictive control for the operation of building cooling systems. IEEE Trans. Control Syst. Technol. **20**(3), 796–803 (2012)
3. Ray, A.L., Couse, D.: Cement plant fan efficiency upgrades. IEEE Trans. Ind. Appl. **53**(2), 1562–1568 (2017)
4. Liao, Z., Dexter, A.L.: An inferential model-based predictive control scheme for optimizing the operation of boilers in building space-heating systems. IEEE Trans. Control Syst. Technol. **18**(5), 1092–1102 (2010)
5. Mcdonald, A.T., Friskney, S.H., Ulrich, D.J.: Thermal model of the dishwasher heater in air. IEEE Trans. Ind. Appl. **25**(6), 1176–1180 (1989)
6. Incropera, F.P., DeWitt, D.P.: Fundamentals of Heat and Mass Transfer, 2nd edn. Wiley, New York (1985)
7. Mantovani, G., Ferrarini, L.: Temperature control of a commercial building with model predictive control techniques. IEEE Trans. Industr. Electron. **62**(4), 2651–2660 (2015)
8. Minakais, M., Mishra, S., Wen, J.T.: Database-driven iterative learning for building temperature control. IEEE Trans. Autom. Sci. Eng. **16**(4), 1896–1906 (2019)
9. Sharma, K.D., Chatterjee, A., Rakshit, A.: Harmony search algorithm and Lyapunov theory based hybrid adaptive fuzzy controller for temperature control of air heater system with transport-delay. Appl. Soft Comput. **25**, 40–50 (2014)
10. Feng, Z., Zheng, W.X., Wu, L.: Reachable set estimation of T-S fuzzy systems with time-varying delay. IEEE Trans. Fuzzy Syst. **25**(4), 878–891 (2017)
11. Liu, T., Li, Y., Zhang, L.: Flow rate predictive control for volumetric type of fuel oil supplying system driven by variable frequency induction motor. Hydraul. Pneum. Seals **31**(2), 52–56 (2011)

12. Probst, A., Magan, M.E., Sawodny, O.: Using a Kalman filter and a Pade approximation to estimate random time delays in a networked feedback control system. IET Control Theory Appl. **4**, 2263–2272 (2010)
13. Hamdy, M., Abd-Elhaleem, S., Fkirin, M.A.: Adaptive fuzzy predictive controller for a class of networked nonlinear systems with time-varying delay. IEEE Trans. Fuzzy Syst. **26**(4), 2135–2144 (2018)
14. Maiti, R., Das Sharma, K., Sarkar, G.: Linear consequence based fuzzy parallel distributed compensation type L_1 adaptive controller for two link robot manipulator. IEEE Trans. Circuits Syst. I: Regul. Pap. **66**(10), 3978–3990 (2019)
15. Cao, C., Hovakimyan, N.: Design and analysis of a novel L_1 adaptive control architecture with guaranteed transient performance. IEEE Trans. Autom. Control **53**(2), 586–591 (2008)
16. Hovakimyan, N., Cao, C.: L_1 Adaptive Control Theory Guaranteed Robustness with Fast Adaptation. SIAM, Philadelphia (2010)
17. Intel: Intel® Core™ 2 Duo Processor, Intel® Pentium® Dual Core Processor, and Intel® Celeron® Dual-Core Processor: Thermal and Mechanical Design Guidelines (2008)
18. Maiti, R., Das Sharma, K., Sarkar, G., Chandra, S.: Modelling and control of delayed, nonlinear, uncertain, disturbed air heater employing fuzzy PDC-L_1 adaptive scheme. Communicated (2020)
19. Maxim Integrated: DS18B20: Programmable Resolution (2019)

Part V
Epilogue

Chapter 8
Future Research Directions of Hybrid Controller

Abstract This chapter draws the conclusion of this book. The key features of the methodologies presented in this book are portrayed in this chapter. In this work, lbest HS-L_1 adaptive controller and fuzzy PDC-L_1 adaptive controller are developed and employed on four simulation case studies and three experimental case studies. The performances portray that, the lbest HS-L_1 adaptive and fuzzy PDC-L_1 adaptive controllers perform better than the other control methodologies tested in this work. In future, these controllers may be employed on different large scale industrial process, aircraft systems, etc. Artificial intelligence based adaptive controllers may be implemented for different types of systems.

8.1 Significance of the Methodologies Presented in This Book

In this book, to handle practical system consists of unknown constant, time varying uncertainties, disturbances, nonlinearities and delays, hybrid controllers are designed. The significance of this work can be summarized as follows.

I. A special type of adaptive controller, viz., L_1 adaptive controller is utilized to tackle uncertainties and disturbances. The parameters of L_1 adaptive controller are tuned utilizing stochastic optimization and metaheuristics techniques to obtain optimal control performance. A variant of harmony search (HS) algorithm, viz., local best harmony search (lbest HS) is used to tune the parameter values of L_1 adaptive controller.

(a) It is analytically proved that, the lbest HS based L_1 (lbest HS-L_1) adaptive controller provides satisfactory exploration and exploitation phenomenon.
(b) Stability of the metaheuristic optimization technique along with controller en bloc, viz., lbest HS-L_1 adaptive controller is guaranteed in terms of spectral radius convergence.
(c) Satisfactory transient and steady state performance of this method are guaranteed analytically.

R. Maiti et al., *Hybrid L_1 Adaptive Control*, Studies in Systems, Decision and Control 422, https://doi.org/10.1007/978-3-030-97102-1_8

II. Fuzzy parallel distributed compensation (PDC) logic is augmented with L_1 adaptive controller to tackle nonlinearities, delays along with uncertainties and disturbances.

(a) All the components of L_1 adaptive controller are designed with the adaptive laws formulated from the fuzzy Lyapunov stability criteria.
(b) Overall stability of the fuzzy parallel distributed compensation type L_1 (fuzzy PDC-L_1) adaptive controller is guaranteed.
(c) Transient and steady state performance of the system with this controller are analyzed to show fast transient performance with guaranteed robustness and satisfactory steady state tracking performance.

III. The control methodologies are validated on four simulation case studies and three experimental case studies.

(a) In simulation case studies, at first, chaotic Duffing's oscillatory system and nonlinear spring mass damper system are investigated to examine the effectiveness of the lbest HS-L_1 adaptive and fuzzy PDC-L_1 adaptive controllers in tackling systems with disturbances, uncertainties, nonlinearities. Then, a non-minimum phase, inherently unstable inverted pendulum with cart system is controlled and stabilized. System with cross-coupling and higher order nonlinearities, viz., twin rotor MIMO system (TRMS) is also investigated.
(b) In experimental case study, enormously used electrical actuator with load disturbance is controlled. System with unknown dynamics, nonlinearities, viz., two link robot manipulator (TLRM) is controlled through the designed controllers satisfactorily. The air heater system with large delay is also controlled to show the effectiveness of the designed controllers in tackling delay.
(c) The controllers are designed for the systems with small amount of disturbance or no disturbance. Then, they are subjected to the large disturbances. The results show that, the lbest HS-L_1 adaptive controller performs better than the basic L_1 adaptive controller whereas, the fuzzy PDC-L_1 adaptive controller outperforms all other control methodologies tested in this book.

The future research perspectives of the present work are provided in the subsequent section.

8.2 Future Research Directions

In future, the works presented in this book can be extended in different directions. Some of which are discussed as follows.

I. In this book, the exploration and exploitation capability of the lbest HS algorithm is analysed mathematically. Many other optimization techniques, viz., particle swarm optimization [1], ant colony optimization [2], firefly algorithm

[3], cuckoo search algorithm [4], etc., are used in literatures. In future, the stability of local best PSO (lbest PSO) [5, 6] by considering different behavior of particles in different regions can be investigated. Investigation of earliest fast hitting time from the local convergence may be of great interest. Stability analysis of firefly algorithm with variable population size [7] can be carried out owing to its high convergence speed. Hence, in future, intensification and diversification of these optimization techniques and their varients may be formulated analytically.

II. Fractional order differential equations may be used to model systems with unmodelled dynamics, nonlinearities [8]. Different types of fractional order controllers, viz., fractional sliding mode controller [9], optimal fractional order controllers [10], fractional order PID controllers [8] were utilized in different literatures. Fractional order controllers improve closed loop response and robustness of time delay systems [8]. Owing to the advantages of fractional order controller, fractional order L_1 adaptive controller may be developed in future. The uncertainties, disturbances present in the system may have fractional relationship with the system model. Therefore fractional order adaptation of their values may provide better estimation of them. Other fractional order controllers, viz., fractional order backstepping controller, fractional order internal model controller may be augmented with basic L_1 adaptive controller to improve the overall controller performance.

III. Applications of artificial intelligence in designing controllers increase gradually for handling situations where human intelligence is required but physical presence is difficult. Artificial intelligent controllers can take decisions and perform works based on its intelligence and learning [11, 12]. In future, artificial intelligence based L_1 adaptive controllers, artificial intelligence based fuzzy controllers, etc., may be developed for different types of practical applications in hazardous situations, large scale industrial process, etc.

IV. The designed hybrid controllers may be implemented on different practical systems, viz., large scale industrial manipulator systems, aircraft systems, under water vehicles, etc. The application of theses controllers may be extended in power systems, smart grid systems, system biology, etc.

References

1. Das Sharma, K., Chatterjee, A., Rakshit, A.: A PSO–Lyapunov hybrid stable adaptive fuzzy tracking control approach for vision-based robot navigation. IEEE Trans. Instrum. Meas. **61**(7), 1908–1914 (2012)
2. Liao, T., Socha, K., Montes de Oca, M.A., Stützle, T., Dorigo, M.: Ant colony optimization for mixed-variable optimization problems. IEEE Trans. Evol. Comput. **18**(4), 503–518 (2014)
3. Dekhici, L., Redjem, R., Belkadi, K., Mhamedi, A.E.: Discretization of the firefly algorithm for home care. Can. J. Electr. Comput. Eng. **42**(1), 20–26 (2019)

4. Chrouta, J., Chakchouk, W., Zaafouri, A., Jemli, M.: Modeling and control of an irrigation station process using heterogeneous cuckoo search algorithm and fuzzy logic controller. IEEE Trans. Ind. Appl. **55**(1), 976–990 (2019)
5. Schmitt, M., Wanka, R.: Particle swarm optimization almost surely finds local optima. In: Genetic and Evolutionary Computation Conference. ACM (2013)
6. Bonyadi, M.R., Michalewicz, Z.: Analysis of stability, local convergence, and transformation sensitivity of a variant of the particle swarm optimization algorithm. IEEE Trans. Evol. Comput. **20**, (3), 370–385 (2016)
7. Li, F.X.F., Hou, Z.: A firefly algorithm with self-adaptive population size for global path planning of mobile robot. IEEE Access **8**, 168951–168964 (2020)
8. Biswas, D., Das Sharma, K., Sarkar, G.: Stable adaptive NSOF domain FOPID controller for a class of non-linear systems. IET Control Theory Appl. **12**(10), 1402–1413, (2018)
9. Davijani, N.Z., Jahanfarnia, G., Abharian, A.E.: Nonlinear fractional sliding mode controller based on reduced order FNPK model for output power control of nuclear research reactors. IEEE Trans. Nucl. Sci. **64**(1), 713–723 (2017)
10. Vinagre, B., Feliu, V.: Optimal fractional controllers for rational order systems: a special case of the wiener-hopf spectral factorization method. IEEE Trans. Autom. Control **52**(12), 2385–2389 (2007)
11. Wang, D., He, H., Liu, D.: Intelligent optimal control with critic learning for a nonlinear overhead crane system. IEEE Trans. Industr. Inf. **14**(7), 2932–2940 (2018)
12. Lan, X., Liu, L., Wang, Y.: ADP-based intelligent decentralized control for multi-agent systems moving in obstacle environment. IEEE Access **7**, 59624–59630 (2019)

Appendix A
Norms of Vector and Matrix

The norms of vector and matrix are discussed in this appendix.

A.1 Vector Norms

The norm of vector or matrix is a real valued function denoted as $\|\cdot\|$ with the following properties for all vectors and matrices $x, y, \rho \in \Re$.

1. $\|x\| > 0$ if $x \neq 0$ and $\|x\| = 0$ iff $x = 0$.
2. $\|x + y\| \leq \|x\| + \|y\|$.
3. $\|\rho x\| = |\rho| \; \|x\|$.

Different types of norms of any vector $w = [w_1, w_2, \ldots, w_n] \in \Re^n$ is defined as follows.

1. p-norm of a vector is defined as:

$$\|w\|_p = \left[\sum_{i=1}^{n} |w_i|^p\right]^{1/p} \tag{A.1}$$

2. 1-norm of a vector is defined as:

$$\|w\|_1 = \sum_{i=1}^{n} |w_i| \tag{A.2}$$

R. Maiti et al., *Hybrid L_1 Adaptive Control*, Studies in Systems, Decision and Control 422, https://doi.org/10.1007/978-3-030-97102-1

3. 2-norm of a vector is defined as:

$$\|w\|_2 = \sqrt{\sum_{i=1}^{n} |w_i|^2} = \sqrt{w^T w} \tag{A.3}$$

4. ∞-norm of a vector is defined as:

$$\|w\|_\infty = \max_{1 \leq i \leq n} |w_i| \tag{A.4}$$

A.2 Matrix Norms

Let, a matrix $A \in \Re^{n \times m}$ is observed as an operator to map the space of m dimensional vectors into the space of n dimensional vectors.

1. Then, the operator norm or the induced p norm of a matrix is defined as:

$$\|A\|_p = \sup_{x \neq 0} \frac{\|Ax\|_p}{\|x\|_p} = \sup_{x \neq 0} \left\| A \frac{x}{\|x\|_p} \right\|_p = \sup_{\|x\|_p = 1} \|Ax\|_p \tag{A.5}$$

From the above definition of induced p norm, other induced norms are defined as follows.

2. The induced 1-norm of a matrix is defined as:

$$\|A\|_1 = \max_{1 \leq j \leq m} \sum_{i=1}^{n} |a_{ij}| \tag{A.6}$$

3. The induced 2-norm of a matrix is defined as:

$$\|A\|_2 = \sqrt{\lambda_{\max}(A^T A)} \tag{A.7}$$

where $\lambda_{\max}(\cdot)$ denotes the maximum eigenvalue of matrix A.

4. The induced ∞-norm of a matrix is defined as:

$$\|A\|_\infty = \max_{1 \leq i \leq n} \sum_{j=1}^{m} |a_{ij}| \tag{A.8}$$

Matrix norms have these following properties.

1. $\|A\| = \|A^T\|$.

2. $\|Ax\| \leq \|A\| \; \|x\|$.

where x and A are arbitrary vector and matrix respectively.

3. All the norms are equivalent. if $\|\cdot\|_p$ and $\|\cdot\|_q$ are two different norms of any arbitrary vector or matrix X, then:
$c_1\|X\|_p \leq \|X\|_q \leq c_2\|X\|_p$. where $c_1 > 0$ and $c_2 > 0$ are two constants.

Appendix B
Spectral Radius Convergence

Let, M be the iterative matrix of Jordon form $M = Z\,J\,Z^{-1}$.

where $J = diag(J_0, J_1)$.

Then, for 'itr' number of iteration it can be expressed as $M^{itr} = Z\,J^{itr}Z^{-1}$.

If the spectral radius of iterative matrix is $\rho(M)$, then it follows that, $\rho(M) = \rho(J)$.

Now, If $\rho(M) = \rho(J) < 1$ then, J^{itr} converges to 0 and as well as M^{itr} converges to 0 (viz. $\lim_{n\to\infty} \left\| M^{itr} \right\| = 0$).

Again, if $\rho(M) = \rho(J) > 1$ then, $(J^{itr})_{i,i} = \lambda^{itr}$,

where λ is an eigenvalue with $|\lambda| > 1$.

Now, if ξ is the ith row and ξ' is the ith column of Z^{-1}, then

$$\xi' M^n \xi = \xi' Z\,J^n Z^{-1} \xi = \lambda^n.$$

Therefore, $\lim_{n\to\infty} \left|\xi' M^n \xi\right| = \infty$ and $\lim_{n\to\infty} \|M^n\| = \infty$.

Mathematically, it is expressed as:

$$\text{if } \rho(M) < 1 \text{ then } \lim_{n\to\infty} \left\| M^n \right\| = 0 \tag{B.1}$$

and

$$\text{if } \rho(M) > 1 \text{ then } \lim_{n\to\infty} \left\| M^n \right\| = \infty \tag{B.2}$$

Therefore, the iterative matrix M will convergence if its spectral radius will less than 1, viz., $\rho(M) < 1$.

R. Maiti et al., *Hybrid L_1 Adaptive Control*, Studies in Systems, Decision and Control 422, https://doi.org/10.1007/978-3-030-97102-1

Index

R. Maiti et al., *Hybrid L_1 Adaptive Control*, Studies in Systems, Decision and Control 422, https://doi.org/10.1007/978-3-030-97102-1